Communication and Cybernetics 11

Editors: K. S. Fu W. D. Keidel H. Wolter

Structure and Process in Speech Perception

Proceedings of the
Symposium on Dynamic Aspects of Speech Perception
held at I.P.O., Eindhoven, Netherlands,
August 4–6, 1975

Edited by
A. Cohen and S. G. Nooteboom

Springer-Verlag
Berlin Heidelberg New York 1975

Professor Dr. Antonie Cohen
Dr. Sibout G. Nooteboom
Institute for Perception Research, Eindhoven, Netherlands

The Symposium was sponsored by the Institute for Perception Research, Eindhoven, and held under the auspices of the *Koninklijke Nederlandse Akademie van Wetenschappen* (Royal Netherlands Academy of Arts and Sciences) and the *Nederlandse Organisatie voor Zuiver Wetenschappelijk Onderzoek* (Netherlands Organization for the Advancement of Pure Research).
The Planning Committee of the Symposium was formed by:
A. Cohen, M. P. Haggard, S. G. Nooteboom, S. E. G. Öhman, and D. B. Pisoni.

With 62 figures

ISBN-13: 978-3-642-81002-2 e-ISBN-13: 978-3-642-81000-8
DOI: 10.1007/978-3-642-81000-8

Library of Congress Cataloging in Publication Data. Symposium on Dynamic Aspects of Speech Perception, Eindhoven, 1975. Structure and process in speech perception (Communication and Cybernetics; 11). Sponsored by the Institute for Perception Research, Eindhoven. Bibliography: p. Includes index. 1. Speech perception – Congresses. I. Cohen, Antonie. II. Nooteboom, S. G., 1939 –. III. Instituut voor Perceptie Onderzoek. IV. Title. [DNLM: 1. Speech – Congresses. 2. Auditory perception – Congresses. W1 C0427X v. 11 / BF455 S986s 1975] BF251.S95 1975 153.7 75-35692.

Softcover reprint of the hardcover 1st edition 1975

PREFACE

The purpose of the Symposium was to provide a meeting place for those working in the field of speech perception, whose main interest is in the study of the perceptual processes in the decoding of connected speech, hence the title Dynamic Aspects of Speech Perception.

It was felt, after the meeting of the 8th ICA in London and the 2nd Speech Communication Seminar in Stockholm, 1974, that there should be an opportunity for an exchange of ideas on this topic with the emphasis on discussion and interpretation, rather than on the presentation of experimental results.

The initiators set themselves up as a planning committee and asked the present editors to organize a symposium in Eindhoven at the Institute for Perception Research, which has a well established tradition in perceptual work.

The present proceedings contain papers by contributors invited by the planning committee, as well as discussions. Most of the papers were circulated well in advance of the Symposium. We arranged the material in five sections:

I. Theoretical Issues in Speech Perception Research
II. Prosody in Speech Perception
III. Some Storage Properties of Speech Perception
IV. Short-term context effects, Dichotic listening, Speech Perception in the very young
V. General discussion and tentative conclusions

The order of papers is not the same as at the Symposium. Two papers are printed that were not introduced orally by their authors, viz. the papers of Darwin and Carterette, who could not be present for reasons beyond their control. Darwin's paper was introduced by Haggard. Carterette's was not presented orally at all.

We have attempted to include as much of the relevant discussions as possible. To this end we asked participants to write out their contributions to the discussions during the conference. These written versions were typed out immediately and presented to the discussants for approval. Since we set great store by early publication the last day of the conference was the dead-line for submitting written contributions to the discussions. This had as a result that relatively few of the last day's discussions are included in the proceedings.

Section V contains the written versions of a contribution made by Klatt and one by Levelt and Flores d'Arcais to the general discussion, dealing respectively with recent developments in speech understanding systems and reflections from a psycholinguistic point of view on the conference. The section ends with a general statement by the editors in an attempt to give as fair an assessment of the present research interests as could be gathered from the intensive contacts during and between the sessions, and particularly during the general discussion.

The organization of the Symposium would not have been possible without the support of the IPO its director Prof.Dr. C.A.A.J. Greebe, and in particular the secretarial devotion of Linda Savenije who also took a large share in the preparation of the proceedings together with Wilma Blom. We are grateful to Jan Brokx and Ko de Rooij whose contribution in collecting and preparing material for the discussions was of invaluable help to us. We also thank Ben Lopes Cardozo for his editorial advice.

We greatfully acknowledge the willingness of the Royal Netherlands Academy of Arts and Sciences and The Netherlands Organization for the Advancement of Pure Research to act as patrons to the Symposium.

Finally we thank Springer-Verlag for their friendly cooperation.

August 1975

A. Cohen

S.G. Nooteboom

CONTENTS

PARTICIPANTS

P.J. Bailey, The Queen's University of Belfast, Department of Psychology, Belfast BT7 1NN, Northern Ireland

A.P. Benguerel, The University of British Columbia, Faculty of Medicine, Division of Audiology and Speech Sciences, Vancouver V6T 1W5, Canada

C.I. Berlin, Louisiana State University Medical Center, Department of Otorhinolaryngology and Biocommunication, Kresge Hearing Research Laboratory, 1100 Florida Avenue, New Orleans, Louisiana 70119, U.S.A.

P. Bertelson, Université Libre de Bruxelles, Laboratoire de Psychologie expérimentale, Avenue Adolphe Buyl 117, 1050 Bruxelles Belgium

R. Carlson, Royal Institute of Technology (KTH), Department of Speech Communication, Speech Transmission Laboratory, S-100 44 Stockholm, 70 Sweden

* E. Carterette, University of California, Department of Psychology, Los Angeles, California 90024, U.S.A.

A. Cohen, Institute of Phonetics, University of Utrecht, Oudenoord 6, Utrecht, The Netherlands

R. Collier, Institute of Applied Linguistics, Department of Linguistics, Blijde Inkomststraat 21, B-3000 Leuven, Belgium

* C.J. Darwin, Psychology Department, University of Sussex, Brighton, Great Britain

E. Fischer-Jorgensen, Kongestien 45, Virum, Denmark

G. Flores d'Arcais, Rijksuniversiteit Leiden, Psychologische Funktieleer, Stationsplein 10, 2e etage, Leiden, Holland

H. Fujisaki, University of Tokyo, Department of Electrical Engineering, Bunkyo-Ku, Tokyo 113, Japan

J.H.V. Gilbert, The University of British Columbia, Faculty of Medicine, Division of Audiology and Speech Sciences, Vancouver V6T 1W5, Canada

B. Granström, Royal Institute of Technology (KTH), Department of Speech Communication, Speech Transmission Laboratory, S-100 44 Stockholm, 70 Sweden

M.P. Haggard, The Queen's University of Belfast, Department of Psychology, Belfast BT7 1NN, Northern Ireland

J.'t Hart, Instituut voor Perceptie Onderzoek, Den Dolech 2, Eindhoven, The Netherlands

A.W.F. Huggins, Massachusetts Institute of Technology, Dept. of Electrical Engineering and Research Laboratory of Electronics, Cambridge, Mass. 02139, U.S.A.

D.H. Klatt, Massachusetts Institute of Technology, Dept. of Electrical Engineering and Research Laboratory of Electronics, Cambridge, Mass. 02139, U.S.A.

I. Lehiste, The Ohio State University, Department of Linguistics, 1841 Millikan Road, Columbus, Ohio 43210, U.S.A.

W.J.M. Levelt, Katholieke Universiteit, Psychologische Funktieleer, Erasmuslaan 16, Nijmegen, The Netherlands

J.G. Martin, University of Maryland, College Park, Maryland 20742, U.S.A.

D.W. Massaro, The University of Wisconsin, Department of Psychology, Psychology Building, Madison, Wisconsin 53706, U.S.A.

S.G. Nooteboom, Instituut voor Perceptie Onderzoek, Den Dolech 2, Eindhoven,The Netherlands

S.E.G. Öhman, Uppsala University, Department of Linguistics, Kungsgatan 37, Uppsala 853 21, Sweden

D.B. Pisoni, Indiana University, Department of Psychology, Psychology Building, Bloomington, Indiana 47401, U.S.A.

L.C.W. Pols, Instituut voor Zintuigfysiologie, Kampweg 5, Soesterberg, The Netherlands

Q. Summerfield. The Queen's University of Belfast, Department of Psychology, Belfast BT7 1NN, Northern Ireland

A.J.W.M. Thomassen, Katholieke Universiteit, Psychologisch Laboratorium, Erasmuslaan 16, Nijmegen, The Netherlands

A. Wingfield, Brandeis University, Department of Psychology, Waltham, Massachusetts 02154, U.S.A.

* Absent

I. Theoretical Issues in Speech Perception Research

UNDERSTANDING SPEECH UNDERSTANDING

Keynote address by

Mark Haggard
Department of Psychology
Queen's University
Belfast, N Ireland

The perception of segments of speech one or two phonemes long is now quite well understood. This is because listeners rather readily accept the sort of experimental situation where they are asked simply to identify brief utterances, not always meaningful words in their language. A variety of powerful paradigms using variants of simple identification (for example reaction times and dichotic presentation) have been developed to specify factors involved in identifying these short segments of speech. Those not trained in the intellectual tradition of analytical experimental science are easily dissatisfied with the formal as opposed to superficial relevance that these tasks have to the perception of speech in real life situations, and those of us who have been studying perception at the phonemic level for many years have a kind of obligation not to rebuff their demands that we should be concerned with chunks of material of size that could actually constitute an utterance in a communication situation. Hence the title of this symposium "Dynamic Aspects of Speech Perception" is deliberately vague to avoid the implication that we are exclusively concerned with prosody, syntax, pragmatics or whatever. We are concerned with continuous speech perception.

It became fashionable recently in the philosophy of science to say that facts are inevitably theory-laden and that even investigators who profess not to have or to want theories have some theory, implicit in the sorts of experiments they do and the way that they make interpretations. Now there never has been any set of fundamental assumptions that one could call a theory of speech perception and I am sure that many decades will pass before there is, but there are many theoretical ideas implicit in what we do that recur in the writings of different authors and which probably do not change very much over the years. In what follows I have set myself a dual task. Firstly I want to try and point out the sort of ideas that we are likely to be led to by the exhortation to become interested in the dynamic aspects of speech perception. Secondly I propose to give examples of how theoretical ideas enter interpretations of experiments and hope to be able to show that we can monitor the quality of our scientific descriptions in a productive way.

Structure and Process in Speech Understanding

The old distinction between competence and performance has served those interested in the study of language well in the past and should continue to do so providing that we conceive of performance not merely as a set of unfortunate limitations that obscure our picture of the beauties of linguistic competence but rather as that systematic set of factors which govern how people can and will do things both in real life and laboratory situations. I should not want to say that we should be concerned exclusively with performance even in this sense because at the present time I find it profitable to make a distinction between two sorts of approach to psychological problems that rather cross-cuts the distinction. The dominant tendency in contemporary mainstream academic psychology is to be concerned with (hopefully) basic processes (such as various forms of memory representation) and with (hopefully) general types of task-solving strategy which could in principle come into play in a very wide variety of laboratory tasks and hence are likely to be playing some role in any situation where a human operator has to process information quickly and accurately. There is some concern (Newell 1973) as to just how task-specific the results obtained may be but the results are of general interest and give us a far more detailed understanding of what may be happening in a laboratory experiment than we had 10 years ago. This whole approach may be contrasted with that which I take to be the approach of the majority of contributors here. Here we limit our problems by saying that we know what the information to be processed in the stimulus is and by giving the listener a task which relates at least simply to something which he has generally been doing for a decade and a half before he enters our laboratory, namely understanding what others say. By accepting this limited definition of the problem we avoid the psychological and philosophical concern with the general nature of the human mind, and this may be no bad thing. One advantage that we get is the formal description of the structure of the stimulus which is likely to prove at least as powerful a predictor as the logical structure of our experimental design of the outcome of our experiment.

This is true whether we use the mathematics of speech production theory or the categories of a generative grammar to construct our stimulus sets. Although I would be very glad to hear this view questioned I am here taking the view that our objective should be to construct some sort of process model. We want to understand what goes on when people perceive speech; description of the linguistic, phonetic and acoustical structure is a preliminary to that goal. If we take language as an abstraction of a set of programs that have to run on a particular biological apparatus - the human brain - then we would see linguistics as a branch of cognitive psychology rather than a discipline of its own. I am not here staking an exclusive claim as a psychologist because linguistics could equally be called a branch of general systems theory or of cultural anthropology.

So the formal description of our stimulus is an important starting point and we have to proceed via demonstrations of the importance or psychological reality of the units in that description to the much more difficult task of inferring by manipulations of input and output the nature of the processing through which these units are passed. The need for experimental tasks which make sense of the units in the description of the stimulus has been appreciated for about 15 years. We have however seen the spread of some sophistication in the use of the concept of a 'unit'. For example in the work of Wingfield presented at this symposium, there is an explicit denial that there could be one particular unit size and a reluctance to claim that there is a normal unit size or even a single important unit size related to the particular circumstances. This sophistication is the product of two realisations. Firstly, though we like to draw models on paper that have unidirectional, sequential and hierarchical properties, we realise that it is much more likely that the real process of speech understanding proceeds by simultaneous decisions about units at a variety of different levels. The second realisation is that depending upon the particular question that you ask your experimental subject you may be able to tap into these levels experimentally. The psychology of human information processing in general is not sufficiently advanced to provide recipes for tapping into any particular level desired, but a sufficient diversity of results has been obtained to force us to think in a rather open-ended way about the whole problem.

I have made a distinction between structure and process which I think may be useful for understanding the perceptual processing of speech, but before I proceed to give some examples of how present day paradigms are used and interpreted I should like to mention two other uses of the word 'structure' that we should not forget about. The first of these we can for convenience call the structure of the mind. Many psychologists like to summarise contemporary understanding of information processing with a schema or block diagram model in which various boxes are interconnected; approximate properties are postulated for each of the boxes. Though it may be more appropriate to think of each of the boxes as being a process ie a particular type of program running at an undefined time in an undefined place, it is much simpler to represent it as a box connected to a limited number of other boxes. Because of this problem of representation the dictionary distinction between structure and process begins to break down and I have sometimes heard the inter-relationships between these boxes described with positive evaluation as "structural models". This is unfortunate because the contemporary preference is to regard the mind as having very little in the way of fixed structure. There is perhaps a valid structure/process distinction between a box diagram model which may be little more than a conceptual schema and a flow-diagram model or code model where the nature of the input-output relationships of the boxes is specified. The other sense of structure is that of physiological structure. It is inappropriate to ascribe a particular type of functioning that we analyse from the behavioural repertoire of a complete organism to a small set of neurons in that organism. However, animal studies and clinical studies continually offer informative data giving new dimensions of variation in performance with tasks whose effects on normal adult listeners we have already explored. The contribution of Berlin and Cullen to this symposium reminds us that the process has to run on a physiological structure and that selective interference with that structure can often have effects upon the processes that are not merely non-random but can be reconciled with laboratory data.

Difficulties in Conceptualising Dynamic Processes in Perception

There seems little doubt that prosody plays an important part in speech perception. Not only are we sensitive to dimensions of variation in the stimulus at a suprasegmental level but the experiments of Darwin and of Dooling (1974) show that when this information is disturbed so too is our perception. Neisser (1967, p262) suggested that rhythmic aspects of a sentence serve as a sort of frame on which the segmental details are hung. One might then see prosody as a way of off-setting a limited resolution of order in some form of short-term memory. If this were its only function then one might speculate that perception should work in a fashion more closely tied to its input in order to avoid memory problems. Rather it is attractive to postulate that prosody has a more creative role; the question is what role, and how do we begin to conceptualise it? This is a problem adumbrated in the papers by Martin and by Darwin but not yet faced up to in detail. It involves considering first what the various roles for prosody might be and then devising ways of submitting them to empirical test.

It has long been thought necessary to conceive of two modes of processing in speech perception: a bottom-up mode and a top-down mode. Advocates of a top-down mode might stress the syntactic and semantic determinants of our perception while advocates of a bottom-up mode would stress relatively inflexible forms of acoustic phonetic processing up to the word level. It is of course a mistake to call these modes theories (for example, active or analysis-by-synthesis, and passive respectively) and to oppose them. It may also be a mistake to think of one mode of processing as predominating even at one moment for a particular class of material as, very probably, both modes are simultaneously engaged in any real act of speech perception. The mistake I think derives from our limited ability to conceptualise any process in a highly parallel network. It is easy to think in terms of analogies derived from flow-charts for solving problems on today's highly serial digital computers. This is where we get the idea of a process starting with a "hypothesis" related to one level of abstraction in a stimulus and then moving in sequence to other levels in order to gather further evidence for that hypothesis before making a linguistic decision to determine output.

Serial thinking is convenient and may continue to serve us for a while as a way of describing what is going on but eventually it must give way to more sophisticated concepts. An example of this way of thinking is given in the proposal for a limited speech recognition system by Newell _et al_ (1973). They were concerned with "Voice CS" that is a system which allows someone to enquire along a telephone channel what computing facilities are available. Revealingly, the problem posed was one of using a rather convenient set of lexical, syntactic and semantic limitations to offset a very approximate and preliminary form of acoustic phonetic analysis. It was suggested that the system should scan the input looking for a particular keyword. When an adequate match for this word was found then a sub-grammar for that word would be looked up which would suggest where in the input related words might be found which would jointly specify one of a limited set of possible queries in a limited set of sentence forms. This engineering solution does only metaphorical justice to human speech perception. For realism we should have to think also of particular stretches of the acoustical input resolving themselves into tentative phonetic transcriptions on the basis of local constraints; they would also call for lexical and semantic analysis to resolve their internal discrepancies so as to ensure a more permanent form of representation. Thus we would have to embody the principle of 'heterarchy' introduced to artificial intelligence by Winograd (1972). This simply refers to a process at one level being able to call a process at any other level without a hierarchical transfer of control. Heterarchy itself is not too difficult to conceptualise. The difficulty comes when conceiving of multiple simultaneous heterarchical calls such that the concept of control begins to lose value altogether. However we have to try and lose 'control' and conceive of the problem as the mutual corroboration of bundles of cognitive and acoustical evidence seeking corroboration for one another at the end-nodes of a phonetic, phonological, syntactic and semantic tree-structure. Such a system has no predominant mode of output, no single scale of 'percept', but can be tapped at many levels. Providing this sort of search process is progressive and not self-erasing, search time can be reduced by the order of one square root by advancing the search from both ends. Neuroeconomically, therefore, we must postulate both bottom-up and top-down processing.

The above discussion has not involved prosody as such, only the long-appreciated need to view speech understanding as more than a process of acoustical analysis. Where might prosody fit in? Prosody carries direct phonetic cues to certain semantic and grammatical classes; it therefore serves to restrict the search processes whereby contact is made between a cognitive representation and acoustic representation. The restriction could take either of two forms, or both. Firstly the role of prosodic information could be additive; that is to say the direct cues to clause boundary, to form class, to contrastive stress could add together with information already present at various levels to make the choice more rapid or more accurate. In most areas of psychology effects of this nature are conceived of as a type of biasing, a shift in the value of evidence required for a decision due to the availability of _a priori_ probabilities derived from some other source.

The second alternative is in some ways more complicated and more interesting. The term "interactive" can be used here although it is sometimes wrongly used for the first additive type of contribution of two sources of information. Here we have to conceive of prosody as influencing the actual allocation of processing capacity in terms of power, temporal location and duration. I think Darwin and Martin envisage just such a control or interfacing function for prosody although they are wisely cautious about interpreting any evidence that we at present have as favouring such a view. By contrast with the preceding type this sort of effect is manifest in an index of sensitivity. When contextual, particularly prosodic, information is used appropriately then control processes should ensure the most efficient solution for a given long-term processing demand. Whether this recruitment of effort is directed to words, phonemes or whatever would have to be determined.

Caution about interpreting existing evidence as favouring such a view is justified on the following grounds. Consider an experiment performed by Anne Cutler (1975). She was interested in the fact that reaction times to the initial phonemes of stressed words tend to be shorter than for unstressed words, and she found that this was true even when the actual target words were not stressed words but the listeners were led to expect them to be so by the prior context. The problem here is determining what the actual experimental task implies for real speech perception; we may eventually find out but we must not jump to the superficial conclusion that a faster response time represents a more highly tuned perceptual apparatus. Let us imagine that we do not know the result of the experiment; what would we predict? We could assume that when a stressed word was coming up this indicated that the speaker was conveying a word of high lexical uncertainty, where it was important to make a correct decision otherwise the interpretation of what followed might be jeopardised. Under such circumstances he would be justified in setting high standards of evidence or slow responding criteria. On the other hand we might have predicted that in languages such as English stressed syllables are a courtesy provided for the listener enabling him to maintain a generally somnolescent state at the price of occasional flurries of perceptual activity. This would predict the obtained result of generally more rapid response. The extra quality of processing might not be restricted to targets in expected positions in expected words but might also be seen in low false positive rates, good discrimination and memory for adjacent phonemes and possibly in effects on secondary tasks. I think this indication that the result is not as obvious as first appears will serve to show that we should be cautious about interpreting tasks in which we ask our subjects to do slightly unnatural things.

New Paradigms for Old

We do not make progress simply by having theoretical ideas and dreaming up the best ways of testing them. Rather theories are suggested by data often gathered when we have been carried away by the sheer beauty of an experimental situation. We have to guard against churning out data from experimental paradigms simply because they are old or simply because they are new. This means that a certain amount of our effort is justified in examining paradigms for their own sake and in setting up criteria for deciding when to regard a paradigm as simply too old or simply too new to give data that are both intelligible and interesting. It is clear that as we switch to an interest in dynamic aspects of speech perception some of our tried old paradigms will fail in their usefulness. Two seem to be established. The technique of rapid monitoring of a particular unit such as a phoneme in the signal and the scoring of identification errors under conditions of selective distortion. We shall have to add to these to obtain the necessary convergence upon perceptual processes and if I knew exactly what to add then I should be doing experiments at a great rate rather than giving a keynote address. I do however have two suggestions to make. Firstly we should not be shy of asking our listeners to make certain judgments about large segments of speech. Of course the situation can become slightly uncontrolled and it may be necessary to establish just which aspects of the speech they are judging. But the experiments of Nooteboom and of Summerfield show that stretches more than a syllable long can have an effect upon perceptual processes which we tap with a phoneme identification. It may be possible (as has been found by those who have succeeded in scaling various subjective quantities) both to train our subjects and ask them the right question in such a way that enables us to make fuller use of their intelligence. We can ask them is a particular stretch of speech natural or unnatural, too fast or too slow, or does it match up appropriately with some other defined stretch?

As an example of this approach, a few years ago I played listeners various versions of a synthetic sentence asking them which of a pair of versions they preferred from the point of view of naturalness. Within a given pairing only the rate of tongue articulation in the articulatory model underlying the synthesis was allowed to vary but different pairings that they were asked about had different syllabic rates. The results showed that there was not a particular set of acoustical values that was preferred across all syllabic rates; rather the listeners preferred the underlying rate of articulation and hence of formant transitions to be constant across the syllabic rates so that in the faster examples certain targets were not reached, bursts became abbreviated and the principle of reduction was manifest. The result is a predictable one but it shows that even with a small and unsophisticated experiment you can get meaningful answers when you ask people questions about the abstract properties of sentences.

The other suggestion that I have involves the use of the concept of priming. This is nicely illustrated in an experiment by Lackner and Garrett (1972) which suggests to me among other things that not all dichotic experiments are bad experiments. They presented to one ear, sentences which were ambiguous at a variety of levels - lexical, surface structure and deep structure. For example, the sentence "the sailors enjoyed the port in the evening" has a lexical ambiguity which can be resolved by a sentence concerning either preferences for fortified wines or the lack of social life on the open sea. Lackner and Garrett presented such disambiguating sentences at -5 to -10 DB on the unattended ear. Subjects were adamant that they had heard nothing of these unattended sentences yet their paraphrases of the target ambiguous sentences were influenced by them. Like some current experiments in visual masking, this result suggests that there can be direct access to the semantic system for information that is never represented at any explicit level in either a phonological or sensory code. It suggests that there are multiple parallel computations of possible semantic readings for an attended message. It suggests that incompleteness of attention is not simply a matter of breakthroughs but rather one of levels of processing, and that there is a variable level of processing which could be deployed dynamically in other ways in real speech perception.

I believe that the study by Lackner and Garrett is important because it shows that if we devise sensitive stimulus material we can show how streams of information interact at many different levels. Full exploitation of the paradigm is based upon the use of the dichotic method whereby different streams of information may gain access to central processing without disturbing one another at a peripheral level. There are a number of interesting parameters to be explored such as the dependence of the effect upon the total processing demand in the attended message, the number and complexity of unattended messages and the critical temporal location of the effective disambiguating information. In the 'priming' paradigm in general I think we have a useful inroad into the theoretically powerful idea of levels of processing.

Interpretation of Speech Experiments

Very often, conclusions are drawn from experiments in rather informal language which may accurately record some of the intuitive motivation for the experiment but really are not justified by the data. One way to guard against this is by listing the alternative interpretations that might have been envisaged. As an example of how widespread this informality of interpretation is, I am choosing an example which is located between mainstream human information processing research and contemporary research on the segmental aspects of speech. Methodologically this is a tight piece of research from a well-reputed author. Wood (1974) was concerned with the reaction times to simple phonetic and acoustic properties of speech signals. In a two-choice task when an irrelevant stimulus dimension varies randomly, reaction times tend to be longer than when the irrelevant dimension is fixed. However, when the irrelevant variation is completely redundant upon the target variation then reaction times are again shorter. Wood found this to be the case with identification of consonant place of articulation and overall pitch (high/low) varied in a correlated fashion. He concluded that phonetic and auditory processing must therefore be in parallel. The inferential model assumes that reductions in reaction time accrue from a sort of OR gate following the parallel processing and the response control mechanism. Because each separate process has a distribution of latencies, the OR gate will take the first available on any occasion. This interpretation is extremely reasonable but the conclusion about parallel processing is likely to be misconstrued by others because the meaning of the word 'auditory' is not in fact clear. At least two alternatives exist. 'Auditory' could mean the common acoustical processing which must precede the systematic perceptual analysis of both speech and non-speech sounds or it could mean non-linguistic processing as a whole in opposition to linguistic processing. It seems a necessary matter not an empirical one that both these senses of 'auditory' correspond to real phenomena, the empirical issue being what experimental methods separate them, and how complicated the parallel non-linguistic processing is or can be. The design of the task allows the OR gate to be at a very late stage before response, so what the experiment tells me is that people are sufficiently flexible to indulge in some sort of attention-sharing between linguistic and non-linguistic tasks when it pays them to do so, a methodologically useful conclusion. That these two forms of processing are in parallel seems on reflexion to be a necessary interpretative assumption and not part of the conclusion. The use of the word 'auditory' may wrongly suggest that the result questions the traditional and necessary sequence of the auditory level and the phonetic level in a conceptual schema for processing such as that of Pisoni. This danger arises through not listing and discounting the first of the two alternative conceptions of 'auditory' which I mentioned. There is room for some philosophical rigour in making usable and legitimate interpretations from experiments; this is not assured by experimental design. A discriminating use of theoretical ideas is what I am urging here, not a bias to depict oneself as a theoretician. One might perhaps chide Cole and Scott (1974) for the latter transgression. These authors have been taken to task sufficiently elsewhere for interpreting experiments with cross-spliced bursts plus transitions as indicating that there are invariants in the burst cues for stop consonants. This chimerical success in the search for the "phantom in the phoneme" (or, more appropriately, the spectre in the spectrum?) led Scott and Cole to speculate that the function of the formant transitions was the prevention of auditory streaming. Auditory streaming is the phenomenon, known at least since J.S. Bach, whereby certain variables especially frequency separation limit the ability of stimulus elements to be heard as a single pattern. The relevance of this phenomenon outside melody problems is indeterminate, but certain components of the speech stream such as fricative and stop bursts may, after prolonged repetition, appear to float off and not cohere any longer with the main source of the information. It is likely that any disruption of the coherence of a signal such as deleting its formant transitions would increase the incidence of stream separation of the components in a repeated speech signal, but if one manipulates properties of synthetic speech rather than repeating natural speech, the quality of the fricative itself is the major determinant of the subjective separation. Either way I know of no data that justify a teleological use of the concept "to prevent streaming" in a theory of perception or, more plausibly, production. Transitions are present in speech because of limited sets of ways in which it is possible to move the articulatory organs and, presumably because of their information value, perceptual mechanisms appear to be rather sensitively tuned to aspects of transitions. The interesting thing about speech pattern processing is that networks are built up precisely to pool different types of information that may not be highly coherent either in time or in frequency, e.g. some bursts and the related transitions. Cole and Scott add to the foregoing suggestions about the roles of burst cues and transition cues Neisser's suggestion about the framing function of amplitude envelope cues. To call this set of suggestions even "Towards a theory of speech perception" seems to me a misuse of the word 'theory' , at least as unfortunate as that in the term 'motor theory'.

I hope I have conveyed with these examples the distinction between a concern with theoretical ideas necessary for interpretation and the eagerness to produce something that people will call a 'theory'. What seems to emerge from the papers presented at this symposium is that we are really not yet concerned with theory as such. While investigators are content to use a number of ideas that have theoretical content, they seem reluctant even to anticipate what the next level of explanatory formalism might be. I think this caution is entirely realistic. Man may be a theoretical animal but his theories do not need to be very explicit, wide-ranging or formal. We can all appreciate why some of the experiments that people do are intuitively appealing at a given point in time. For me the experiments of Darwin have this quality because they link variables of known importance (Relevance), the results are relatively unambiguous (Control) and because the results are neither painfully obvious nor wildly inexplicable (Appropriate hypothesis specifity). One way of framing questions about the dynamic aspects of speech perception would be to do experiments with these properties and to have the data refine the existing theoretical ideas into a position where they prescribe further experiments. Despite my speculations about serial and parallel processes, I think we must proceed in this semi-empirical fashion. However, I hope that my excursion through points other than data points will have both assisted some evaluative discussion by the symposium participants and also helped the less familiar reader of the published proceedings to build up a picture of what workers in this field are attempting to do and what we might reasonably expect them to achieve in the next few years.

REFERENCES

Cole, R.A. and Scott, B. Toward a theory of speech perception. Psychological Review, 1974, 81, 348-374.

Cutler, A. Personal communication, 1975.

Dooling, D.J. Rhythm and syntax in sentence perception. Journal of Verbal Learning and Verbal Behaviour, 1974, 13, 255-264.

Lackner, J.R. and Garrett, M.F. Resolving ambiguity: effects of biasing context in the unattended ear. Cognition, 1972, 1, 359-372.

Neisser, U. Cognitive Psychology. New York: Appleton-Century, 1967.

Newell, A. You can't play 20 questions with Nature and win. In W.G. Chase (ed) Visual Information Processing, 1973, 283-308. Academic Press: New York.

Newell, A. et al. Speech Understanding Systems. Final Report of a Study Group, 1973. North Holland: Amsterdam.

Winograd, T. Understanding natural language. New York: Academic Press, 1972.

Wood, C.J. Parallel processing of auditory and phonetic information in speech perception. Perception and Psychophysics, 1974, 15, 501-508.

Some Stages of Processing in Speech Perception

David B. Pisoni and James R. Sawusch

Indiana University
Bloomington, Indiana 47401 U.S.A.

Introduction

In this paper we consider a number of theoretical issues that have occupied the attention of investigators concerned with human speech perception. Most extant theories of speech perception have been quite general and vague and for the most part not terribly well developed. It is our view that many of the theoretical issues have not been dealt with adequately in the past primarily because of the failure of these theories to make explicit and precise statements that can be tested empirically. Indeed, most of the current theories are formulated in such broad terms that the scope of the theory cannot be delimited in any satisfactory way to deal with existing phenomenon or predict new findings.

The present contribution is not an empirical paper and we do not plan to report any new experimental findings. Rather, we feel that an attempt to deal with some of the fundamental issues in speech perception within an organized framework is warranted at this time, especially at a conference such as this. We consider speech perception in both the broad sense, encompassing syntactic and semantic analysis, and, in a more narrow sense, involving only auditory and phonetic analysis. However, our major emphasis will be on the earliest stages of perceptual analysis of speech. The contents of this chapter are organized into four major sections. In the first section, we briefly summarize what we see as the major theoretical issues in speech perception. Several of the traditional issues involving segmental analysis will be touched upon briefly here such as the lack of invariance and segmentation, the role of articulation in speech perception and units of perceptual analysis. In the second section, a distinction between taxonomic and dynamic models of speech perception is considered along with a brief review of several prominent speech perception theories. In the third section, we describe a fairly general information processing model of speech perception. Although the major emphasis in this discussion is on phonetic perception, the model relies heavily on syntactic and semantic processes which are used to guide earlier stages of perceptual analysis. In the fourth section, we briefly examine the role of higher-level linguistic processes in speech perception.

Our major purpose in preparing this contribution has been two-fold. First, we believe that it is useful to theorize in a general way about the process of speech perception outside the context of the particular experimental paradigms. For the present it seems like a reasonably good strategy to focus on general questions in speech perception rather than particular issues. The second purpose is our hope that the approach we have taken here will prove to be useful not only in thinking about speech perception but also in stimulating further theoretical discussion in the future.

Some Theoretical Issues in Speech Perception

Invariance and Segmentation

Most investigators agree that the two most important problems for speech perception theory deal with the lack of invariance and segmentation of the physical signal. The earliest observations showed that there was generally a lack of one-to-one relation between information in the acoustic signal and segments resulting from linguistic analysis. Examination of spectrographic patterns of real speech indicated that there were, in general, no segments in the speech signal that corresponded uniquely to segments in the message. The early perceptual experiments with synthetic speech indicated that a single segment of the acoustic signal carried information about several successive phonetic segments. Other studies showed that very different acoustic signals were often perceived as identical linguistic segments and identical acoustic signals were often perceived as different linguistic segments depending on the context. The acoustic cues to a given phonetic segment have been shown to be numerous and quite variable depending on the phonetic context, stress, speaking rate and speaker, which are natural consequences of the dynamics of the articulatory apparatus. For our purposes, the most important implication of these findings is that they place fairly rigid constraints on the types of perceptual theories that might be proposed. For example, filter and template matching theories would seem to be poor candidates for phonetic perception since the segments for analysis cannot be defined exclusively by properties of the physical signal. Based on these considerations the only acceptable class of theories are those that employ higher level linguistic information in the earliest stages of recognition.

Articulation in Speech Perception

Theoretical accounts of speech perception have often emphasized some type of articulatory-motor involvement during perceptual processing. There seem to be two primary reasons for this concern with articulation. The first is basically historical. Acoustic phonetics and much of the experimental research on speech perception were guided by the traditional articulatory descriptions developed by phoneticians. Thus, the acoustic cues underlying the perception of different classes of sounds were related to the articulatory gestures and dimensions that distinguished these sounds in production. Early work indicated that speech sounds could be specified precisely in terms of a few simple and relatively independent dimensions which provided the same distinctions in perception that articulatory dimensions provided in production and suggested that perceptual and articulatory processes may be intimately linked.

A second reason for assuming some type of interaction between perception and production is that speech sounds are produced by a sound source which has well defined acoustic constraints for the listener (Fant, 1960; Flanagan, 1972). Recent perceptual work indicates a close articulatory-perceptual match between changes in vocal tract shape and the acoustic signal. For example, Stevens (1972) has found that in certain parts of the vocal tract, a small change in the position of the tongue constriction resulted in a large variation of the formant frequencies; whereas in other regions of the vocal tract which correspond to the "natural" articulatory positions, a large variation in the position of the major tongue constriction had only a small effect on the formant frequencies. Stevens argues that the articulatory positions for particular phonetic segments, especially the consonants, have been chosen at locations where large deviations in articulation will produce only minimal changes in the acoustic signal. The existence of "Quantal" relations of this type between articulation and sound output may place constraints on the possible inventory of articulatory and acoustic properties that may be used as the basis for phonetic features in natural languages. Thus, according to Stevens, an acoustic property that would be a suitable candidate for a phonetic feature would be one that could be generated by the vocal tract without requiring very precise positioning or manipulating of the articulatory apparatus. Sounds produced in these "natural" regions might represent both articulatory and acoustic correlates of a particular phonetic feature. Thus, if knowledge of articulatory constraints is used in perception, the important questions for any theory of speech perception become <u>where</u> and <u>how</u> this knowledge is put to use by the perceptual system. In our view, the failure of the Motor Theory and Analysis by Synthesis approaches as psychological theories stems from their vagueness in specifying how knowledge of articulation would be used in perception.

Units of Perceptual Analysis

A central issue in speech perception has always been the choice of a unit of perceptual analysis. Numerous investigators have argued for the primacy of the feature, phoneme, syllable or word as their candidate for the basic perceptual unit. Other investigators, motivated chiefly by developments in generative linguistic theory, have proposed much larger units for perceptual analysis such as clauses or sentences. Many of the problems about the choice of a perceptual unit could be resolved if a distinction were made as to the level of perceptual analysis under consideration. It should be obvious that the size of the processing or structural unit will vary as a function of the level of processing in the linguistic system. There is now sufficient psychological evidence available to argue for the simultaneous existence of all units of linguistic analysis from features to syllables to clauses. To argue that there is one basic unit in speech perception is to acknowledge that language exists primarily in one form rather than another. However, the major fact about human language is that it exists in many forms, most of which are inaccessible to conscious inspection. Thus, we normally do not listen to phonetic details in perceiving speech. We recognize words, not sounds, and find it difficult to separate our observations from their interpretations. For example, consider the distinction between phonemes and syllables. Several investigators have argued that the syllable is a more basic perceptual or linguistic unit than the phoneme. The major claim here is that the phoneme is more "abstract" than the syllable because syllables exist independently as articulatory or acoustic units and phonemes do not. But most of these arguments have failed to take note of the symbiotic function of phonemes and syllables. As Studdert-Kennedy points out, the syllable serves as the carrier of phonetic information; its purpose is to create contrast for phonetic perception so that the listener can detect sound segments and features for the subsequent operations of the perceptual system. By choosing the syllable as a basic unit, some investigators have assumed that the invariance issue could be resolved. But there are several problems with this line of reasoning too. First, there are difficulties in defining the syllable acoustically. Second, syllables are not indivisible units; they have a complex internal structure which at some point must be recovered by the perceptual system. And finally, the acoustic cues for syllables are affected by context in much the same way as the cues for phonemes. One way to resolve the issue of units is to simply assume that the perceptual mechanism requires information distributed over at least a whole syllable in order to derive minimal information for a reliable phonetic interpretation. We should note, as Fodor, Bever and Garrett (1974) have remarked, that this strategy does not necessarily reflect on the size of the perceptual unit but only on the type of information that this analysis requires.

Syntax and Semantics

It is well known that words presented in sentential context are more intelligible than the same words presented in isolation. The traditional explanation of this finding was that syntax and semantics played only a minor role in perception by narrowing down or constraining the number of possible alternatives available to the listener. This view of speech perception assumed that the process was organized serially and that phonetic and phonemic characteristics of an utterance were recognized more or less directly from a sequence of physical properties in the sound wave. However, much of the research on the perception of connected speech indicates that processing does not proceed serially on the basis of a sequence of independent decisions about individual segments. Indeed, a good part of the information used in the perceptual process is not determined exclusively by the physical properties of the signal. There is good evidence, for example, that it is extremely difficult if not impossible for a linguist to discover the correct phonemic representation of an utterance without the use of higher-level syntactic information (see for example, Shockey & Reddy, 1974). And the limited success of recent speech understanding systems has been brought about by the explicit deployment of syntactic, semantic and conceptual information in the recognition process. This suggests that syntactic and semantic information must play an active and intimate role in any theory of speech perception since this information guides the early stages of phonetic and phonological processing. As we note below, most of the current theories of speech perception have failed to consider syntax and semantics seriously.

Prosody and Speech Perception

Perhaps one of the most neglected topics of research in speech perception is prosody. As Svensson (1974) has recently observed, there is a very wide gap between the work on isolated segments and features and the role of syntactic factors in speech perception. For the most part, prosody and grammar have been ignored in speech perception theories. The way prosodic information might be used in the recognition process has not been considered in any detail. This is surprising since there is some evidence that prosodic features such as pitch, for example, may provide cues to segment the speech stream into constituents that would be suitable candidates for further syntactic analysis (Lea, 1973). We feel that this is an important problem that must be considered in any theoretical account of speech perception. Prosody may serve as the interface between low-level segmental information and higher-levels of grammatical structure in speech.

Paradigms in Speech Perception Research

Several remarks need to be made about paradigms in speech perception. Much of the speech perception research that has been carried out over the last twenty-five years developed around specific experimental paradigms instead of being focused on general issues. Thus, most of the experimental questions that investigators devoted their attention to were defined by the particular types of experiments that were currently in vogue rather than being derived from some specific theoretical perspective. This is, of course, not surprising since relatively few specific predictions could be derived from the currently available theories. For example, numerous categorical perception experiments were carried out, the results of which were used for a long time as the major empirical support for the Motor Theory of Speech Perception. More recently, there has been a strong interest in the use of dichotic listening techniques to study hemispheric differences in speech perception and to detail some of the processes underlying speech perception. Although the dichotic listening findings may have important implications for understanding some aspects of brain function in speech perception, we feel that most of the results have only limited theoretical value. Indeed, we would be willing to argue that much of the dichotic listening work has detracted from the main goal of research in this area, namely, the development of a theory of speech perception. Finally, there has been a great deal of work in the last year or two using selective adaptation procedures to study the functional organization of the speech processing system (see Cooper, 1975 for a review). The early work with this technique raised the issue of the possible existence of feature or property detectors which mediate the perception of various phonetic distinctions. The specific issues are intricate and the arguments complicated. But there seems to be fairly good evidence for the existence of property detectors which respond to certain types of acoustic information in the signal. The evidence for specialized detectors that respond to more abstract phonetic information is somewhat weaker. Nevertheless, work using this technique has proceeded at an extremely rapid rate. It is, however, still too early to tell whether this line of research will produce any long range payoffs. To summarize, we feel that there is clearly a need to focus on basic issues rather than paradigms in speech perception research if we are to make any advances in the field and if we are to develop a strong theoretical framework in which to pursue these questions.

Models of Speech Perception

Taxonomic vs. Dynamic

Historically models of speech perception were derived from the taxonomic principles of segmentation and classification. It was assumed that linguistic structure, especially the phonological level, could be discovered by the use of these procedures applied automatically to a corpus of data. Thus, according to this view, a separate level of phonological structure could be described independently of higher-level grammatical or syntactic considerations. More recent models, developed within the framework of generative linguistic theory, however, have stressed a more active or dynamic view of the process. For example, Chomsky and Halle (1968) have argued quite convincingly that speech perception is an "active process" in which acoustic information is utilized to form hypotheses about the structure of sentences. These hypotheses which are then based in part on expectations and knowledge of the language are subsequently used to generate structures which are compared against the original input. In this way, the listener uses phonological principles to determine the phonetic properties of the stimulus. As Chomsky and Halle have put it, what the listener "hears" is what is internally generated by the rules for the hypothesized structure. This view is, of course, a version of analysis by synthesis which we will discuss below. Although it is a very abstract theory, it emphasizes an important point that has often been overlooked. A listener's interpretation of a speech stimulus may not be determined exclusively by the properties of the physical signal. Since our conscious experience of language almost always follows the semantic intent of the message and not its form, it is difficult to become aware of the physical properties of the stimulus. Similarly, the final semantic interpretation may involve information which has no physical correlate. Thus, what is finally perceived depends on the listener's knowledge of his language as well as numerous extragrammatical factors that determine what the listener expects to hear.

In the remainder of this section we briefly review several of the current theories of speech perception. We do not intend this review to be exhaustive in any sense and only those theories that have played a prominent role in recent research will be considered.

Motor Theory of Speech Perception

The basic assumption of the Motor Theory is that "speech is perceived by processes that are also involved in its production (Liberman, et al., 1967)." This view of speech perception was motivated by the observation that the perceiver is also a speaker and it would be unparsimonious to assume that the speaker-listener might use two entirely distinct processes for language perception and production. As we mentioned earlier, one of the central problems in speech perception and perhaps the major reason for postulating a motor theory is the issue of "encoding" or "lack of invariance" between the acoustic signal and phonemic perception. One possible way of resolving the invariance problem was to assume that the same perceptual response to widely differing acoustical patterns arises because the pattern would be produced by the same articulation or underlying motor commands to the articulators. Similarly, different perceptual responses to fairly similar patterns arise from different articulations or different motor commands.

Although the motor theory has occupied an overwhelmingly dominant place in contemporary perceptual theory, the link between empirical data and theory has not been very strong. Indeed, much of its early support was based on recurrent demonstrations of the same experimental outcome using near identical techniques and a very restricted set of stimuli. For example, the primary data used to support the early versions of the Motor Theory came from categorical perception experiments with synthetic stop consonants and steady-state vowels. Although this particular issue has received a great deal of attention over the last few years we will not elaborate on these findings because of space limitations. Most of the current evidence seems to suggest that the categorical perception results are more a function of coding processes in short-term memory than differences in production between various classes of speech sounds (Pisoni, 1973).

If we set aside the categorical perception findings, there seems to be very little direct empirical support for some active mediation of articulatory knowledge or information during perceptual processing. In our opinion, most of the current arguments for Motor Theory rest on parsimony, logic, or faith rather than a firm empirical foundation. The one exception is the recent finding of Cooper (1974) on perceptuo-motor adaptation who showed that listening to repetitions of a /pi/ utterance produced correlated shifts in the subject's VOT values in speech production. Although the magnitude of the effect was small, it is, nevertheless, the first direct support for the hypothesized link between speech perception and production.

From our point of view, the most serious problem for Motor Theory rests on the failure to specify the level of perceptual analysis where articulatory knowledge is employed in recognition. In the information processing model to be described below, it is possible to specify in detail where articulatory information is employed. However, because the model is not completely dependent on the overlap of perceptual and productive processes and only uses articulatory knowledge in a heuristic way, we think that the model is more general than traditional Motor Theory approaches.

Analysis-by-Synthesis

The Analysis-by-Synthesis model proposed by Stevens and his colleagues is much more explicit than the Motor Theory (Stevens, 1960; Stevens & Halle, 1967; Stevens & House, 1972). The basic assumption of the model is similar to the Haskins position; there exist close ties between the processes of speech production and perception, and there are components and operations common to both. The perceptual process begins initially with peripheral processing of the speech signal to yield a description in terms of auditory patterns. In cases where phonetic features are not strongly context-dependent, the auditory pattern will provide a relatively direct mapping of those features during preliminary analysis. The output of preliminary analysis is a rough matrix of phonetic segments and features which is then transferred to the control system. Thus, recognition of some features is thought to take place by relatively direct operations on the acoustic information output from peripheral analysis. However, when there is no invariant attribute to identify the phonetic feature, additional processing is required. An hypothesis concerning the representation of the utterance in terms of phonetic segments and features (i.e., an abstract feature matrix) is constructed. This representation then forms the input to a set of generative rules which produce candidate patterns which are compared against the original patterns. Results of this match are then sent to a control component. The phonetic description is transferred to higher stages of linguistic analysis. Analysis-by-synthesis is simply a more carefully specified version of the Motor Theory except that the invariance problem is presumably resolved at the neuroacoustic level rather than the neuromotor level. The Analysis-by-Synthesis Theory, like the Motor Theory, is quite abstract and there has been little direct experimental evidence to support this view.

Fant's Auditory Theory

Fant's theory is not terribly well developed. His major objection to the "active" motor-type theories is simply that the evidence used to support these theories is not conclusive (Fant, 1967). We agree with this statement. Fant claims that all the arguments brought forward in support of the motor theory would fit just as well into sensory-based theories where the decoding process proceeds without postulating the active mediation of speech motor centers. The basic idea in Fant's model is that the motor and sensory functions become more and more involved as we proceed from the peripheral to central stages of analysis. He assumes that the final destination is the concept of a "message" which comprises brain centers common to both perception and production. According to Fant, there are separate sensory (auditory) and motor (articulatory) branches, although he leaves the possibility of interaction between these two blocks open. Auditory input is first processed by the ear and subject to primary auditory analysis. These incoming auditory signals are then submitted to some kind of direct encoding into distinctive auditory features. Auditory features are then combined together in some as yet unspecified way to form phonemes, syllables, morphemes and words. Although much of Fant's concern has been on continued acoustical investigations of distinctive features his model obviously leaves much to be desired. It is at present much too gross to be tested and the problems of invariance and segmentation which we noted earlier still remain to be resolved.

Bondarko's Stage Theory

In line with a number of recent formulations, Bondarko et al., (1970) have proposed a model that has a series of hierarchically organized stages of perceptual analysis between the acoustic signal and intended message. The following stages are proposed: (a) auditory analysis, (b) phonetic analysis, (c) morphological analysis, (d) syntactic analysis and (e) semantic analysis. Although the model is quite general, it has several properties that are similar to our own views. We consider only the first two stages in their model. The first stage involves auditory analysis of the signal and provides a description of the stimulus in terms of "auditory features." Some type of spectral analysis is assumed to take place at this stage but the presence of possible auditory property detectors which may respond differentially to specific acoustic attributes is also considered. Thus, some type of low-level auditory feature analysis may take place relatively close to the periphery. The second stage involves phonetic analysis. The output of this stage is abstract and may be either an acoustic or articulatory representation of the speech event. This representation is thought to be based on phonetic segments or distinctive features. Provision is also made for the use of various types of normalization routines so that some of the variability from different talkers can be reduced for subsequent processes of recognition. The invariance problem can be resolved in this model by assuming that multiple decisions are made on the basis of outputs from a number of auditory feature detectors. Thus, this aspect of the model is designed not to focus on isolated auditory features (i.e., acoustic cues) but to examine multiple cues simultaneously and in parallel. Segmentation is accomplished at higher levels of analysis. The importance of this approach to speech perception should be emphasized. The model is based, in part, on a number of assumptions about the flow of information within the organism. Memory and coding processes are given strong emphasis and decisions are assumed to take place at many levels of perceptual analysis. The model, in its simplest form, incorporates many of the ideas that have proved fruitful in recent information processing approaches to perception and memory, to which we now turn.

Information Processing in Speech Perception

In recent years the study of speech perception has begun to adopt the aims and methods of information processing models that have been employed in the study of visual and auditory perception. The basic approach views perception as a hierarchically organized sequence of events involving structures for storage and processes for the transformation of information over time. During these stages information is transformed, reduced, elaborated, stored, recovered and used to make various types of decisions.

Three Assumptions

There are three basic assumptions in current information processing models. First, perception is not immediate but is the outcome of distinct operations distributed over time. One goal of information processing models is to attempt to specify in as much detail as possible the operations which occur from the onset of a stimulus to the response of an observer. A flow chart is typically used to represent the various stages of information processing. The second assumption is that there are "capacity limitations" at various stages of analysis during the processing of this information. Because the nervous system cannot maintain all aspects of sensory stimulation there are limits on the capacity to store and process sensory data which require that information be recoded into different forms. One goal of research in information processing is to identify where these capacity limitations occur and the nature of the recoding operations. Finally, the third assumption is that perception necessarily involves various types of memorial processes since recoding and retention of information must occur at all stages of information processing. The study of speech perception, therefore, necessarily entails the study of perceptual memory.

Stages of Processing

Several recent accounts of speech perception have begun to emphasize process and to divide this process into a hierarchy of stages: auditory, phonetic, phonological, syntactic, semantic, etc. (see for example; Studdert-Kennedy, 1974; 1975).

Figure 1 shows some of the processes which are assumed to take place between the initial acoustic signal and its final conceptual representation in the mind of the listener. According to this view, the speech signal undergoes a series of successive transformations whereby information is recoded into more and more abstract forms of representation. On one hand, the stages of processing are thought to be partially successive since spoken language is inherently a temporal phenomena. However, in real-time operation, decisions at various stages must also occur in parallel to permit information from higher more abstract levels of analysis to be employed in processes at lower levels. Figure 2 shows the type of approach we have in mind here. Phonological, syntactic and semantic decisions can all take place simultaneously in order to derive a phonological representation for a sentence. Rather than elaborate on this formulation, it seems sufficient to remark that it is consistent with the types of arguments we made earlier about the role of higher-level linguistic processes in perception. Wè will return to this again below.

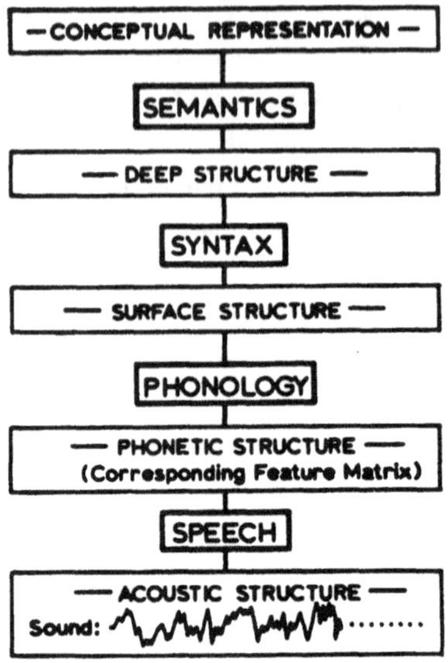

Fig. 1 Stages of Processing in Speech Perception. (Adapted from Liberman, 1970)

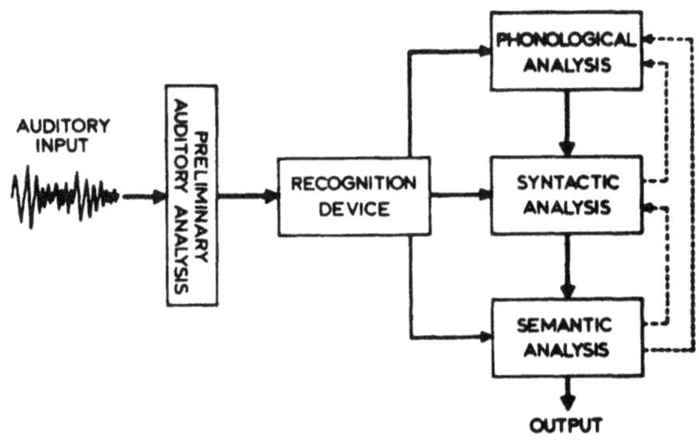

Fig. 2 Functional Organization of Speech Perception Processes.

An Information Processing Model

In this section we briefly sketch the structure of an information processing model, particularly the earliest stages of analysis that deal with recognition. Figure 3 shows a block diagram of the components. Auditory input enters the system

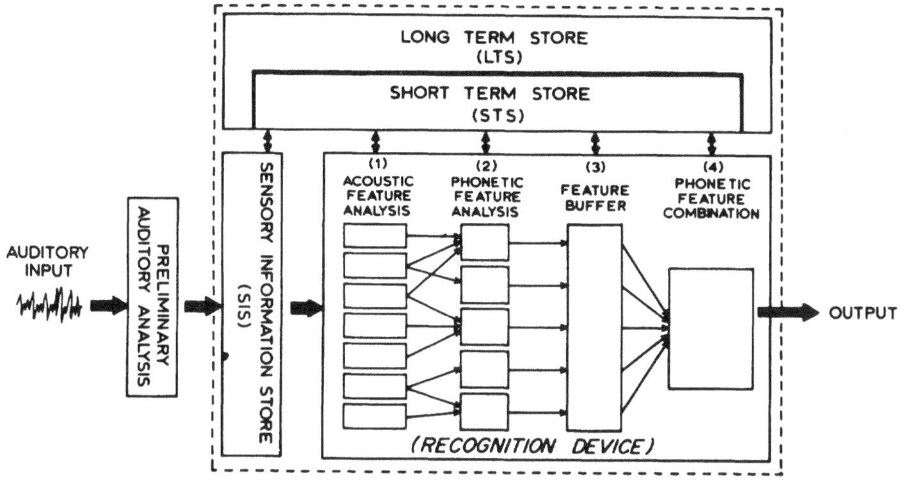

Fig. 3. Components and Organization of the Speech Recognition Process

and is processed in progressive stages. The output of Preliminary Auditory Analysis is assumed to be some type of spectral display in terms of frequency, time and intensity. Sensory input is processed automatically (i.e., not under the control of the subject) through several levels of analysis without the operation of conscious selective attention. Sensory information is maintained in a relatively gross unanalyzed form in the Sensory Information Store (SIS). Information is further processed by the "recognition device" which is shown as four distinct stages in Figure 3. Information from any or all of these stages of processing is placed in Short-term Store (STS) where the subject can selectively rehearse, encode, or make conscious decisions about it. Information in Long-term Store (LTS) is also assumed to be used in the recognition process. In keeping with recent formulations, STS may be thought of as simply that portion of LTS that is temporarily activated (see Shiffrin, 1975; Bjork, 1975).

Automatic processing by the recognition device is assumed to take place as follows. In Stage 1, Acoustic Feature Analysis, we suggest that auditory features of the speech signal are recognized by a system of individual auditory property detectors (Stevens, 1973). For example, consider the case of a simple CV syllable. We assume that there are detectors which will respond selectively to at least some of the following types of auditory information in the signal: (1) presence or absence of the fundamental frequency from the beginning of the stimulus; (2) abrupt rise in the frequency of the fundamental at the transition from consonant to steady-state vowel; (3) presence or absence of noise from the beginning of the syllable; (4) frequency of the spectral maximum of burst; (5) bandwidth of the noise; (6) duration and intensity of the noise; (7) presence or absence of a rapid change in the spectrum; (8) direction, extent and duration of a change in the spectrum.

In Stage 2, Phonetic Feature Analysis, we propose that a set of decision rules is used to map the multiple auditory features extracted from Stage 1 into phonetic distinctive features. These decision rules can be thought of as having knowledge of the articulatory constraints on speech production in as much as they affect the nature of the acoustic signal although it is not necessary for the model. We would also rather not assume that a "phonetic processor" exists as a distinct physiological mechanism. Instead, we would prefer to describe its function simply in terms of decision rules at the present time. We assume that decisions about features and segments within a syllable, for example, are made on the basis of information distributed across the whole syllable. As we remarked earlier, this does not commit us to arguing for some special status for the syllable.

These features are subsequently maintained in Stage 3, the Feature Buffer which is a storage mechanism to maintain decisions about the feature composition of a particular syllable. The reason for a feature buffer is two-fold. First, not all phonetic features are assumed to be processed (i.e., recognized) at the same rate. Secondly, some type of storage is needed to preserve feature information more or less independently for subsequent stages of phonological analysis. Feature information is then used in Stage 4, Phonetic Feature Combination, where individual features can be recombined to form discrete phonetic segments. Information about the feature specification of these phonetic segments is in the form of a rough distinctive feature matrix which is then made available to higher levels of processing for phonological and syntactic analysis.

One may ask at this point whether the approach we have advocated here has any relevance to the perception of connected speech in dynamic context. We think it does for several reasons. First, although much of the early research on speech perception used isolated CV syllables, these studies were aimed at specifying the minimal and sufficient cues for perception of a specific phonetic distinction. Thus, these cues may be thought of as representing the lower bound on the information contained in the waveform which is necessary for a decision assuming no higher level information was available. We would anticipate better or more accurate performance on a given task if higher-level grammatical information were available. Second, we feel that the characterization of these minimal acoustic cues that were found in isolation must be in part correct. Various synthesis by rule programs for generating connected speech have been written which formalize the use of these isolated cues in a systematic way and which produce fairly intelligible speech

(Mattingly, 1968; Kuhn, 1973).

To summarize briefly, low-level analyses are assumed to be relatively automatic processes of the perceptual system. Higher-level syntactic and semantic processes are assumed to operate in a heuristic way in perception. One implication of this approach is that the perceptual system makes available to higher stages of processing only a "crude" analysis of the input. We assume that there is considerable ambiguity at these early stages of analysis and only very tentative decisions are made about the phonological structure of the input. Before proceeding, several remarks should be made about the role of the memory system in information processing approaches.

Sensory, Short-term and Long-term Memory

A basic concept of information processing analyses is the notion of an iconic or preperceptual store (Sperling, 1960; Massaro, 1972). This is thought of as a very temporary storage medium which preserves all of the stimulus information in a literal or veridical form for several hundred milliseconds so that pattern recognition processes can operate on the input. During this time period, the sensory information is converted into more persistent and abstract memory codes for representation in short-term storage. Short-term store (STS) or "working memory" is thought to have a very limited capacity from which information is rapidly lost unless active control processes are operating such as rehearsal. The contents of STS may be thought of as a person's conscious memory. Long-term store (LTS) on the other hand, is assumed to be a permanent repository for information with a relatively unlimited capacity. However, the availability of information in LTS is constrained by various search and retrieval operations (Shiffrin & Atkinson, 1969). LTS receives information from STS. Rehearsal of information in STS is considered to be an "active" process which regenerates decaying memory traces and also causes information in STS to be entered into LTS. Recognition is assumed to be a process whereby the sensory input or some derived version of it "makes contact" with a stored representation in long-term memory or some type of representation that has been constructed or synthesized by rules in long-term memory. The information present in short-term memory is thought to consist of a combination of information from both the sensory input and information currently activated from long-term memory. Information that is in the sensory information store is not simply transferred to STS but is "recoded" while still being maintained at the earlier stage of processing. It is generally assumed that the earliest stages of the recognition process occur automatically and without conscious control by the subject. A good part of the information from the earliest stages of processing is lost by decay and only a relatively abstract representation of the input is maintained in STS. Although SIS, STS, and LTS are typically referred to as "structural" properties of the memory system, no claim is made as to whether they correspond to distinct neurological systems. Indeed, recent formulations simply represent different memory stores as different phases of activation of a single neurological system (see Shiffrin, 1975).

Coding Processes

In this section we briefly consider the nature of coding processes in speech perception. One of the problems with current information processing models is that they say very little about the exact form in which information is represented at various stages of processing or the types of recoding operations that are assumed to take place. Thus, for example, no claims are made with regard to the physiological reality of any coding scheme. For the most part, this is an arbitrary question. However, with regard to speech perception we think the situation is somewhat different. Speech sounds are not an arbitrary set of acoustic signals in the environment. That is, they are not symbols but rather signs and they form a class of natural categories in much the same way as other sign stimuli in the environment (Mattingly, 1972). It seems reasonable to suppose then that the non-arbitrariness of speech sounds derives in part from the production system and the way in which they are produced. Thus, some aspects of the coding processes in memory may ultimately be linked to articulation. Although this has been a topic of considerable debate, its resolution does not seem to be very close in hand. It is sufficient to note here that the properties of memory codes for speech sounds and the nature of recoding processes would appear to be central theoretical issues in speech perception. It seems to us that contemporary information processing models have failed to come to grips with these types of issues.

Higher Level Processing

In this section we consider some aspects of higher-level analyses in speech perception. Three stages of processing will be distinguished; phonological, syntactic and semantic. These are not conceived of as separate entities but rather as labels for the type of encoding and form of information from long-term memory that is used in processing. Together, these stages form an integrated system whose net product is comprehension of the acoustic input. However, before treating the integration of these three processes, they should be examined separately.

Phonological processing consists of integrating segmental and prosodic information from the recognition device. The result of this analysis is a tentative string of phonetic segments, including some word and morpheme boundaries. This tentative string is fed to the syntactic processor, along with prosodic information directly from the recognition device.

The purpose of the syntactic processor is to construct a preliminary parse tree of the input. Traditionally, this process has been conceived of as the application of a sequence of rules, the net result of which is a parse tree or parse graph. We propose that the theoretical syntactic processor follows real-time limitations on memory and processing capacity that are known to occur in humans (see Fodor, Bever & Garrett, 1974). One approach to this appears to be the augmented transition network (ATN) proposed recently by Woods (1970, 1973). The ATN is neutral with regard to the type and number of rules used in syntactic processing. The choice of rules can follow transformational grammar (Chomsky, 1965) or systemic grammar (Halliday, 1961; 1967) or any other system of rules. What is important is that the ATN makes specific use of memory in syntactic processing and provides a processor for applying syntactic rules, whatever their origin (see Kaplan, 1972).

The semantic processor works on information from the syntactic processor in conjunction with prosodic information from the recognition device. The semantic processor does not have to wait for the construction of a full parse tree since it is intimately involved with all aspects of the syntactic processing. We assume that the semantic processor draws upon knowledge of context, specific lexical meaning, and other forms of semantic information stored in long-term memory. This is not the place to propose a specific model of semantic information processing. For the present, we simply note that there are more than a few such theories available (Anderson & Bower, 1973; Bierwisch, 1971; Kintsch, 1974; Quillian, 1968; Rumelhart, Lindsay & Norman, 1972; Schank, 1972; Winograd, 1972). However, the psychological evidence distinguishing these theories is still inconclusive.

Up to this point, we have not mentioned lexical insertion. In this present framework, it is not necessary to conceive of a specific stage of lexical insertion. Instead, each of the three stages may look-up and insert correct lexical entries as necessary. Lexical search and insertion is a property of higher level analysis rather than a specific stage of processing. Another point about the organization of these three processes should be noted. These are not sequential stages, one following the other. They operate basically in parallel. Phonological analysis operates on smaller segments of the output from the recognition device than the syntactic or semantic processors. As a result, some aspects of its processing are faster than those of syntax and semantics. The net result is that, while information is processed in parallel, it gradually converges upon more abstract processors.

Two final points should be mentioned about higher level processing. The first of these concerns feedback from higher levels of processing to lower levels and the second concerns the existence of a central processor or "executive" to oversee the whole process. As shown in Figure 2, there are feedback routes from semantic and syntactic levels. However, there is no direct feedback to the recognition device. As we mentioned earlier, the recognition device is assumed to operate automatically and it provides only a crude analysis of the acoustic input. However, the higher levels of processing, including phonological, syntactic and semantic, must be capable of exchanging information and revising previous decisions.

In the present conception, there is no central executive that oversees the operations of the total system. Rather, each level operates on information as it becomes available. The semantic processor has the final responsibility of accepting or rejecting the results of previous analysis. Since the goal of processing an acoustic input is to recover the speaker's intended message, the speech perception process must ultimately be directed by a semantic system.

Concluding Remarks

We began this paper by considering a number of contemporary theoretical issues in speech perception. Our main purpose in reviewing these problems was to show how they place constraints on the possible class of theories in speech perception. We then described several of the prominent theories and attempted to point out some of their strengths and weaknesses. We tried to show that most of the current theoretical efforts have failed to consider seriously the role of higher-level linguistic processes in perception, particularly syntax and semantics and the influence of prosodic factors. In our view, it seems that most of these theories have tacitly endorsed the traditional taxonomic assumption that an autonomous level of phonological structure can be derived from an input without the contribution of higher-order linguistic information.

We also described some aspects of the information processing approach that underlies our own thinking about speech perception and sketched the type of model that we have in mind. One of the points that we emphasized was the role of coding processes in speech perception, especially the types of recoding operations and the properties of memory codes at different stages of analysis. In the final section, we briefly examined some aspects of higher-level analyses in speech perception and how these might form an integrated system. We tried to emphasize that the earliest stages of perceptual processing provide only a rough and tentative analysis of the acoustic input while larger units are constructed and subsequently made available for syntactic and semantic analysis.

As we mentioned earlier, one of our goals in this paper was to focus on general questions in speech perception rather than particular issues in narrowly defined experimental contexts. One reason for doing this was our feeling that work in speech perception might eventually have implications for deciding theoretical questions in linguistics, particularly phonetic and phonological issues (see Lieberman, 1970). Thus, for example, issues dealing with the possible inventory of universal phonetic features or various types of phonological processes might be motivated on perceptual grounds as opposed to strictly formal linguistic criteria (see for example, Bever and Langendoen, 1971). Indeed, several investigators have recently argued that phonetics and phonology should play a more predictive role in linguistic theory rather than its current interpretative function (Lindblom, 1971; Ohala, 1971).

However, these sorts of goals may never be reached until the theoretical work in speech perception becomes formalized within some unified framework. We hope that some of the issues we have considered in this paper will stimulate further theoretical discussion in the future.

Acknowledgments

Preparation of this paper was supported in part by USPHS grants MH-24027-02, NS-12179-01 and MH-11219-06 to Indiana University.

References

Anderson, J. R. & Bower, G. H. Human associative memory. Washington, D. C.: V. H. Winston, 1973.

Bever, T. G. & Langendoen, D. T. A dynamic model of the evolution of language. Linguistic Inquiry, 1971, 2, 433-463.

Bierwisch, M. On classifying semantic features. In D. D. Steinberg & L. A. Jakobovits (Eds.), Semantics. Cambridge, G. B.: Cambridge University Press, 1971.

Bjork, R. A. Short-term storage: the ordered output of a central processor. In F. Restle, R. M. Shiffrin, N. J. Castellan, H. Lindman, & D. B. Pisoni (Eds.) Cognitive theory: Volume I. Potomac, Maryland: Erlbaum Associates, 1975.

Bondarko, L. V., et al. A model of speech perception in humans. Working Papers in Linguistics No. 6. Computer & Information Science Research Center, Ohio State University, Columbus, Ohio. Technical Report 70-12, 1970.

Chomsky, N. Aspects of a theory of syntax. Cambridge, Mass.: MIT Press, 1965.

Chomsky, N. & Halle, M. Sound Pattern of English. New York: Harper & Row, 1968.

Cooper, W. E. Perceptuomotor adaptation to a speech feature. Perception & Psychophysics, 1974, 16, 2, 229-234.

Cooper, W. E. Selective adaptation to speech. In F. Restle, R. M. Shiffrin, N. J. Castellan, H. Lindman & D. B. Pisoni (Eds) Cognitive theory: Volume I. Potomac, Maryland: Erlbaum Associates, 1975.

Fant, G. Acoustic theory of speech production. The Hague: Mouton, 1960.

Fant, G. Auditory patterns of speech. In W. Wathen-Dunn (Ed.) Models for the perception of speech and visual form. Cambridge, Mass.: MIT Press, 1967.

Flanagan, J. L. Speech analysis, synthesis and perception. (Second Edition) New York: Academic Press, 1972.

Fodor, J. A., Bever, T. G. & Garrett, M. F. The psychology of language. New York: McGraw-Hill, 1974.

Halliday, M. A. K. Categories of the theory of grammar. Word, 1961, 17, 241-292.

Halliday, M. A. K. Notes on transitivity and theme in English. (Part I). Journal of Linguistics, 1967, 1, 37-81.

Kaplan, R. M. Augmented transition networks as psychological models of sentence comprehension. Artificial Intelligence, 1972, 3, 77-100.

Kintsch, W. The representation of meaning in memory. Hillsdale, N. J.: Erlbaum Associates, 1974.

Kuhn, G. M. A two-pass procedure for synthesis-by-rule. Journal of the Acoustical Society of America, 1973, 54, 339 (A).

Lea, W. A. An approach to syntactic recognition without phonemics. IEEE Transactions on Audio and Electroacoustics, 1973, AU-21, No. 3, 249-258.

Liberman, A. M. The grammars of speech and language. Cognitive Psychology, 1970, 1, 301-323.

Liberman, A. M., Cooper, F. S., Shankweiler, D. S. & Studdert-Kennedy, M. Perception of the speech code. Psychological Review, 1967, 74, 431-461.

Lieberman, P. Towards a unified phonetic theory. Linguistic Inquiry, 1970, 1, 3, 307-322.

Lindblom, B. E. F. Numerical models in the study of speech production and speech perception: some phonological implications. Papers from the Institute of Linguistics, University of Stockholm, October, 1971, Publication No. 7.

Massaro, D. W. Preperceptual images, processing time, and perceptual units in auditory perception. Psychological Review, 1972, 79, 2, 124-145.

Mattingly, I. G. Synthesis by rule of general American English. Supplement to Haskins Laboratories Status Report on Speech Research, April, 1968.

Mattingly, I. G. Speech cues and sign stimuli. American Scientist, 1972, 60, 3, 327-337.

Ohala, J. J. Physical models in phonology. Paper presented at the VII International Congress of Phonetic Sciences, Montreal, 1971.

Pisoni, D. B. Auditory and phonetic memory codes in the discrimination of consonants and vowels. Perception & Psychophysics, 1973, 13, 2, 253-260.

Quillian, M. R. Semantic memory. In M. Minsky (Ed.), Semantic information processing. Cambridge, Mass.: MIT Press, 1968.

Rumelhart, D. E., Lindsay, P. H. & Norman, D. A. A process model for long-term memory. In E. Tulving & W. Donaldson (Eds.), Organization of memory. New York: Academic Press, 1972.

Schank, R. C. Conceptual dependency: a theory of natural language understanding. Cognitive Psychology, 1972, 3, 552-631.

Shiffrin, R. M. Short-term store: the basis for a memory system. In F. Restle, R. M. Shiffrin, N. J. Castellan, H. Lindman, and D. B. Pisoni (Eds.), Cognitive theory: Volume 1. Potomac, Maryland: Erlbaum Associates, 1975.

Shiffrin, R. M. & Atkinson, R. C. Storage and retrieval processes in long-term memory. Psychological Review, 1969, 76, 179-193.

Shockey, L. and Reddy, R. Quantitative analysis of speech perception: results from transcription of connected speech from unfamiliar languages. Paper presented at the Speech Communication Seminar, Stockholm, Sweden, August 1-3, 1974.

Sperling, G. The information available in brief visual presentations. Psychological Monograph, 1960, 74, (11, Whole No. 498).

Stevens, K. N. Toward a model for speech recognition. Journal of the Acoustical Society of America. 1960, 32, 47-55.

Stevens, K. N. The quantal nature of speech: evidence from articulatory-acoustic data. In E. E. David, Jr. and P. B. Denes (Eds.), Human communication: A unified view. New York: McGraw-Hill, 1972.

Stevens, K. N. The potential role of property detectors in the perception of consonants. Paper presented at the Symposium on Auditory Analysis and Perception of Speech, Leningrad, USSR, August, 1973.

Stevens, K. N. & Halle, M. Remarks on analysis by synthesis and distinctive features. In Wathen-Dunn, W. (Ed.), Models for the perception of speech and visual form. Cambridge: MIT Press, 1967.

Stevens, K. N. & House, A. S. Speech perception. In J. Tobias (Ed.), Foundations of modern auditory theory: Volume II. New York: Academic Press, 1972, 1-62.

Studdert-Kennedy, M. The perception of speech. In T. A. Sebeok (Ed.), Current trends in linguistics, Vol. XII, The Hague: Mouton, 1974.

Studdert-Kennedy, M. Speech perception. In Lass, N. J. (Ed.), Contemporary issues in experimental phonetics, Springfield, Illinois: C. C. Thomas, 1975.

Svensson, S. G. Prosody and grammar in speech perception. Monographs from the Institute of Linguistics, University of Stockholm, 1974, No. 2, Institute of Linguistics, Stockholm, Sweden.

Woods, W. A. Transition network grammars for natural language analysis. Communications of the ACM, 1970, 13, 10, 591-606.

Woods, W. A. An experimental parsing system for transition network grammars. In R. Rustin (Ed.), Natural language processing. New York: Algorithmics Press, 1973.

Winograd, T. Understanding natural language. New York: Academic Press, 1972.

DISCUSSION of "Some stages of processing in speech perception"
PISONI

GILBERT:

It would seem to me that in the development of a theory of speech perception, which DAVID PISONI advocates, we have to take cognizance of recent animal studies on syllable discrimination, the fact that individuals with zero hearing have been observed to use verbal language and finally, that there is at least one literature reference to a child who learned language without having the motor ability to produce speech sounds. If individuals can learn language-speech, without either hearing or motor control, then obviously our definition of speech needs revision.

What is it that we perceive when we perceive speech?

Sven Öhman

Department of Linguistics
Uppsala University

There is a manner of reasoning in contemporary experimental work on speech perception that may be described in somewhat simplified form as follows.

The investigator constructs an acoustic signal by means of a speech synthesizer so as to produce the auditory impression of a piece of speech (usually a mono- or bisyllabic nonsense word). Certain physical parameters of the signal are allowed to vary in specified ways, and native listeners are asked to identify the signal as a (nonsense) word of their language. The typical mode of reasoning associated with this kind of experiment is implicit in a certain hope which might be formulated as follows: one is interested in determining just what sorts of acoustic patterns will cause the perception of a certain phoneme.

The crucial idea involved here seems to be the assumption that perception is an input/output mechanism in which the acoustic signal figures as the input and the phoneme as the output. The acoustic vibrations go in, and the phonemes come out.

Needless to say, this paradigm raises a number of very difficult questions of which it will be possible to discuss only a few here. The main question is perhaps this one: what sorts of things *are* these phonemes that are the 'outputs' of speech perception?

Three answers seem distinguishable in the literature. The oldest of these states that the output phoneme is an abstract entity whose characteristic properties are 1) its being in some sense maximal among the units that *do not systematically carry* any meaning, and 2) its being in some sense minimal among the units that *may systematically differentiate* between meanings.

I will not consider here the critical analysis of this notion that has recently been performed by several linguists. The reason is that although most linguists seem to posit phonemes in one form or another, they do *not* seem to think of these phonemes as being necessarily anything other than elements with which *representations* of utterances can be constructed (more or less naturally or adequately re. some purpose). Now, if someone produces an utterance, say a syllable, of my language, and if I happen to perceive it, then I refuse to believe that I have actually perceived a *representation* of the utterance rather than the utterance itself. I refuse to believe this for the same reason that I refuse to believe that in seeing e.g. Gunnar Fant I do not in fact see Gunnar Fant but merely a representation of him. Thus it seems to me that those who say that there are phoneme outputs in speech perception must mean something else by 'phoneme' than the linguists mean by that word.

When somebody utters a syllable and I perceive it, I definitely do not perceive something abstract or mental. Rather, I perceive a concrete event developing in the vocal apparatus of the speaker. The syllable that I perceive is this event, not its phonemic representation.

The phoneme theorists of speech perception will perhaps argue against this by pointing out that when I thus perceive the articulatory event, something specific must happen _in me_, e.g., certain neural circuits in my brain will be brought into certain states, and that _this_ is the output of the perceptual mechanism. I find this argument beside the point, however.

For one thing, it merely confuses matters to use the same term to refer _both_ to elements of representations of articulatory events in a speaker _and_ to certain neural events taking place in the brain of a listener. Secondly, there is nothing to show that the latter neural events, although perhaps adequately representable (re. some purpose) by phonemes, are at all well _described_ in terms of phonemes. Finally, and most importantly, the talk about neural brain events would be directly relevant to a theory of the neurophysiology of speech perception. But it has only peripheral relevance to a psychological theory of speech perception; i.e., we must discuss perception first of all in terms of the subjective experience of perception. Discussion in terms of the physiological events taking place in the perceiver _during_ perception will have to come later, in fact, _cannot_ come until later.

There is one more possible objection that should be answered. This objection states that, in rejecting that we perceive representations, I am forgetting that an act of perception may be characterized by several levels of abstraction, and that I am trying to chain speech perception to a single level only, and, moreover, to a level which is much too low. Listening to speech as speech is not the same thing as listening to it as just any old sound. The former mode of listening would be a more abstract, more inwardly directed, and grammatically-lexically dependent sort of listening, whereas the latter type would be more concrete, outgoing and empirically unbiased. Moreover, the former type, being more abstract, would be an abstraction _from_ the data given by the latter; i.e., the 'primary', 'sound percept' would be a 'substratum' or 'substance' carrying the 'secondary' linguistic percept.

This I find utterly difficult to believe. Consider by analogy any visible object, say my own car. I can see this car as one of the many vehicles regularly trafficking the streets of my city. Or, I can see it as one of the several causes of my mediocre economy. Is the former car a more concrete substratum for the latter? This question is of course ridiculous. Any object perceived is perceived in some aspectual context; i.e., any thing, A, can be perceived as an x, as a y, a z, ...etc. The x, y, z, ... define aspects under which A may be perceived, and there may be systematic relations between these aspects. But this does not entail that the thing I perceive when I perceive A now as an x, now as a y, and so on, varies in concreteness, or that one of these things is abstracted from the other. My car seen as a source of bad economy is just as concrete as my car seen as a city vehicle, and neither is abstracted from the other. They are in fact the same thing regarded as functioning in different contexts. Consequently, there is also no reason to believe that a certain vocalization when perceived as a syllable is more abstract etc., than when it is perceived as, say, a yawn.

Thus, shifts of aspectual context do not by themselves entail any differences in the level of perceptual abstraction. It might be proposed, however, that it should be possible to keep the aspectual context fixed, yet experimentally vary the richness of detail or structure perceived in an object. There could be more or less casual listening and more or less attentive listening. And listening to speech might be characterized by a peculiar combination of casualness and attentiveness within the framework of a fixed aspectual context.

Although there is some truth in this, it certainly does not help the abstract phoneme theory of perceptual outputs. For, the greater or lesser attentiveness in perception should be an attentiveness directed to something concrete in the external environment, e.g., to the shape of the front window of my car, or to the expression of somebody's face, or to the mouth position held by someone who articulates a funny consonant. These things do not seem to be the kinds of things thought of as outputs in the abstract phoneme theory of speech perception.

This leads me to a second type of answer to the question concerning the nature of the 'outputs' of speech perception. This answer states that the outputs are primarily not phonemes, but distinctive features of subjectively delimited sound segments. The phonemes are configurations or bundles of such distinctive features.

My reaction to this position depends on the interpretation given to the term 'distinctive feature'. If it is understood merely as an entity entering into the systematic construction of representations of utterances, the criticism must be analogous to that levelled against abstract phonemes. We do not perceive representations of things. If, again, the term is taken to refer to a pattern of some kind which is present in the input, my reaction is one of doubt.

To clarify this point let me draw up an analogy with vision. Suppose, for instance, that I perceive a bus that is driving down a street. Obviously, this bus does not have any distinctive features all by itself, but only relative to some fixed aspectual frame of reference. When my perception started to react to the signals emitted by the bus, but before I knew that what I saw was in fact a bus, my frame of reference may have been the class of objects likely to be expected in a street. Under this aspect the distinctive features of the bus are presumably partly a set of features that it has in common with other vehicles, and partly a set that differentiates it from the non-bus vehicles.

Several other frames of reference are possible, however. For instance, I may expect to see a bus or a taxi (knowing for instance that at this time of the day only busses and taxis are likely to appear). In this situation, the distinctive features of the bus may include a set of features differentiating different bus lines.

My problem here is this: isn't it possible to perceive the bus without identifying it as 'something particular', as a this or a that? And isn't it so that in perceiving the object we alternate between a number of aspectual frames, relatively randomly chosen, some more general and others more particular? For each choice, various features of the object are noted, some distinctive and others non-distinctive. At any moment in this process the features previously noted are associated with a percept which increases with time in its richness of structure. The distinctive features of the object under _any one_ of the possible aspectual contexts would then merely represent a selection from among the features available. I cannot see, therefore, how it can be certainly true or even meaningful to say that perceiving the object consists in perceiving its distinctive features, _simpliciter_. Certainly, in perceiving the object, I may note or search for its distinctive features re. some fixed frame of reference. But this seems to presuppose, rather than _be_, perception of the thing _per se._

There is a third answer to the question concerning the nature of the perceptual outputs which will be the last one to be discussed here. This answer states that neither abstract phonemes, nor distinctive features are perceived, but rather the sound itself.

On the best version of this theory, there is associated with every word of a language a specific musical pattern, viz. _the sound_ of the word. Knowing the language means among other things knowing how each word of the language should sound. In speaking a word the speaker, by moving his vocal organs appropriately, sets the air into vibration. And these vibrations actualize in the listener the sound of the word in question. It is an empirical fact, it is said, that the sound of the words - whether actualized or not - can be considered as objects *sui generis*; they can be compared, decomposed, and analyzed in various ways. The phonemes are systematically established components of the word sounds. And it is these components that are the outputs of perception.

What I admire most in this theory is the undeniability of the facts that it argues from. The notion of 'the sound of a word' seems immediately comprehensible, and it can hardly be doubted that it refers to something real. But *what* is it that it refers to? Is it really something perceptual?

To illustrate the reasons for my doubts in this matter, let me consider the following sort of situation: From where I am standing, let us suppose, I can hear (but not see) her moving around in the kitchen. Now I hear her opening the refrigerator, bringing the whisky bottle out, etc.

In the way I am listening, I listen to these events. I *do not* listen to *the sounds* of the events. I *could* listen to the sounds of the events, however, if I wanted to. I would then listen to them as a sort of concrete music, disregarding their physical meaning. This latter sort of listening is what I take to be meant by 'auditory sensation'. It consists in an immediate awareness of the developing states of my auditory sense. As such it is a form of perception, viz. perception of the states of my own body.

Thus, there are two kinds of perception, each of which consists in an immediate awareness of something. Ordinary perception is an immediate awareness of the developing states of an external physical system (e.g., a woman moving around in a kitchen). Perceiving in this way, my mind 'constructs' the external system-sate configurations.

Sensation, on the other hand, is an immediate awareness of the developing states of the physical system that I myself am. Perceiving in this way, my mind 'constructs' internal system-state configurations.

I will call the former type of perception _objective_ and the latter _subjective_. Apparently, the objective percept, i.e. the external system-state configuration constructed by the perceiver, is constructed on the basis of the subjective percept _plus_ previous knowledge of the world at large. And the subjective percept is avoided, sought for, etc. in view of properties of the objective percept and previous experience. There seems to be an intricate sort of oscillation between the two.

In a manner of speaking, objective perception is a process by which the subjective percept is 'projected' out into a posited physical system. This projection can be more or less successful, i.e., one may succeed in projecting only a certain part of the subjective percept, thus leaving the remaining part uninterpreted. Most of us are familiar with this from the experience of hearing a sound the objective percept for which does not develop until long after the subjective percept was felt. Conversely, we sometimes seem to be able to disconnect our objective perception from the subjective phase, and concentrate on the latter, as in listening to music.

The question therefore arises, whether the phonetic percept in speech perception is mainly subjective or mainly objective. What seems to argue against the former proposal is the high degree of coarticulation and reduction commonly observed in speech. It is difficult to understand how the notoriously variable speech signal can be 'decoded' except by projection into a highly constrained physical system, viz. the vocal apparatus of the speaker as regulated by the phonological laws of the language.

On the other hand, the projection process would be time consuming and psychically laborious. It might therefore be suspected that the phonological pattern of a language is always so constructed that speech can be correctly perceived without much reference to the objective system that generates it.

If the latter proposal is correct, the task of speech perception research should be to determine the acoustic invariants carried by the speech signals and to discover how subjective perception goes about extracting them.

If, on the other hand, the former and, in my opinion, much more likely proposal is the correct one, then speech perception research should strive to determine the nature of the listener's subjective model of the speech <u>production</u> system. In particular, it should be investigated what physical states of the production system the listener considers to be its possible states.

A second problem for the objective percept theory, is that of discovering the dynamics of the perceptual process. One wants to know how the subjective percept is elicited by the sound, and how the objective percept is constructed from this data. The listener's model of the production system and the laws regulating its operation must obviously enter as an integral part into this process. The phonetic percept could then be described as the state sequence attributed by the perceiver to his model of the speaker's production system. Moreover, under this theory, our model of speech perception should explain how the acoustic signal evokes a subjective percept and by what mechanisms this data is used to assign a specific state sequence to the production model.

An advantage that this approach seems to have is the possibility of describing the perceiver's subjective notion of the production system (as well as his phonetic percepts) <u>in objective physical terms</u>. The perceptual model would partly take the form of a production model associated with a set of physical states so defined as to reflect the listener's way of perceiving speech. These states should be definable in terms of the physical concepts (such as 'source location', 'filter shape'. etc.) with which the acoustic theory of speech production is currently formulated.

This is not enough to explain the listener's *linguistic identification* of the signals, however; i.e., it does not explain how the listener succeeds in perceiving the phonetic event *as* some element of his language. To this end, a further stage seems to be needed: in particular, a stage in which certain crucial or distinctive states (or approximations to such states) are detected in the objective percept.

It is not obvious that the linguistic identification of the signal is tantamount to a determination of its full phonological structure, whether phonemic or feature, however. Phonological identification is clearly necessary for someone who is learning to write and spell properly. But in normal speech perception the all important thing is to be able quickly to perform the relevant cognitive-semantic operations that call forth the *Vorstellungen* intended by the speaker. We do not know much about these operations, nor about the size and nature of the phonetic expression units that call them up. In view of the contrast between the difficulty one has in learning to use even an essentially phonemic alphabet for writing purpose, on the one hand, and the ease with which the analphabet uses spoken language, on the other hand, it seems quite likely that the expression units, which are directly tied to the cognitive-semantic operations, are phonetic Gestalts much larger in size than phonemic segments. Hence, the phoneme output assumption for speech perception needs independent proof; i.e., it is not clear *a priori* that phonemes figure in speech perception at all.

We have proposed that the listener interprets his sensory data in terms of a subjective model of the production systems. How does this model relate to the listener's own actual production system?

Given that the sound of a word is defined subjectively as a temporally distributed complex of distinctive states of a private production model, it seems rather reasonable to assume that one is able to regard one's own actual articulatory apparatus as a case to which the model is applicable; to produce a word is then a matter of bringing one's speech organs through the desired state sequence, an act which requires the application of definite (privately well-known) physiological operations.

Does this mean that speech perception involves simulating the input signal through covert operations on one's own actual production system? No, of course not. Perceiving a state sequence in an external system is not the same thing as perceiving the operations, if any, that that system applies to itself in order to generate the state sequences in question. In fact, in most cases of objective perception it is not even meaningful to talk about such 'self-operations'. What seems generally to be the case, however, is that objective perception is the perception of physical system-state histories. It is hard to see what necessitates the proposition that speech perception should be an exception to this rule. In any case, this rule seems to be entirely compatible with the notion that one's private model of the human speech production system serves as a common psychological basis for both the perception and the production of speech.

In summary, then, a simple-minded input/output model of speech perception, assuming phonemic output, is beset with conceptual difficulties. A more constructive theory allowing for a mutual interaction betweeen the subjective (sensory) and the objective (physical) percepts through the activation of a private production model, would seem to be more in consonance with our immediate experience with the process in question. Also, the assumption that the phonetic percept is identified linguistically as a linear phoneme string in normal speech perception should be questioned.

References: In writing down the above thoughts, I was inspired by J. Gibson's views as expressed in his 'The Senses Considered as Perceptual Systems' (Houghton Mifflin, Boston 1966). I also had many statements by A.M. Liberman, K.N. Stevens and M. Halle in mind. For a review of recent work on speech perception by these and other authors, see M. Studdert-Kennedy, Speech Perception, to appear in 'Contemporary Issues in Experimental Phonetics', ed. N.J. Lass (Springfield, Ill.: CC. Thomas).

II. Prosody in Speech Perception

HOW A FULL ACCOUNT OF SEGMENTAL PERCEPTION DEPENDS ON PROSODY AND VICE VERSA

Quentin Summerfield

ABSTRACT

The synthesis time-base of a synthetic precursor phrase is one factor determining the position of the phoneme boundary on a continuum of synthetic CV target syllables, which vary in the durational cue Voice Onset Times (VOT) and are introduced by the precursor. Reducing the time-base, thereby increasing the rate of speech in the precursor, increases the probability of the consonant in the target being perceived as voiceless. The results of such perceptual experiments are compared to those of a production study in which six adult male speakers of British English produced examples of CV syllables composed of the consonants /b,p,g,k/ and the vowels /i,a/ in a sentence frame. VOTs in productions of /p/ and /k/ were 20 milliseconds shorter at fast than at slow rates of speech, warranting a normalisation of VOT duration in perception like that actually obtained.

VOTs were longer in /ki/ than in /ka/. It has been suggested that voicing onset is retarded before high vowels because the oral constriction is reduced more slowly than before lower vowels, thus delaying the attainment of the transglottal pressure drop required for voicing. However, VOTs in /pi/ were shorter than those in /pa/. In the case of voiceless bilabials the aerodynamic factor appears to be outweighed by a mechanical influence of anticipatory co-articulation of tongue position which leads to larynx elevation and less vocal cord abduction before high vowels. The production result for velars, but not that for bilabials, is paralleled in perception where longer values of VOT are required for stops to be perceived as voiceless before /i/ compared to /a/ at all places of production. This failure of the perceptual process to follow exactly the constraints in production provides an illustration of a heuristic rather than an algorithmic perceptual strategy presumably designed to allow fast decisions while tolerating some loss of accuracy in exceptional cases.

The 'precursor target' paradigm used in these experiments could be extended to examine prosodic influences on other segmental distinctions, but also, to determine the perceptual substrates of prosodic variables such as "rate of speech" and to measure the precision of perceptual expectations for vowel and consonant durations in different sentential and syntactic environments.

INTRODUCTION

In this paper I shall compare some production data with some perceptual data which, taken together, expose a limit on the precision of the articulatory model which the perceptual process utilises when processing running speech and from which inferences can be made about the 'heuristic' processing of a particular acoustic feature. In the perceptual experiments a synthetic target syllable drawn from an acoustic continuum is introduced by a synthetic precursor phrase on which constrained acoustic variations are imposed. In general, if variation of a particular acoustic aspect of a precursor influences the decision as to the value of a phonetic feature in a target, then we may impute to that aspect a contextual influence on the processing of the feature. I believe that this technique forms a useful complement in our armoury of experimental paradigms to the now popular battering ram of perceptual adaptation. Moreover, by stepping smartly to one side, we may turn the weapon upon itself. If a phonetic feature decision in a target syllable can be shown to be influenced by only one type of information deriving from a precursor phrase, then we may use the magnitude of such effects to determine how that information is extracted from the precursor.

DATA

Summerfield (1974) determined the positions of phoneme boundaries on the dimensions /ga-ka/ and /gi-ki/ on which syllables were distinguished by variation in Voice Onset Time (VOT). The two most interesting conditions were differentiated by the synthesis time-base of the three syllable precursor phrase which introduced the target monosyllables. The phrase, "Why are you...?", was synthesised with durations of 528 and 704 msec. in the two conditions, a ratio difference in overall duration of 1:1.33. Phoneme boundaries were determined by the adaptive algorithm PEST (Taylor and Creelman, 1967) and two major results emerged:

(i) The mean phoneme boundary on the dimension /gi-ki/ occurred at a VOT 6.6 msec. greater than that on the dimension /ga-ka/.

(ii) Averaging over both dimensions, the mean phoneme boundary position was 5.4 msec. greater when syllables were introduced by the longer, slower precursor compared to the faster, shorter precursor.

Result (i) was interpreted as reflecting perceptual recalibrations based on perceived vowel identity designed to take account of either (a) the slower rate at which the oral constriction is released in /ki/ compared to /ka/ which results in a slower oral pressure drop and consequently later voicing onset before the high front vowel; or, (b) the greater burst amplitude and duration which generally accompany productions of /ki/ compared to /ka/, in whose absence in these synthetic target syllables, a longer VOT was required to cue voicelessness. Lisker (1974) has since shown that a transitionless first formant increasingly cues voicedness as its frequency is lowered. Whether a steady F1 onset frequency is an additional cue to a voiced F1 transition remains to be determined. Lisker suggests that a low F1 at voicing onset cues voicedness directly through its implication of a closed vocal tract, providing a more parsimonious rationale for certain differences (Summerfield and Haggard, 1974) between various vocalic contexts than our own, being based on an acoustic cue rather than on recalibrations resulting from feedback from at least a parametric representation of the vowel.

Result (ii) was interpreted as a potentially appropriate normalisation to variations in the mean VOTs of (in particular) voiceless stops that would accompany changes in rate of speech in production.

Regardless of how the perceptual adjustments implied by such phoneme boundary shifts are achieved, it is appropriate to determine the extent to which they reflect changes in the mean VOTs of voiced and voiceless stops which accompany variations in contextual factors such as vocalic context and rate of speech in production. Summerfield (1975b) persuaded six adult male speakers of English to produce repetitions of the sentence:-

"Why are you a C_1V_1 when you're a C_2V_2?"

C_1 was always the same as C_2 and was one of /b/, /p/, /g/ and /k/.
V_1 always differed from V_2; in each sentence one vowel was /i/ and the other was /a/. Each speaker produced fifteen repetitions of each utterance at each of three rates of speech, nominally identified as "Fast", "Normal" and "Slow". Four measurements were made from oscillograms of each production:-

M1: the time from the start of the utterance to the beginning of the release of C_1.

M2: the VOT of C_1.

M3: the time from the beginning of the release of C_1 to the beginning of the release of C_2.

M4: the VOT of C_2.

We shall concentrate on the data from the productions of voiceless stops. The average values of M1, M2, M3 and M4 for each of /pi/, /pa/, /ki/ and /ka/ are tabulated in Table 1. In Figure 1a values of M2 are plotted against M1 and in Figure 1b values of M4 are plotted against M3. Thus in each figure, average VOT is plotted against the average duration of the four immediately preceding syllables. Table 1 shows that stops in sentence-final syllables were produced with slightly longer VOTs than were stops in sentence-medial syllables. Apart from this, Figures 1a and 1b show essentially the same pattern and illustrate three results:-

(i) As we should expect from Lisker and Abramson's definitive (1967, 1970) spectrographic measurements, /k/s are produced with longer VOTs at each rate of speech and in each vocalic context than /p/s.

(ii) VOTs produced in voiceless stops increase as the rate of speech decreases. This is not surprising, although I do not think that it has been shown before. The pooled data disguise the fact that one of the six speakers did not display this pattern in his bilabial productions which showed no variation in VOT, but the general pattern was displayed in his velar productions and by all the other speakers in both their bilabial and their velar productions.

(iii) The single surprising result is the interchange of absolute VOTs between vowel and place contexts at all rates of speech. For these six speakers, at least, /ki/s were generally produced with longer VOTs than /ka/s while, paradoxically, /pa/s were produced with longer VOTs than /pi/s. An explanation for this result may be found by considering the conditions which must apply for normal voicing to occur. These are that the vocal cords should be adducted and a transglottal

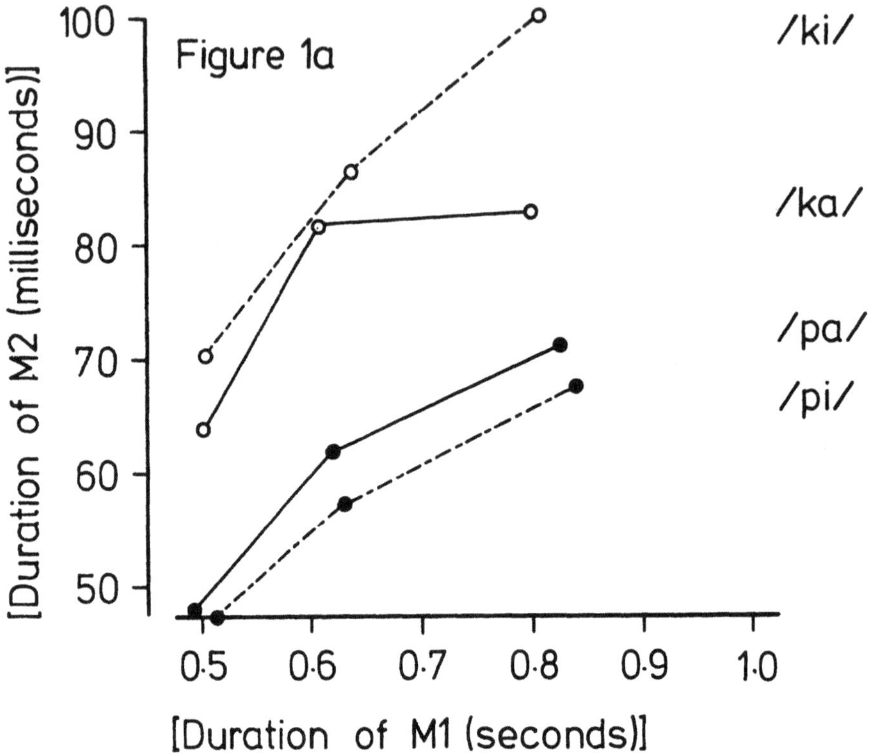

Figure 1a: Plots of data from sentence-medial productions (C_1V_1) of the syllables /pa/, /pi/, /ka/ and /ki/ from the sentence:

"Why are you a C_1V_1 when you're a C_2V_2?"

VOT in C_1V_1 (M2: milliseconds) is plotted against the total duration of the immediately preceding four syllables (M1: seconds) for each syllable at each of three rates of speech. Each point plots the mean of 90 data (15 repetitions by each of 6 speakers). The means are tabulated in Table 1.

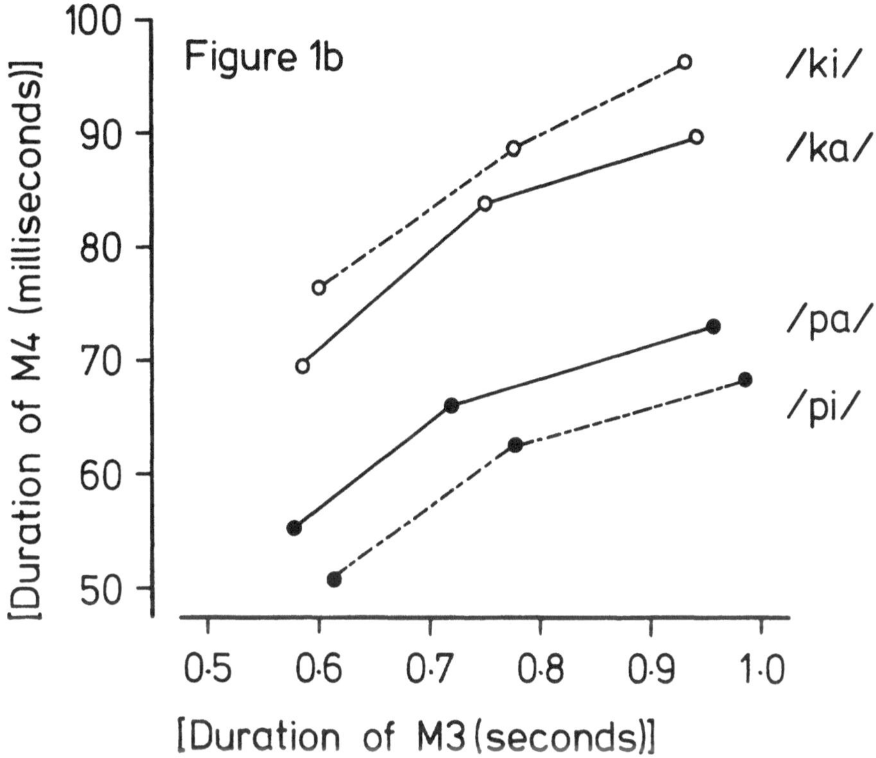

Figure 1b: Plots of data from sentence-final productions (C_2V_2) of the syllables /pa/, /pi/, /ka/ and /ki/ from the sentence:

"Why are you a C_1V_1 when you're a C_2V_2?"

VOT in C_2V_2 (M4: milliseconds) is plotted against the total duration of the immediately preceding four syllables (M3: seconds) for each syllable at each of three rates of speech. Each point plots the mean of 90 data (15 repetitions by each of 6 speakers). The means are tabulated in Table 1.

Table 1 Values in milliseconds of the measures M1, M2, M3 and M4 for each syllable averaged over six speakers.

M1 and M2 refer to sentence-medial productions (C_1V_1). M3 and M4 refer to sentence-final productions (C_2V_2). M1 and M3 are the total durations of the four syllables preceding C_1V_1 and C_2V_2. M2 and M4 are the VOTs of C_1V_1 and C_2V_2.

Sentence Medial Productions (C_1V_1)

RATE of Speech		/pa/	/pi/	/ka/	/ki/
FAST	M1	493	513	503	503
	M2	48.1	47.4	63.8	70.3
MEDIUM	M1	620	630	611	636
	M2	61.8	57.2	81.9	86.4
SLOW	M1	824	841	800	808
	M2	71.2	67.8	83.0	100.2

Sentence Final Productions (C_2V_2)

RATE of Speech		/pa/	/pi/	/ka/	/ki/
FAST	M3	578	614	583	600
	M4	55.3	50.8	69.6	76.4
MEDIUM	M3	718	777	750	777
	M4	66.1	62.8	84.1	88.0
SLOW	M3	955	984	939	931
	M4	73.1	68.4	89.7	96.4

pressure drop should exist. Thus, three factors will influence the timing of voicing onset relative to oral closure release: (i) most importantly, the relative timing of the commands for adduction relative to release; (ii) the physical separation of the vocal cords at the moment when adduction is initiated; and, (iii) the rate at which the oral pressure falls to establish the transglottal pressure drop. Klatt (1973) suggests that factor (iii) underlies the occurrence of longer VOTs in /ki/ compared to /ka/, since, before /i/ the tongue movement at release is almost tangential to the point of closure, while before /a/, it is almost perpendicular. It may also be an important determinant of the development of the VOT differences between the three places of production although these are presumably maintained actively in adults through factor (i). Kim (1970) has shown that factor (ii) accompanies the variations in VOT in the three categories of stops in Korean, at least in his own speech. It is this factor which is responsible for shorter VOTs in /pi/s compared to /pa/s despite aerodynamic considerations favouring the reversed pattern. With bilabials, in running speech, the tongue position for the vowel can be anticipated during the stop closure. With velars, much less anticipatory coarticulation can occur because the tongue body is involved in the closure gesture itself. Summerfield (1975b) argued that the high front position of the tongue for /i/ raises the hyoid which results in increased vocal cord tension. The same effect here results in less complete glottal opening and faster adduction in /pi/ compared to /pa/. In consequence, the aerodynamic factors which delay voicing onset in /ki/ are outweighed by mechanical factors which advance voicing onset in /pi/ relative to /pa/.

While this phenomenon and its explanation are of some interest, per se, the effect is small. It is important, however, because such a mechanical influence is only likely to appear in the production of stressed /p/ before a high vowel, while the aerodynamic factors should dominate for the other places of production and for all stops, including /p/, in consonant clusters. It is precisely the sort of 'exceptional' situation which a heuristic perceptual procedure would not allow for. If, as Lisker has suggested, one of the influences of a low frequency F1 at the onset of voicing is to cue voicedness through the implication of a more closed vocal tract, then we may test whether this cue is processed algorithmically by examining its effect in different vocalic contexts at each of the three places of production. If the closure cue is processed algorithmically then we should expect phoneme boundaries to occur at shorter VOTs on the continuum /bi-pi/ compared to /ba-pa/. If, on the other hand, the cue is processed heuristically, as seems likely, then we should expect the reverse pattern so that the boundary will always occur at a longer VOT in /i/ compared to /a/ context, just as both the present production data and the perceptual data of Summerfield (1974) find for velars.

Accordingly, Summerfield (1975a) determined phoneme boundaries on the nine synthetic VOT continua formed by the combination of three places of production and the vocalic contexts /a/, /ɜ/ and /i/ with steady-state F1 frequencies of 664, 504 and 277 Hz respectively. To demonstrate the generality of the syllabic rate influence on voicing perception, target stimuli were again presented under two conditions differentiated by the duration of the steady-state portions of precursor phrases. Two precursors were used. The longer, slower precursor had an overall duration of 1832 msec. and the shorter, faster precursor an overall duration of 1464 msec. The ratio of durations was, therefore, slightly smaller than in the previous experiment, being 1:1.26. In Table 2, the average phoneme boundaries of the sixteen subjects and the average shifts in the positions of these boundaries

Table 2 Average phoneme boundary positions and average phoneme boundary shifts between two rates of speech on nine VOT continua in milliseconds.

CONTINUUM	AVERAGE PHONEME BOUNDARY POSITION	AVERAGE PHONEME BOUNDARY SHIFT	DIFFERENCE BETWEEN /i/ AND /a/ BOUNDARIES	No. OF SUBJECTS
/ba-pa/	15.5	2.53		16
/bɜ-pɜ/	22.8	2.33	6.4	16
/bi-pi/	21.9	4.31		16
/da-ta/	17.1	1.90		16
/dɜ-tɜ/	23.3	1.94	9.2	16
/di-ti/	28.3	3.00		13
/ga-ka/	25.3	1.45		16
/gɜ-kɜ/	29.8	1.89	8.3	16
/gi-ki/	33.6	4.45		13

are tabulated in milliseconds of VOT.

We may note firstly that for every place of production the phoneme boundary in /i/ context occurred at a longer VOT than that in /a/ context. Table 2 tabulates the differences between /a/ and /i/ contexts which averaged 8.0 msec. and are statistically equivalent to one another. This suggests that the F1 closure cue is assessed heuristically in the sense that its interpretation is not influenced by higher order decisions concerning vowel identity as Summerfield (1974) had presumed. Secondly, Table 2 shows that in each place/vowel context, the effect of increased syllabic rate in the precursor is to cause a greater number of voiceless percepts; i.e., the phoneme boundary shifts, appropriately, toward lower VOT values.

While the 'heuristic/algorithmics' distinction is of some interest, for the purposes of the present symposium the demonstration of a prosodic influence on the perceptual interpretation of a segmental feature which parallels an equivalent variation in production, is more important. It is a moot point whether the size of the boundary shifts found in these perceptual experiments would be large enough to normalise differences in VOT occurring in production. In Figures 1a and 1b the ratio of the durations of the four syllables immediately preceding the CV syllables between the "Fast" and "Slow" speaking rates is 1:1.62. The difference in average voiceless VOT was about 20 msec. If variation in rate of speech does not significantly affect the duration of VOTs produced in voiced stops, which was the case here, in so far as it could be assessed, then we should expect a shift in the boundary position of half the shift in mean voiceless VOT; i.e., a shift of about 10 msec. in phoneme boundary should accompany a change in on-going syllabic rate of 62%. We have already noted that Summerfield (1974) found that an average shift of 5.5 msec. resulted from a syllabic rate variation of 33%, which is of the right size, therefore. However, the sizes of the boundary shifts tabulated in Table 2, which resulted from a syllabic rate variation of 26%, are rather small. I do not think that any hard and fast inference can be drawn from these comparisons. However, it would clearly be interesting to couple synthetic targets to real speech precursors spoken at different rates where I should expect the production/perception correlation to be more precise.

DISCUSSION

The precursor/target paradigm can obviously be extended and, I predict, would demonstrate speech rate influences on most other feature distinctions subsumed by durational cues. It would be of greater interest to determine whether the value of a particular duration cue in a precursor could act as the dominant criterion-setting influence for the same cue in the target. However, given that phoneme boundary shifts on the VOT dimension can indicate changes in the overall perceived rate of speech, the paradigm can be turned upon its head and then exploited in three ways.

Firstly, the paradigm could be used to determine the perceptual substrates of what I have rather vaguely been calling "syllabic rate" or "rate of speech". One component may indeed be the rate of syllables over time. Another could be the rate of fundamental frequency changes over time, though we have yet to test this notion. Before doing so, it is necessary to establish whether other contextual variables carried in precursor phrases, apart from syllabic rate, can influence the perception of voicing in the target. So far, I have only examined one such possibility. Summerfield (1975a) manipulated the vocal tract lengths implied by precursor phrases and target syllables in addition to the syllabic rate of the precursors.

I found that neither the position of the phoneme boundary on the dimension /bi-pi/, nor the size of the shift in its position which resulted from variation in precursor syllabic rate, were influenced either by variation in the vocal tract length implied by the precursor or by the vocal tract length relationship between the precursor and the target. The result implies either that the voicing cues in the target syllable are not interpreted in relation to the current perceptual vocal tract length reference, or that the target syllable itself supplies that reference. In either case, it seems that syllable timing information in the precursor does influence voicing perception in the target while other articulatory variables do not. The result encourages the use of the paradigm in investigating the perception of the parameters of 'syllabic rate' since it suggests that only these parameters influence the perception of target voicing.

Secondly, a parametric study which titrated the duration of the portion of the precursor abbutting the target on which duration variations were imposed, could determine the "time-window" over which syllabic rate is assessed. Summerfield (1975a) carried out a pilot study directed toward this end. It appears that the effective time window is less than one second in duration since synthesising the first half of a 2 second precursor at one rate (a) and the second half at another rate (b) produced exactly the same effect on the perception of voicing in the target as did synthesising both halves at the second rate (b).

Thirdly, and finally, the paradigm could be used to determine what perceptual expectations apply to syllabic durations in different sentential and syntactic environments. In the production data reviewed here, it was found that some speakers produced consistently longer VOTs in sentence-final syllables compared to sentence-medial syllabes. If the same precision of perceptual expectation applies here as in the situations reported by Klatt and Cooper (1975) then I should expect phoneme boundaries to occur at longer VOTs in sentence-final compared to sentence medial stop-vowel syllables. Supporting this hypothesis, Summerfield (1971) found that increasing the duration of the vowel increased the probability of the stop which preceded it being perceived as voiced. In natural speech vowel steady-states are the main site of pre-pausal lengthening (Lindblom, 1968; Oller, 1973). In the same way, I should expect even longer vowel durations to be required to cue voicedness post-vocalically in sentence-final syllables than in sentence-initial and sentence-medial syllables.

In all these situations, considerable interest lies in determining the precision of the perceptual expectation for variations in duration, which, given the demonstrably low variance of timing variation in production (Kozhevnikov and Chistovich, 1965; Lackner and Levine, 1975), can be expected to be high. Nevertheless, as with the effects of vocalic context reported here, the nature of perceptual processing may be powerfully exposed by a few exceptional cases, where perceptual effects and production constraints do not coincide.

REFERENCES

KIM, C.W. (1970). A theory of aspiration. Phonetica, 21, 107-116.

KLATT, D.H. (1973). Voice onset time, frication and aspiration in word-initial consonant clusters. M.I.T. Research Laboratory of Electronics, Quarterly Progress Report No. 109, 124-136.

KLATT, D.H. and COOPER, W.E. (1975). Perception of vowel duration in sentence contexts. Paper presented to the 89th meeting of the Acoustical Society of America, Austin, Texas.

KOZHEVNIKOV, V.A. and CHISTOVICH, L.A. Speech Articulation and Perception. JPRS 30, 543. Washington: U.S. Department of Commerce, 1965.

LACKNER, J.R. and LEVINE, B.K. (1975). Speech production: evidence for syntactically and phonologically determined units. Perception and Psychophysics, 17, 107-113.

LINDBLOM, B. (1968). Temporal organisation of syllable production. Speech Transmission Laboratory Quarterly Progress Report, 2/3, 1-5. (Royal Institute of Technology, Stockholm.)

LISKER, L. (1974). Is it VOT or a first formant transition detector? Paper presented to the autumn meeting of the American Association of Phonetic Sciences.

LISKER, L. and ABRAMSON, A.S. (1967). Some effects of context on Voice Onset Time in English stops. Language and Speech, 10, 1-28.

LISKER, L. and ABRAMSON, A.S. (1970). The voicing dimension: some experiments in comparative phonetics. Proceedings of the 6th International Congress of Phonetic Sciences. Prague, 1967. Prague: Academia.

OLLER, D.K. (1973). The effect of position in utterance of speech segment duration in English. Journal of the Acoustical Society of America, 54, 1217-1247.

PETERSON, G.E. and BARNEY, H.L. (1952). Control methods used in a study of the vowels. Journal of the Acoustical Society of America, 24, 175-184.

SUMMERFIELD, A.Q. (1971). Some tests of an information-flow model of speech perception. Unpublished Bachelor's Thesis, University of Cambridge.

SUMMERFIELD, A.Q. (1974). Towards a detailed model for the perception of voicing constrasts Speech Perception No. 3, 1-26. (Progress Report, Department of Psychology, The Queen's University of Belfast.)

SUMMERFIELD, A.Q. (1975a). Cues, contexts and complications in the perception of voicing contrasts. Speech Perception No. 4, (in press). (Progress Report, Department of Psychology, The Queen's University of Belfast.)

SUMMERFIELD, A.Q. (1975b). Aerodynamics versus mechanics in the control of voicing onset in consonant-vowel syllables. Speech Perception No. 4, (in press). (Progress Report, Department of Psychology, The Queen's University of Belfast.)

SUMMERFIELD, A.Q. and HAGGARD, M.P. (1974). Perceptual processing of multiple cues and contexts: effects of following vowel on stop consonant voicing. Journal of Phonetics, 2, 279-295.

TAYLOR, M.M. and CREELMAN, D.D. (1967). PEST: Efficient estimates on probability functions. Journal of the Acoustical Society of America, 41, 782-787.

DISCUSSION of "How a full account of segmental perception depends on prosody and vice versa"
SUMMERFIELD

KLATT:

Your physiological explanation for the slightly shorter VOT in /pi/ than in /pa/ in the production data is not consistent with other data. It is probably true that the larynx is more raised in the /pi/ environment and that the vocal folds are thereby stiffened, but this seems to inhibit voicing rather than facilitate an earlier voicing onset. Evidence includes the devoicing of Japanese high vowels in these environments, studies of laryngeal preparations, and theoretical arguments (Stevens, VIIIth Congress of Phon. Scs., Leeds 1975).

SUMMERFIELD:

I agree. The /pi/ - /pa/ inversion is paradoxical. However, the intention of my paper was not really to provide a detailed physiological explanation for such paradoxes in production but, rather, to suggest the existence of a continuously estimated perceptual duration, or rate, parameter. Not only could this be useful in maintaining a rate-dependent phonetic distinction as here, but it could interact with knowledge of the number of syllables in a word to predict the level of that word in the parsing tree or vice versa. While the perceptual paradigm appears to provide a very sensitive measure of this parameter's extraction (as also in SIEB NOOTEBOOM's paper), it would be interesting to cross-calibrate it against a prosodic measure.

LEHISTE:

I would like to suggest an explanation for the fact that you got such a short VOT for voiceless plosives in isolated syllables. It is possible that people expected prevoicing for voiced plosives produced in syllables presented in isolation.

SUMMERFIELD:

This question relates to some data which I presented verbally at the Symposium but which was not included in the text above. I had determined the positions of phoneme boundaries on a synthetic /gi-ki/ VOT continuum both when the target stimuli were introduced by precursor phrases and when they were presented in isolation. The rate of speech of the precursors was varied over a considerably greater range than the variations referred to in the experiments reported above and exposed the limits on the size of phoneme boundary shift that could be obtained by such speech rate variation. Surprisingly, for 5 of my 6 subjects, boundaries estimated in isolation coincided with boundaries estimated when the same target continuum was introduced by the precursor phrase at the _fastest_ rate of speech. The 6th subject's isolated boundary coincided with her boundary estimated at the _slowest_ rate of speech. While I realise that stop consonants have different VOT characteristics when produced in isolation as opposed to sentence contexts. I think that this result reflects a major difference between the _perceptual_ processing of speech relevant durations in and out of sentence context as well as a dramatic difference between individual perceivers, rather than a specific perceptual expectancy for prevoicing.

FISCHER JØRGENSEN:
I am not sure that the relation you have found between the VOT-values of _pi_ and _pa_ is universal, and I doubt whether your physiological explanation is correct. If it is, _pu_ should also have a shorter VOT-value than _pa_, but according to my experience _pu_ has normally a longer VOT-value than both _pa_ and _pi_.
I also think that the difference between _k_ and _p_ have aerodynamic and physiological explanations: the smaller subglottal cavity and the slower movement of the tongue compared to the lips will have the effect that it takes longer time for the supraglottal air-pressure to come down to a level permitting vocal chord vibrations.

Perception of Segment Duration in Sentence Contexts

DENNIS H. KLATT

Research Laboratory of Electronics
and Department of Electrical
Engineering and Computer Science

and

WILLIAM E. COOPER

Department of Psychology

Massachusetts Institute of Technology
Cambridge, Mass. 02139

Abstract

A category judgment technique has been used to quantify the preferred duration for phonetic segments in words placed in different syntactic environments. Sentences containing the stressed vowel /i/ of the word "deal" and the postvocalic fricative /š/ of the word "fish" were modified by deleting or duplicating portions of the waveform from the phonetic segment in question. Results indicate that listeners adjust their expectations of segment duration depending on the syntactic position of the word in the sentence. In particular, the segmental lengthening that is seen in the production of a phrase-final syllable is represented in a listener's perceptual scheme. The data were further examined in order to determine the just-noticeable difference (JND) for a change in duration to a single segment, in a randomized set of different sentences. The JND was found to be 25 msec or more. JND differences between sentence environments suggest the following constraints on the perception of durational information: (a) sensitivity to changes in duration is best for phonetic segments in non-word-final syllables, (b) sensitivity is influenced by a backward masking effect of any following words, and (c) the offset time of an utterance-final fricative is particularly difficult to perceive.

Introduction

The temporal organization of speech is based on a complex system of rules involving a number of linguistic and nonlinguistic variables. The duration of a phonetic segment is influenced by linguistic factors that include aspects of the semantic, syntactic and phonetic environments of the segment in a spoken utterance. For example, in order to predict the duration of a vowel in American English, one would have to include rules to account for the following observations (see Klatt, 1975 for data and literature citations):

(1) Each vowel type has a different inherent duration.
(2) Vowel duration depends on features of the postvocalic consonant (a large effect is seen only in phrase-final syllables).
(3) Unstressed vowels are of shorter duration than stressed vowels.
(4) Vowels in word-final syllables are slightly longer than in other positions.
(5) Vowels in phrase-final syllables are longer than in other positions within a phrase.

The perceptual relevance of these effects can be estimated using any of several psychophysical techniques. Nooteboom (1973) employed synthetic speech and a method of adjustment to show that in Dutch, the difference in duration between inherently long and short vowels is a part of the listener's internal representation for these segments. Fry (1958) has shown that changes in the relative durations of vowels in a two-syllable word can change the perceived stress pattern for the word. The changes in vowel duration induced by features of the following consonant have been shown to carry a functional load in speech perception in the sense that a change in the duration of the vowel will change the perception of the voicing feature of the following consonant (Denes, 1955; Raphael, 1972; Nooteboom, 1974).

The fourth durational effect involving the position of a syllable within a word has been shown by Nooteboom (1973) to influence vowel perception. He determined the preferred durations for vowels as a function of number of syllables and syllable position in a word, using synthetic speech and a method of adjustment. He showed that preferred vowel durations are shorter as a function of the number of syllables left to be produced in the word. The amount of shortening was similar to production data in Dutch, Swedish (Lindblom and Rapp, 1973) and in English (Oller, 1973; Klatt, 1975). Nooteboom (1974) has also shown that the perceptual boundary along a synthetic continuum between phonemically long and short vowels in Dutch decreases by approximately 15 msec if a second syllable is added to the word. He concludes that a categorical decision concerning vowel duration must be delayed until at least the end of a word.

The fifth durational effect, involving the position of the syllable within a phrase, has not been related directly to perceptual expectations of listeners, but Lehiste (1973) has shown that the last metric foot of a spoken sentence must be of longer duration than earlier metric feet or else listeners will perceive the final metric foot as too short.

Fluent pauses ofter occur at phrase boundaries. O'Malley et al. (1973) have shown that the locations of pauses can be used by naive listeners to disambiguate spoken algebraic expressions. Segment lengthening occurs at phrase boundaries whether or not a physical pause is present (Klatt, 1975). It is not known whether this lengthening is used by listeners to determine the constituent structure of an utterance, nor is it known whether the lengthening is expected when listening to a fluently spoken sentence. The experiments described here were designed in order to examine the latter question.

A set of sentences has been constructed in which the morpheme "deal" appears (a) in one and two-syllable words, (b) as a noun, verb or adjective, and (c) in phrase-final and non phrase-final positions. Subjective estimates of the preferred duration for the vowel /i/ have been obtained from listeners by modifying the duration of /i/ in natural recordings of the sentences.

The duration of English consonants may also be influenced by the locations of word boundaries and phrase boundaries (Oller, 1973; Klatt, 1974). Accordingly, a second experiment was run using sentences containing the postvocalic fricative /š/ of the morpheme "fish". A fricative was chosen because the duration of stops appears to be less influenced by the locations of word and phrase boundaries.

The experimental design permits calculation of a just-noticeable difference (JND) for segmental duration as a function of position within a sentence. The JND data from these experiments are of interest because the experiments involve a randomized set of 6 or 7 sentences, which producs a more natural listening situation than in previous JND studies in which the same segment (Abel, 1972; Creelman, 1962; Small and Campbell, 1962), word (Fujisaki et al., 1973; Nooteboom, 1973), or sentence (Huggins, 1972a) was repeated over and over again.

Stimulus Preparation

The sentences listed in Table 1 were recorded several times by one of the authors. The most natural sounding recording of each sentence was selected for computer processing. The waveforms were digitized at 10,000 samples per second, using a high-quality linear-phase 5000 Hz lowpass filter to remove alias components. The digitized waveforms were then digital-to-análog converted, low-pass filtered again, and rerecorded onto audio tape. Broadband spectrograms (Presti, 1966) were made to check for artifacts and to help in the location of good segmentation points. A spectrogram of one of the digitized utterances is shown in the lower half of Figure 1.

Experiment 1: Vowel Duration

A waveform editing program (Huggins, 1969) was employed to define the beginning and end of each sentence and to place time markers at selected points in the sampled waveform. For the sentences involving the vowel /i/ of the word "deal", time markers were placed at nearly the same sample in each of five selected pitch periods in the approximately steady-state part of the vowel, as determined from the formant pattern

Table 1. Sentences used in the experiments involving the words "deal" and "fish".

DURATION OF /il/ (MSEC)	SENTENCE
165	Bill will play dealer's choice.
195	The deal situation is bad.
210	Ken was the next dealer.
210	Henry will deal the next hand.
220	The deal rotates to the left.
310	He passed up the deal.
340	Deal.

DURATION OF /I/ (MSEC)	DURATION OF /š/ (MSEC)	SENTENCE
55	95	Bill played near the fishing hole.
85	105	The fish market is open.
80	110	Ken wants to go fishing.
90	100	Henry will fish until dark.
135	160	The small fish were biting.
200	200	He cleaned the fish.

seen in the spectrogram. An example of time marker placement in the vowel portion of the waveform for the sentence with the shortest vowel duration is shown in Figure 2. As can be seen in the figure, a durational increment of two pitch periods was selected for the sentences of the first experiment on the basis of an informal pretest.

Nine stimuli were constructed from each of seven sentences involving the word "deal". In stimulus number 1, waveform segments numbered 1, 2, 3 and 4 were deleted from the digitized recording, thereby yielding the sentence with the shortest duration vowel. (In Figure 2, waveform segment number 1 extends from time marker number 1 to time marker number 2, etc.) Stimulus number 5 was the original recording, and stimulus number 9 was generated by duplicating waveform segment numbers 1 through 4. Spectrograms of the vowels excised from stimuli constructed from one of the sentences are shown in the upper part of Figure 1.

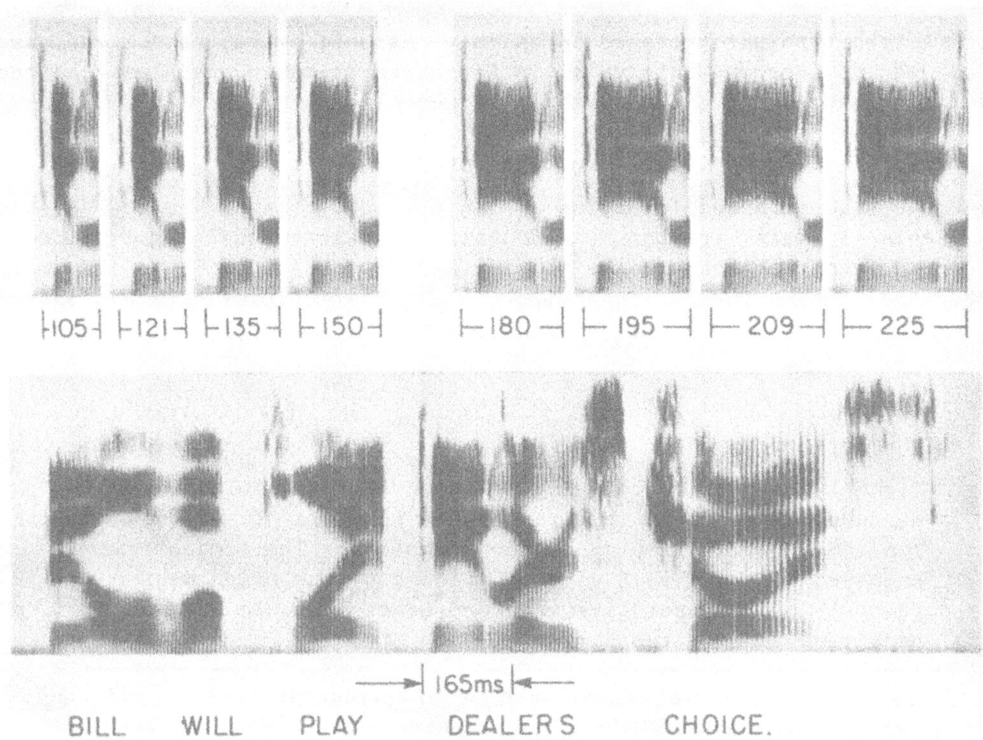

Figure 1. A broadband spectrogram is shown of the original recording of the sentence "Bill will play dealer's choice". Upper: spectrograms of the word "deal", as excised from the eight other sentences that were constructed from this recording are shown.

The composition and measured duration of each syllable nucleus in "deal" for the sentence "Bill played dealer's choice" is shown in Table 2. The duration measurement for the original recording (stimulus number 5) is taken from the instant of release of the /d/, as indicated by the onset of the plosive burst, to the instant of release of the tongue tip after the /l/, as indicated by the sudden upward shift of the first and second formants, and the concomitant discontinuous increase in waveform amplitude. This measurement corresponds to the duration of the plosive burst, the /i/ and the postvocalic /l/. Measured durations for /il/ in each of the original "deal" sentences are presented in Table 1. The /i/ duration appears to be about half of this value in each of the sentence contexts.

The fundamental frequency of the vowel in "Bill played dealer's choice" was about 133 Hz. Therefore successive stimuli differ in duration by about 15 msec for this sentence. The durational increment was slightly different in other sentences because of changes in fundamental frequency.

A test tape was prepared in which 63 stimuli were randomized and recorded at 5-second intervals. A few extra stimuli were placed at the beginning and end of the randomized sequence.

Experiment 2: Postvocalic Fricative Duration

The six sentences involving /š/ in the word "fish" that are listed in Table 1 were treated in an analogous fashion. Nine stimuli were prepared from each sentence by placing markers at 10, 15 or 20 msec intervals during the /š/ frication noise. The interval size was chosen

Figure 2. A computer display is shown of the sampled waveform corresponding to a portion of the vowel /i/ taken from the sentence "Bill will play dealer's choice". The time markers numbered from 1 to 5 were positioned by the experimenter so that any of 4 consecutive waveform segments could be played back zero, one or two times.

Table 2. The number of times each waveform segment was played to generate nine stimuli for the sentence "Bill will play dealer's choice." Also presented are the durations for the resulting syllable nuclei /il/ in each of the stimuli.

	SEGMENT NUMBER				DURATION OF
STIMULUS	1	2	3	4	/il/ (MSEC)
1	0	0	0	0	105
2	1	0	0	0	120
3	1	0	0	1	135
4	1	1	0	1	150
5	1	1	1	1	165
6	1	2	1	1	180
7	1	2	2	1	195
8	1	2	2	2	210
9	2	2	2	2	225

on the basis of a pretest. Markers were always placed near an up-going zero crossing of the waveform in order to minimize the possibility of generating spurious clicks. Spectrograms of the original recording and the 8 modified versions of /š/ are shown in Figure 3 for the sentence containing the shortest original /š/ duration. Measured durations for the /š/ and the vowel /I/ of "fish" are presented in Table 1 for each of the sentences. A test tape was then prepared in which 54 randomized sentences were recorded with a few extra stimuli at the beginning and end of the sequence.

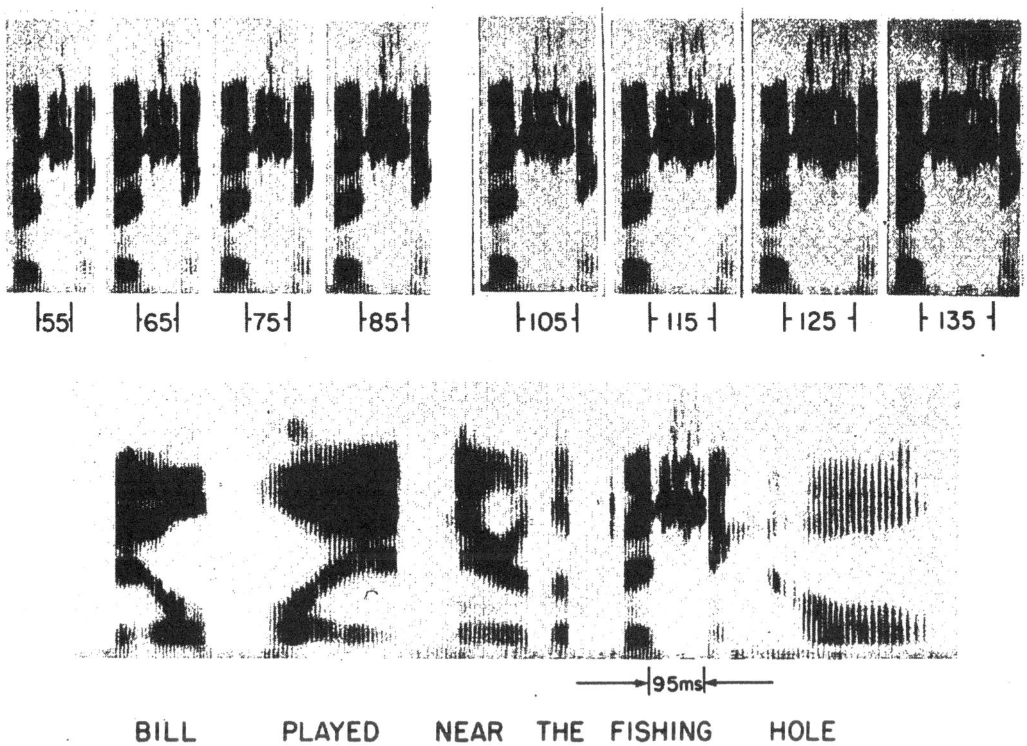

Figure 3. A broadband spectrogram is shown of the digitized recording of the sentence "Bill played near the fishing hole". Upper: spectrograms are shown of the the fricative /š/, as excised from the 8 other sentences that were constructed from this recording.

Test Procedures and Results

The experimental tapes were played to subjects binaurally at a comfortable listening level; an Ampex PR-10 tape recorder and matched Telephonics headphones were used. Ten subjects listened to each test tape a total of 3 times without feedback. Each tape was heard on a separate day, and one of the subjects participated in both experiments. The task of the subjects was to estimate the magnitude of the duration of the vowel /i/ in the word "deal" (Experiment 1) or the fricative /š/ in "fish" (Experiment 2), using the integers 1 - 9. The task represents a special case of the magnitude estimation task and is typically referred to as a category judgment task or an absolute identification task (Braida and Durlach, 1972).

Subjects were played three stimuli before the test: the shortest, natural, and longest duration versions for the third sentence in each group, as listed in Table 1, and they were told that these examples represented duration magnitudes of 1,5, and 9. It was explained that number 5 should be used as a response for sentences in which the vowel had a natural duration relative to the particular sentence context. No more examples were given in order not to bias their judgments concerning what a natural duration might be in other sentence contexts.

Responses were organized in the form of a 9-by-9 confusion matrix for each sentence type. An example of a confusion matrix of pooled responses from 10 subjects is shown in Figure 4. Responses from one of the best subjects are shown in Table 3. It can be seen that this subject is very adept at distinguishing amoung the 9 different durations for the /i/ in "Bill will play dealer´s choice", but he is unwilling or unable to distinguish similar durational changes in the sentence "The deal rotates to the left."

Stimulus \ Response	1	2	3	4	5	6	7	8	9
1	8	13	7	2					
2	1	8	14	4	2	1			
3		4	10	8	7	1			
4		1	2	8	9	7	3		
5				3	15	10	1	1	
6			1	1	3	10	12	3	
7					1	9	9	9	2
8							12	14	4
9							3	10	17

Figure 4. A confusion matrix is shown consisting of the pooled responses of 10 subjects to the 9 stimuli involving the sentence "Ken was the next dealer."

Table 3. Category judgment responses of one subject to 9 versions of the sentence "Bill will play dealer´s choice are shown in row (a) and his responses to the sentence "The deal rotates to the left" are shown in row (b).

	STIMULUS NUMBER								
	1	2	3	4	5	6	7	8	9
RESPONSE TO (a)	1	1	3	3	4	5	6	7	8
DURATION IN MSEC	105	120	135	150	165	180	195	210	225
RESPONSE TO (b)	2	4	4	4	5	3	5	5	6
DURATION IN MSEC	162	177	191	206	220	234	249	263	278

Original Segment Duration

The data in column 1 of Table 4 indicate that the durations of the /il/ of "deal" and the /š/ of "fish" in the original sentence recordings follow a pattern similar to that observed previously. Both segments are of longest duration in sentence-final position. Segments are longer in phrase-final positions. The difference between durations in the verb and in internal phrase-final positions is smaller than expected. The reason may be that each sentence was relatively short and was spoken in

Table 4. Summary of results.

ORIGINAL DURATION OF /il/ (MSEC)	PREFERRED DURATION OF /il/ (MSEC)	PERCEPTUAL MAGNITUDE OF 10 MSEC CHANGE	JND	SENTENCE
165	173	.4	26	Bill will play dealer's choice.
195	203	.25	51	The deal situation is bad.
210	201	.46	22	Ken was the next dealer.
210	213	.26	59	Henry will deal the next hand.
220	225	.23	50	The deal rotates to the left.
310	268	.29	36	He passed up the deal.
340	309	.33	40	Deal.

ORIGINAL DURATION OF /š/ (MSEC)	PREFERRED DURATION OF /š/ (MSEC)	PERCEPTUAL MAGNITUDE OF 10 MSEC CHANGE	JND	SENTENCE
95	114	.56	28	Bill played near the fishing hole.
105	111	.46	37	The fish market is open.
110	104	.6	25	Ken wants to go fishing.
100	98	.43	30	Henry will fish until dark.
160	-	.19	67	The small fish were biting.
200	142	.21	98	He cleaned the fish.

isolation; all segment durations in phrase-final syllables were longer in a connected discourse examined previously by one of the authors (Klatt, 1975).

The segments are shortest in the first syllable of a polysyllabic word. The effect of syllable position within a word is small in these data, presumably because of the relative incompressibility of vowels that have been shortened by the influences of other shortening factors (Klatt, 1973).

Preferred Duration

The pooled data were analyzed to determine the preferred (i.e. most natural sounding) duration for the syllable nucleus or fricative in each syntactic environment. The average category judgment was plotted as a function of stimulus number for each sentence type. Examples of several plots are shown in Figure 5. A straight line was fitted to the data points for each experiment, using the best least-squares fit as the criterion for placement of the line. The stimulus duration

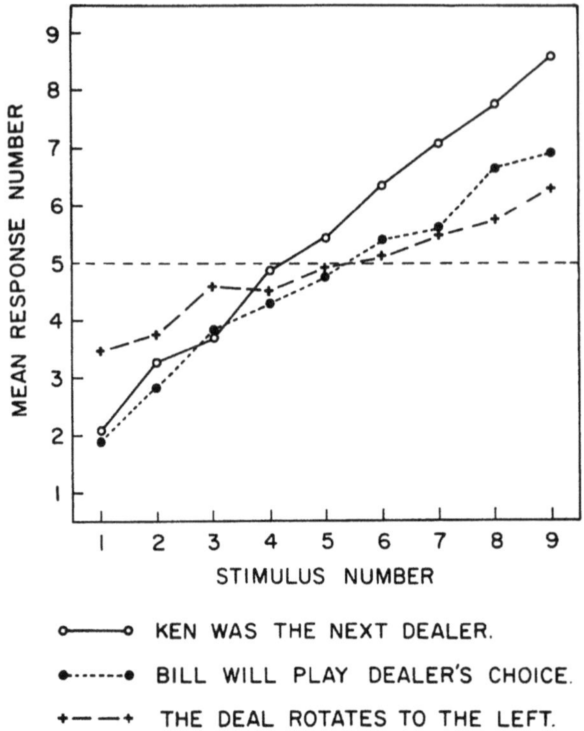

Figure 5. The mean category judgment response is plotted as a function of stimulus number for 3 sentences of the first experiment.

corresponding to the place where this line crosses a response of 5 indicates the preferred stimulus duration. Preferred durations for each of the sentences are presented in column 2 of Table 4.

Agreement between the segment duration in the original sen:ence and the preferred duration is good over a duration range of almost an octave in the vowel data of Experiment 1. A rank order comparison of the original and preferred durations produces a very high positive correlation with no reversals in rank order. There is a slight tendency for preferred durations to be more restricted in total range of variation with sentence context.

The fricative data of Experiment 2 are similar, except that the /š/ of "The small fish were biting" was judged longer than normal for every stimulus, including stimulus 1 where 60 msec had been excised from the /š/. The average response is plotted in Figure 6 as a function of stimulus number for this sentence, and the best-fitting straight line is also shown. A possible explanation for this result is discussed below.

Comparative Perceptual Magnitude of a 10-Msec Change in Duration

The slopes of the curves shown in Figures 5 and 6 indicate the average change in response for a given change in segment duration. In order to make comparisons across all experimental conditions, the slope of each best-fitting straight line has been converted into the change in response magnitude corresponding to a 10-msec change in segment

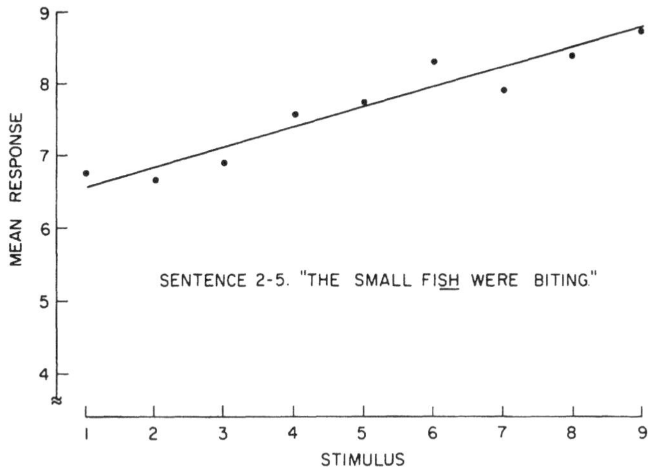

Figure 6. The mean category judgment is plotted as a function of stimulus number for the sentence "The small fish were biting." Note that the /š/ was judged longer than normal even in stimulus 1, which was obtained by excising 60 msec from the original /š/ duration.

duration. The results are presented in column 3 of Table 4. It can be seen that a 10 msec change induces the greatest subjective change in duration if the word contains a following second syllable. Generally, a 10-msec change in the /š/ is perceived to be slightly greater than a corresponding change to /i/, except in sentence-final position, where a 10-msec change to /š/ is judged to be very small.

Just Noticeable Difference in Segment Duration in Sentence Contexts

The perceptual magnitude values cited above are related to the just-noticeable difference in segment duration. In order to compute a meaningful JND, it is necessary to know the variance in the category judgment data, which depends somewhat on the sentence position of the segment. Since data from individual subjects were insufficient to make any quantitative claims, we used pooled data from all ten subjects. The pooled data allow comparisons of relative sensitivity across conditions, but it should be noted that JNDs obtained from pooled responses may underestimate individual discrimination abilities if subjects have different preferred durations. Aspects of the data from individual subjects will be discussed later.

It is assumed that the behavior of the subjects, as reflected by the pooled confusion matrix for each sentence, can be described by the following standard decision model (Green and Swets, 1960). The stimulus duration D of a segment results in a sensation magnitude x. Because of internal noise, external noise and other factors, the distribution of sensation magnitude x resulting from the physical duration D is normally distributed with a mean $\mu(D)$ and a standard deviation σ. The stimuli I=1,2,...,9 have different durations D_I which result in different average sensation magnitudes $\mu(D_I)$, but it is assumed that the standard deviations of the x distributions are not a function of I.

The situation is illustrated in Figure 7 which is based on data obtained from the sentence "Bill will play dealer's choice." A subject behaves as if he compared the sensation magnitude x of an unknown stimulus with eight response critera thresholds that are specific to the sentence in question, and determined his response R on the basis of the number of thresholds that are exceeded plus one. The critera threshold locations of the model are optimized to take into account response biases of the average subject.

From the confusion matrix data, it is possible to find the 8 critera thresholds and the 8 differences between adjacent sensation magnitude means that best fit the data. An example of the results of these calculations is indicated in Figure 7. The most important

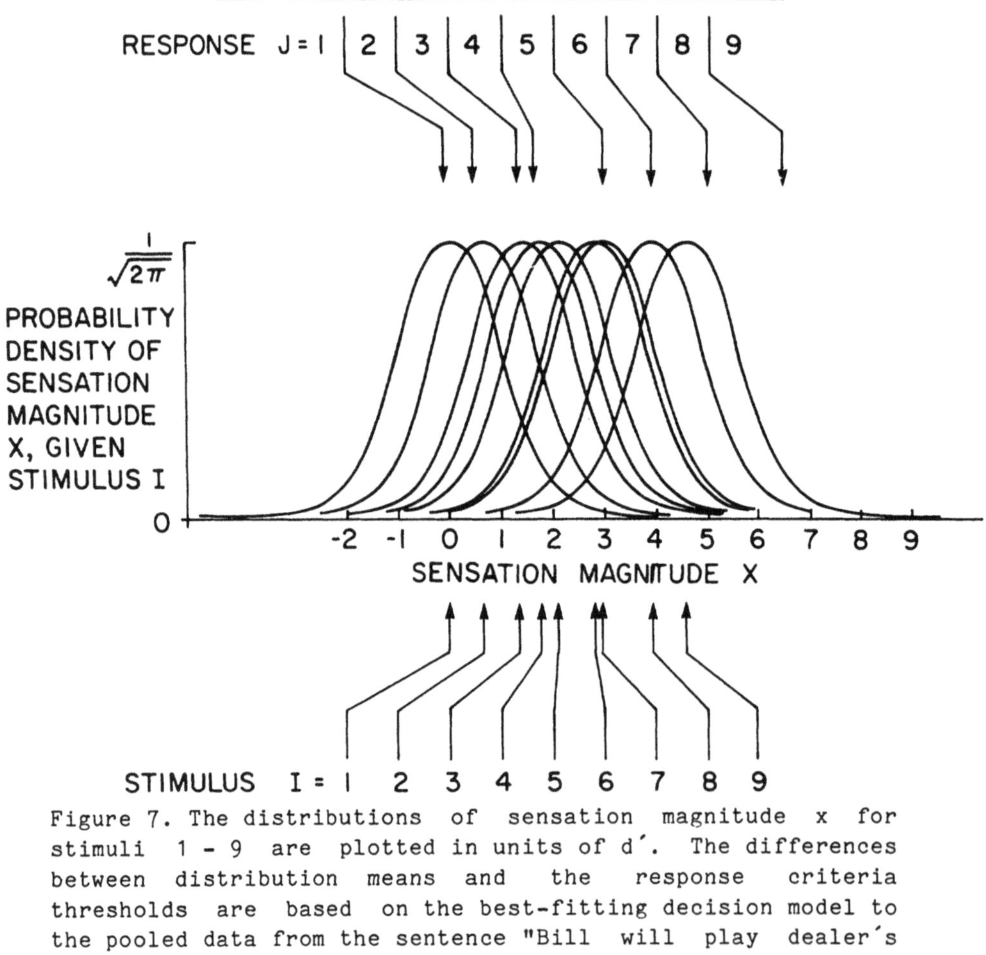

Figure 7. The distributions of sensation magnitude x for stimuli 1 - 9 are plotted in units of d′. The differences between distribution means and the response criteria thresholds are based on the best-fitting decision model to the pooled data from the sentence "Bill will play dealer's choice."

characteristic of the model is that, if σ is normalized to be 1.0, the difference between sensation magnitude means is equal to d′, the standard psychophysical measure of stimulus discriminability. For example, d′ for stimuli 7 and 8 in the figure is approximately 1.0. Therefore these stimuli could be discriminated as different about 75 percent of the time in an appropriately designed paired discrimination test.

The cumulative d′ values associated with going from stimulus 1 to stimulus 9 are a reasonable measure of the average discriminability of segment duration along the stimulus continuum for each sentence. These values have been converted to a just-noticeable difference in msec by computing the change in duration corresponding to a d′ of 1.0. The JNDs are presented in column 4 of Table 4.

The data shown in Figure 7 do not appear to fit the model as well as we might hope. It is surprising and counterintuitive to discover that sensation magnitude is not a smooth function of stimulus duration. Intuitively, d′ should be approximately equal for adjacent pairs of stimuli. There were only 30 responses in each confusion matrix row, whereas the model is usually applied in experiments involving considerably more data from only one subject. Thus some caution must be used in drawing conclusions from the JND values presented in Table 1.

Data from individual subjects were compared by making the following approximation. All trials for a given sentence type can be combined to yield 27 responses (9 category judgments three times). The responses to adjacent pairs of stimuli can be compared to see if the rank order of the responses is such that the stimulus of longer duration received (a) a larger response number than the smaller-duration stimulus of the pair, (b) both received the same response number, and (c) the response numbers were reversed. There are 24 pairs of stimuli for which this comparison can be made for each sentence for each subject. Assuming that these comparisons were the responses of a single two-category judgment experiment, we computed the d′ for each subject. Results did not fit the two-category judgement model very well in all cases, but some general statements can be made. The JNDs thus obtained were in agreement with those presented in Table 4, and the best subject is nearly twice as sensitive to changes in stimulus duration as the worst subject.

Production Data from the Subjects

A few weeks after the perception tests, 4 subjects returned to read the sentences in a random order. Spectrograms were made of the recordings and segmental durations were measured. The /il/ durations in the "deal" sentences were found to be somewhat longer than those of the original speaker. However, the rank ordering of /il/ durations across sentences is nearly perfect when each subject is paired against the original speaker. Only two of 24 rank-order comparisons failed to agree.

In the sentences of the second experiment, the duration of the vowel in "fish" was found to be short, except in sentence-final position. There is some variability among speakers, but a perfect rank order correlation exists between the median data from the four subjects and the data from the original speaker. The durations of the fricative /š/ display more variability and disagreements between all speakers. The median data from the four subjects indicates that an utterance-final or nearly final fricative is longer, and that fricatives are shortened when adjacent to other consonants (Klatt, 1974). The overall picture suggests that phrase-final lengthening within a sentence is optional in a short phrase, depending on whether the reader wishes to call attention to the syntactic boundary.

Discussion

The preferred durations in the perception of /il/ in sentences from Experiment 1 agree well with the measured durations from the original recordings and with theoretical predictions (Klatt, 1975). As we have noted above, the differences in /il/ duration across sentence environments seen in both the original recordings and in the preferred durations are somewhat smaller than those observed previously for the same speaker reading a meaningful connected discourse (Klatt, 1975). It may be that internal phrase boundaries are marked by duration increases to a greater extent in longer, more complex sentences or when sentences form a meaningful part of a connected discourse. Alternatively, subjects in the present experiments may have been making absolute duration magnitude estimates to some extent, even though the instructions were to rate the duration relative to an expected natural duration for a particular sentence. This possibility is suggested by the fact that responses clearly indicate a tendency to judge stimuli from sentences with a normally short duration segment as too short, and segments in utterance-final positions as too long.

The preferred durations for the fricative /š/ in sentences from Experiment 2 also agree with measured durations and theoretical predictions, except in phrase-final positions. The two sentences involving a noun-phrase final use and an utterance final use of the word "fish" produced data indicating poor sensitivity to durational changes, and shorter preferred durations than expected. These data are intriguing, indicating either an error in stimulus preparation or a significant deviation from the expected pattern. Having checked the stimuli carefully, we offer a tentative explanation.

Consider the hypothesis that the end of a fricative in utterance-final position is difficult to perceive because of the gradual decrease in sound intensity with falling subglottal pressure. In sentence-medial positions, a phonetic segment with a rapid voicing onset would normally follow the /š/. Voicing onset is a well-defined acoustic event that can be interpreted as the end of the fricative. Therefore, the fact that the JND is almost 4 times smaller in the utterance-final word "fishing" than in the utterance-final word "fish" can be explained in part by the absence of a clear acoustic cue to the termination of the frication energy in the latter word. It is not surprising that perceptual strategies would evolve to make use of event onset times to determine fricative terminations, since fricative duration is an important cue to the voicing feature for fricatives in English (Klatt, 1975), and room noises and reverberation often mask the true offsets of weak speech sounds.

Given this hypothesis, an explanation can also be offered for the unusual rating behavior of subjects to the sentence "The small fish were biting." Perhaps this /š/ was judged long even when 60 msec was excised from the /š/ duration of the original recording because subjects were not responding to /š/ duration directly, but rather to sentence rhythm, as reflected in the times between stressed vowel onsets (Huggins, 1972b; Allen, 1972). There was no sudden voicing onset following the /š/ in this sentence because the next segment, /w/, was partially aspirated and had a gradual onset of voicing. Thus changes to the duration of the /š/ in this sentence could only be perceived insofar as they produced a perceptible change in the rhythm of the sentence, as reflected in the time of release of the /b/ of "biting".

The JND data from the present experiments are important because they were obtained from fairly natural sentence materials. The JND results indicate that sensitivity to changes in the duration of a vowel or postvocalic fricative is a function of at least three variables: (a) the absolute duration of the segment, i.e. there was a general tendency for the JND data to conform to Weber's law, (b) syllable position within the word, i.e. the JND was smaller for segments in the first syllable of a two-syllable word than for segments of comparable duration in word-final syllables, and (c) word position within the sentence.

An unexpected result of Experiment 1 suggests that backward masking may play a role in determining JND values. The JND for /i/ in sentences involving the single-syllable word "deal" is smaller in utterance-final position even though the absolute duration of the /i/ is longest. There must be no difficulty in determining vowel duration (by using either the timing of the formant transiton cues in /il/ or the offset of voicing at the end of the utterance.) This result suggests that some sort of backward masking effect is present; i.e., if other words follow "deal", the JND increases. It is surprising to find indications of backward masking on a variable such as segment duration. The effect spans too long a time period to be related to peripheral masking (Elliott, 1962), but it may bear some relation to backward masking that has been observed when subjects have to make categorical decisions based on information contained in precategorical storage (Massaro, 1975).

As we have mentioned, the situation for utterance-final /š/ is just the opposite. The JND is very large, thereby indicating the inability of subjects to determine frication offset in an utterance-final fricative.

The present data may be used to estimate the accuracy with which segmental durations must be computed in, for example, a speech synthesis by rule program. If all other segment durations are produced with good accuracy, an error can be tolerated in the duration of a segment such as a vowel or postvocalic fricative consonant of one or more JNDs. Thus the minimum accuracy that would be required in this situation is of the order of 25 to 100 msec. This is a surprisingly large tolerance limit, given some previous experimental results (Ruhm et al., 1966; Nooteboom, 1973), but our results probably reasonable reflect the difference between laboratory measurements made on isolated sounds and human performance in more natural speech contexts. (The Ruhm study also employed an unorthodox definition of JND.)

Huggins (1972a) measured the JND for different types of phonetic segments embedded in a single sentence. He found that a 40-msec or greater range of durational changes were accepted as normal (not long or short) by his subjects. He also investigated the matter of whether subjects would change their judgments of normality of a given segment duration if an adjacent segment were lengthened or shortened (Huggins, 1972b). He found a small negative correlation between the changes to two adjacent segments if and only if the change to the second segment would reestablish isochrony, i.e. restore the onset time of the next stressed vowel to its proper relationship with the previous stressed vowel onset. For example, there was no interaction between members of a stressed consonant-vowel sequence, but interactions did occur across word boundaries and within unstressed syllables.

Fujisaki et al. (1973) studied the discrimination of changes in segmental duration in Japanese vowels and consonants. Their data are of interest because in Japanese there exist minimal word pairs differentiated primarily on the basis of the duration of a vowel or a consonant. The authors found that the JND near a phoneme boundary for a two-syllable word spoken in isolation or placed in a carrier sentence was about 10 msec (for a segment duration of about 100 msec) for vowels, fricatives, plosives and nasals. The somewhat better performance that they report may be due to improved discrimination at a phoneme boundary, or to the fact that each experiment involved only a single word or carrier sentence.

In addition to implications for speech synthesis by rule, results of the present study open up the posssibility that the perception of segment duration is used actively in the decoding of constituent structure during normal listening. This possibility has received little attention in previous studies of sentence perception (Fodor, Bever and Garrett, 1974, Chapters 5 and 6).

The experimental paradigm used in the present study can be applied to this problem in several ways. As a first step, however, a more detailed comparison must be provided between the perception of segment durations in different syntactic environments in which the possible influence of factors other than syntactic environment per se (factors such as the phonetic environment and the number of words or syllables in the phrase or sentence) is neutralized. The sentence materials and data base in the present study were not entirely appropriate for such comparisons. Possibilities for future work include comparing the perception of segment durations in the following environments: (a) other head noun vs verb environments, as in "The deal rotates to the left" vs "Henry will deal the next hand," and (b) head noun immediately dominated by sentence node vs head noun immediately dominated by verb-phrase node, as in "The deal rotates to the left" vs "Tom understands the deal thoroughly." It is hoped that a study of systematic comparisons and additional work along similar lines (e.g., perception of "garden path" sentences) will allow us to determine the extent to which the perception of phonetic segments is used in the decoding of constituent structure.

Acknowledgment

This research was supported in part by N.I.H. Grants NS 04332 and MH 26612. We thank Richard Lippmann for providing a computer program for the estimation of d′ and Steven Lapointe for assistance in running the experiments.

References

Abel, S.M., Duration Discrimination of Noise and Tone Bursts, J. Acoust. Soc. Am. 51, 1219-1223 (1972).

Allen, G.D., The Location of Rhythmic Stress Beats in English: An Experimental Study I, Language and Speech 15, 72-100 (1972).

Braida, L.D. and N.I. Durlach, Intensity Perception II. Resolution in One-Interval Paradigms, J. Acoust. Soc. Am. 51, 483-502 (1972).

Creelman, C.D., Human Discrimination of Auditory Duration, J. Acoust. Soc. Am. 34, 582-593 (1962).

Denes, P., Effect of Duration on the Perception of Voicing, J. Acoust. Soc. Am. 27, 761-764 (1955).

Elliott, L.L., Backward Masking: Monodic and Dichotic Conditions, J. Acoust. Soc. Am. 34; 1108-1115 (1962).

Fodor, J.A., T.G. Bever and M.F. Garrett, The Psychology of Language: An Introduction to Psycholinguistics and Generative Grammar, Chapters 5 and 6, McGraw-Hill, New York (1974).

Fry, D.B., Experiments in the Perception of Stress, Language and Speech 1, 126-152 (1958).

Fujisaki, H., K. Nakamura and T. Imoto, Auditory Perception of Duration of Speech and Non-Speech Stimuli, Proc. Symposium on Auditory Analysis and Perception of Speech, Leningrad, (August, 1973).

Green, D.M. and J.A. Swets, Signal Detection Theory and Psychophysics, Wiley, New York (1960).

Huggins, A.W.F., A Facility for Studying Perception of Timing in Natural Speech, Quarterly Progress Report Number 95, Research Laboratory of Electronics, Massachusetts Institute of Technology, Cambridge, Mass., 81-83 (1969).

Huggins, A.W.F., Just-Noticeable Differences for Segment Duration in Natural Speech, J. Acoust. Soc. Am. 51, 1270-1278 (1972a).

Huggins, A.W.F., On the Perception of Temporal Phenomena in Speech, J. Acoust. Soc. Am. 51, 1279-1290 (1972b).

Klatt, D.H., Vowel Lengthening is Syntactically Determined in a Connected Discourse, J. Phonetics 3, 161-172 (1975).

Klatt, D.H., The Duration of [s] in English Words, J. Speech and Hearing Research 17, 51-63 (1974).

Klatt, D.H., Interaction between Two Factors that Influence Vowel Duration, J. Acoust. Soc. Am. 54, 1102-1104 (1973).

Lehiste, I., Rhythmic Units and Syntactic Units in Production and Perception, J. Acoust. Soc. Am. 54, 1228-1234 (1973).

Lindblom B. and K. Rapp, Some Temporal Regularities of Spoken Swedish, Papers from the Institute of Linguistics, University of Stockholm, Publication 21 (1973).

Massaro, D.W., Preperceptual Auditory Storage in Speech Recognition, Proc. Symposium on Dynamic Aspects of Speech Perception, Institute for Perception Research, Eindhoven, Springer Verlag (August, 1975).

Nooteboom, S.G., The Perceptual Reality of Some Prosodic Durations, J. Phonetics 1, 25-45 (1973).

Nooteboom, S.G., Contextual Variation and the Perception of Phonemic Vowel Length, Proc. Speech Communication Seminar, Stockholm, Vol. 3, 149-154, Almqvist and Wiksell, Upsalla (1974).

Oller, D.K., The Duration of Speech Segments: The Effect of Position in Utterance and Word Length, J. Acoust. Soc. Am. 54, 1235-1247 (1973).

O'Malley, M.H., D.R. Kloker and D. Dara-Abrams, Recovering Parentheses from Spoken Algebraic Expressions, IEEE Trans. Audio and Electroacoustics AU-21, 217-220 (1973).

Presti, A.J., High-Speed Sound Spectrograph, J. Acoust. Soc. Am. 40, 628-634 (1966).

Raphael, L.J., Preceding Vowel Duration as a Cue to the Voicing Characteristics of Word-Final Consonants in English, J. Acoust. Soc. Am. 51, 1296-1303 (1972).

Ruhm, H.B., E.O. Mencke, B. Milburn, W.A. Cooper Jr. and D.E. Rose, Differential Sensitivity to Duration of Acoustic Signals, J. Speech and Hearing Research 9, 371-384 (1966).

Small, A.M., and R.A. Campbell, Temporal Differential Sensitivity for Auditory Stimuli, Amer. J. Psychol. 53, 329-353 (1962).

DISCUSSION of "Perception of segment duration in sentence contexts"
KLATT and COOPER

'T HART:
It seems that a major syntactic break following the target word causes a decrease in perceptual acuity with respect to duration as can be seen on page 9; "Henry will deal the next hand" and "the deal rotates to the left" as well as "Henry will fish until dark" and "the small fish were biting" produce high j.n.d.'s.
From the last part of your discussion I got the impression that it refers to that effect, but I am not sure about that. Could you comment on this?

KLATT:
If a major syntactic boundary follows the word whose duration has been manipulated, subjects seem to expect segment lengthening as an optional way of marking syntactic structure for the listener. The data show that subjects categorize duration changes as smaller in these environments (i.e. no extremely long or extremeley short segments were heard) and, more interestingly, subjects are less accurate in discriminating durational changes.
One possible explanation for the larger j.n.d. is that subjects cannot use sentence rhythm as one criterion for assessing the magnitude of a durational change. There is no regular pattern to the timing of stressed syllable onsets across a major syntactic boundary because of the optional lengthening of segments to the left of the boundary.

MARTIN:
In your statement that vowel types have inherent duration, do you mean that they can be distinguished from each other on the basis of duration, or that their distributions do not overlap, or what?

KLATT:
There are large differences in vowel durations that are measured in a connected discourse of several paragraphs duration (Klatt, 1975, J. Phonetics 3, P.172). A factor analysis of the stressed vowel data revealed that over half of the variance is explained by assuming different inherent durations for the 15 stressed vowel types of English. Other factors accounting for much of the remaining variance include phrase-final lengthening, word-final lengthening and the influence of the postvocalic consonant.

LEHISTE:
You found particularly small j.n.d.'s for the disyllabic words dealer and fishing. I would like to suggest a reason for this phenomenon. Disyllabic words are perceived as a whole - what is perceived is something like the ratio between the syllables. It takes a smaller change to disturb this ratio.

LEVELT:
You state in your paper that "it is surprising and counter intuitive to discover that sensation magnitude is not a smooth function of stimulus duration". However, there are at least two possible explanations for your findings the first of which is my preference, but the second could have special phonetic interest. The first is that your computational model is based on the assumption that sensation magnitudes are normally distributed. This is a strong assumption; it is quite likely that there exists substantial skewness at the two extremes of the scale. The true sensation magnitude distribution could still be a nice smooth function of stimulus duration if we drop the normality assumption. A second possible explanation is that perception is categorical, in the sense that in the middle range all durations lead to equivalent perceptions, whereas this equivalence breaks down at the extremes of the scale increasing sensitivity to durational differences.

FUJISAKI:
I would like to make three comments concerning the apparent discrepancy between my data on discriminability of segmental duration (Fujisaki, H. et al., Auditory perception of duration of speech and non-speech stimuli, Auditory Analysis and Perception of Speech, Academic Press London 1975) and the data presented in this paper.
In the first place, my investigation was concerned with discriminability of segmental durations that are phonemic in Japanese, i.e. presence/absence of an elongation phoneme, while the duration investigated by KLATT is not phonemic but only carries prosodic information.
In the second place, the method of category judgment is expected to yield greater values of DL than those obtained by the standard paradigm of discrimination by paired comparison, since the criteria for magnitude estimation are ad hoc for all the subjects and are not considered as quite stable, and add to the variability of the auditory representation of stimuli themselves.

Thirdly, pooling of the data of individual subjects and of individual sentences will also result in an apparent increase in DL. I might also add that my data on discriminability are expressed in terms of the standard deviation of the discrimination curve, and is equal to 0.625 times the conventional 75%-DL in the absence of the time-order error.

PERCEPTION OF SEGMENTAL DURATION

Rolf Carlson and Björn Granström
Department of Speech Communication
Royal Institute of Technology (KTH)
S-100 44 Stockholm 70, Sweden

Human speech is characterized by sloppy and precise articulatory movements with the acoustical speech wave as result. The listener has to pin-point the crucial events in this wave as well as to neglect unimportant distortions in order to perceive the message. The durational structure of a sentence has been studied on different levels. These studies show astonishingly small variance in certain portions as well as marked variation in others. This paper deals with three questions within the area of speech perception. a) How accurately can a listener perceive segment duration? b) Is this process influenced by segment type as well as the duration of neighboring segments? c) If a segment duration is changed, what compensatory effects does a listener expect?

In the classical study Speech:Articulation and Perception, KOZHEVNIKOV and CHISTOVICH(16) presented evidence that the durational structure of a sentence is preplanned and that the open syllable is an important unit in this process. The articulatory process is changed by sequential commands at preplanned times. This would result in compensatory effects in speech production giving rise to a lengthening of one segment when the following one is shortened or vice versa. Similar results have been presented by LEHISTE (17). These findings have been put into question by OHALA (21) claiming that a "chain model" ("make things after each other") might be as relevant as "the comb model" ("keep the preplanned structure"). Furthermore OHALA discusses the influence of imprecise segmentation, speaking rate and the size of analyzed portions when considering the advantages and disadvantages of these models. If the "comb model" is the most basic one we would expect compensatory effects in speech perception in order to minimize the influence of local disturbances during the decoding of the message. If the opposite model is true there is no immediate need for such a behavior. As should be shown we find evidence for both models in our data and the purpose is not to reject one or the other.

Reasonable models of vowel duration have been proposed by several research groups. These models are based on analysis and production experiments (5, 7, 18, 19, 20), and have been shown to convey perceptually important cues, signalling position and stress within the word and the sentence (4). In a similar way the structure of clusters and syllables has been studied (9, 10, 15, 22). However, the question still remains: how accurate does the descriptive model have to be if the perception sets the limits?

We know from psychoacoustical experiments that the duration DL increases when the duration increases (1, 2, 3, 8). However, it is not a simple linear relation. Pause duration is harder to perceive than the duration of a sinusoid. Does this finding imply that the perception of variation in the duration of a voiceless occlusion is less acute than the perception of variation in the duration of a sound? As shown by FUJISAKI et al. this is not true (8). In most psychoacoustical experiments two simple stimuli are compared so that estimations are made on the basis of little, if any, information given by surrounding sounds. However, in speech we do not seem to estimate the duration of each segment as an isolated part. In tables presented by FUJISAKI et al. we find evidence that the DL is even better than expected from psychoacoustical experiments when the duration is important for phoneme classification. The pause/sound difference has disappeared. To conclude, psychoacoustical experiments cannot be used in general to explain the perception of speech segment duration.

In several experiments HUGGINS(11, 12, 13, 14) has shown that compensatory changes in certain segment durations could be made, and utterances were still classified as "normal". His results indicate that segment durations are not all important in the perception if certain conditions are fulfilled. However, in one position a compensatory change could not restore the "normal" responses. A consonant followed by a stressed vowel was not a sequence in which compensation could help. HUGGINS summarizes his results by putting forward the vowel onset hypothesis, i. e. the vowel onsets of stressed vowels have to be correctly timed and in the interval inbetween a compensation may occur. A distortion in a thus defined sequence could not be compensated by a change in the preceding or following. Experiments with speech production synchronized by external tapping may be interpreted as support of these results (24).

Questions and results described above will be tested and discussed below together with other problems concerning the perception of segmental duration.

Discrimination and identification tests

Using different techniques we have tried to estimate the perceptual need for accuracy in speech production as well as speech synthesis. Most of the experiments reviewed in the introduction shed light on the boundary condition. Far more has to be done to illuminate the acceptability and discriminability of segmental durations. In the following three experiments will be described that were designed to explore this white spot on the phonetic map.

The material

In all experiments ordinary Swedish words were used. The words carried accent II, i. e. primary stress on the first vowel and secondary on the second one. The vowels were open, phonologically short /a/ segments.

The words were recorded in an anechoic chamber and manipulated with a "tape splicing" computer program. In this program the speech signal could be inspected and marked so that e. g. certain parts of the signal could be deleted or doubled by commands from the operator. By means of this method test tapes with manipulated versions of the original were produced.

The stimuli were presented at a comfortable sound pressure level to listeners wearing head sets (HD 414). Words, segment durations, manipulated segments, number of subjects, number of responses for each stimulus and so on are presented for each experiment in Table 1.

Experiment 1

In this experiment two words were tested (plasta, plasma) (see Table 1). Segment durations were changed in order to study discriminability. The durations of the "s" and the "t" in /plasta/ were furthermore changed keeping the total duration of the cluster constant. The experiment was designed as an ordinary ABX test where X could be either A or B. The subjects listened to triads and had to write down if X sounded like A or B. The reference occurred at least once in each triad. The test was balanced so that no within-the-triad sequential effects could influence the result, i. e. each manipulated stimulus resulted in four triads (ABA, ABB, BAA, BAB). The triads were randomized and presented in sequences of 100 with a pause after 50. The first and last two triads were rejected. The result is presented in Fig. 1.

The graphs show the number of incorrect responses. Random responses at 50 % The 5 % and 1 % levels are also indicated. All comparisons are made inbetween the reference, marked in the middle of each graph and a distorted stimulus marked along the X-axis. Four points should be made about the result.

(a) The vowel is clearly more discriminable than the consonants (Fig. 1c). That is correct even in relation to the voiced nasal (Fig. 1e).

(b) A compensatory change of the phoneme boundary inbetween s and t is harder to perceive than a change of a single consonant duration (Fig. 1a, 1b), i. e. compensatory effects exist at the level of discrimination.

(c) Some subjects complained describing the test as "boring" and "difficult".

(d) How much does the test design influence the result? The short term memory of the subjects has to be used during the similarity estimation and three stimuli might introduce memory restrictions.

Experiment 2

In this experiment segments in five test words were studied. Three words "lassa", "platta", "ramma" containing phonologically long consonants were used and compared to a single word "astma" containing all three consonants as a medial cluster. An expanded test on the word "plasta" was included.

Because of the remarks (c) and (d) in experiment 1 the test design was changed to an ordinary AX procedure. The subjects listened to stimulus pairs and marked

	Test type	Word	Variation in single segment	Compensatory variation between segments	Number of subjects	Answers/manipulated stimulus	Answers/reference stimulus	Figure
Experiment 1	ABX	plasta	s, t	st	2	16	0	1a-b
	ABX	plasma	a, s, m	–	4	16	0	1c-e
Experiment 2	AX	lassa	s	–	7	14	112	2a
	AX	platta	t	–	7	14	112	2b
	AX	ramma	m	–	7	14	112	2c
	AX	astma	a, s, t, m	st, tm	8	16	336	2a-f
	AX	plasta	a, s, t	as, st	13, 14	26, 28	536	3d-f
Experiment 3	X	plasta	s	as, st	14	42	336	3a-c

Table 1

on the answering sheet if they perceived them as same or different. This type of test demands a fair portion of equal pairs in order to balance the two alternative responses. The increase of stimuli is a drawback. However, the subjects do not give a 100 % "similar" on equal pairs and the responses on these could be used for estimating the chance level and setting of significance levels. Fig. 2 and 3d-f present the result. Significance levels could be related to Fig. 1. The result is summarized below and will be more generally discussed in the conclusions.

(a) As expected from experiment 1 the perception of the vowel (Fig. 2d, 3d) is more critical than the consonants.

(b) Compensation within a cluster decreases the possibility of perceiving duration variation (Fig. 2e, 3e, 3f). This is also the case for the unvoiced stop voiced nasal boundary which in this case is a syllable boundary (Fig. 2f).

(c) No general trend could be found for the single consonant/consonant cluster relation. The "s" seems to be rather unaffected by the position despite the change of the actual reference duration. This is in disagreement with psychoacoustical findings. The "t" follows these findings but the "m" is contrary to them.

(d) Compensation inbetween the vowel and the fricative (Fig. 3d) does not influence the perception. This might be interpreted as a manifestation of the perceptual need for "correct" vowel length, a need stronger than perceiving the duration balance between syllables. This will be studied more in the production experiment below.

(e) Finally, some comments on peculiar points in Fig. 3e. The two leftmost points in the compensated word give contradicting information. Comparing the same points in experiment 1 (Fig. 1a, 1b) we find evidence to suspect that the second leftmost point in Fig. 3e is misleading. A more detailed analysis gives the result that one of two stimuli carries 80 % of the dissimilar responses for this point.

Experiment 3

This experiment was designed to test the acceptability of stimuli used in some of the discrimination tests. Manipulated versions of the word "plasta" were recorded and presented to the subjects. The acceptability was estimated on a scale from zero ("acceptable") to ten ("not acceptable"). More details are presented in Table 1. The results are presented in Fig. 3a-3c and should be compared to the result of the discrimination test in Fig. 3d-3f.

Conclusions can be summarized in four points.

(a) As expected from earlier experiments (HUGGINS) the duration of the vowel influences the acceptability more than that of the consonants.

(b) The compensated manipulation of the "st" cluster doesn't at least lower the acceptability level. However, a positive change is hard to defend with our data.

(c) The discrimination method is far more sensitive to durational change but the close agreement inbetween the two experiments is obvious.

(d) Astonishingly great duration variation is accepted.

Production experiments with synthetic speech

As mentioned in the introduction we have tried a production paradigm to look into the durational interaction within a word. An advantage with a production experiment is that it gives a direct figure of preferred length and also an accuracy

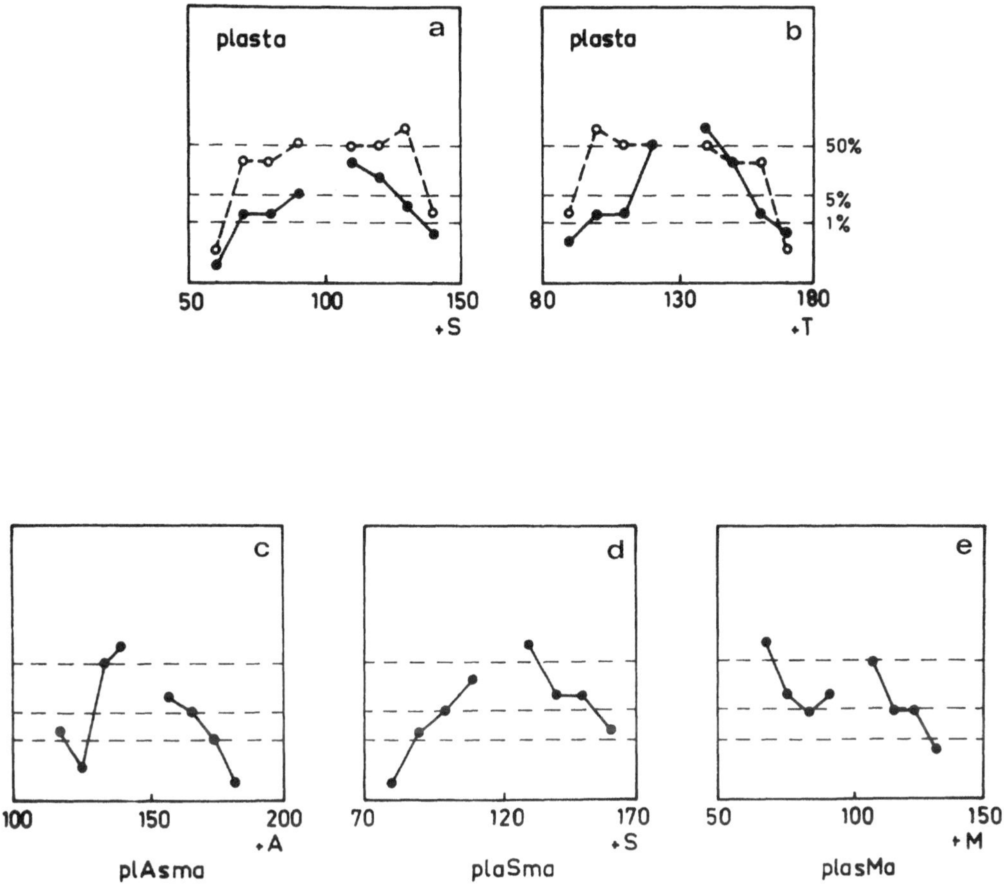

Fig. 1. Results from the ABX test in experiment 1. In the triads one of the first two stimuli was always unaltered. The horizontal axis shows the duration in msec of the manipulated segment indicated below the right side of the squares. The number in the middle is the original length of the segment. The vertical axis shows the number of incorrect responses. The 50 % line indicates chance level. Points below the 5 % and 1 % line respectively are significantly different at these levels. Solid lines refer to manipulations of one segment only. Dotted lines in Fig. "a" and "b" correspond to stimuli where the total length of the cluster "st" was kept constant (compensation).

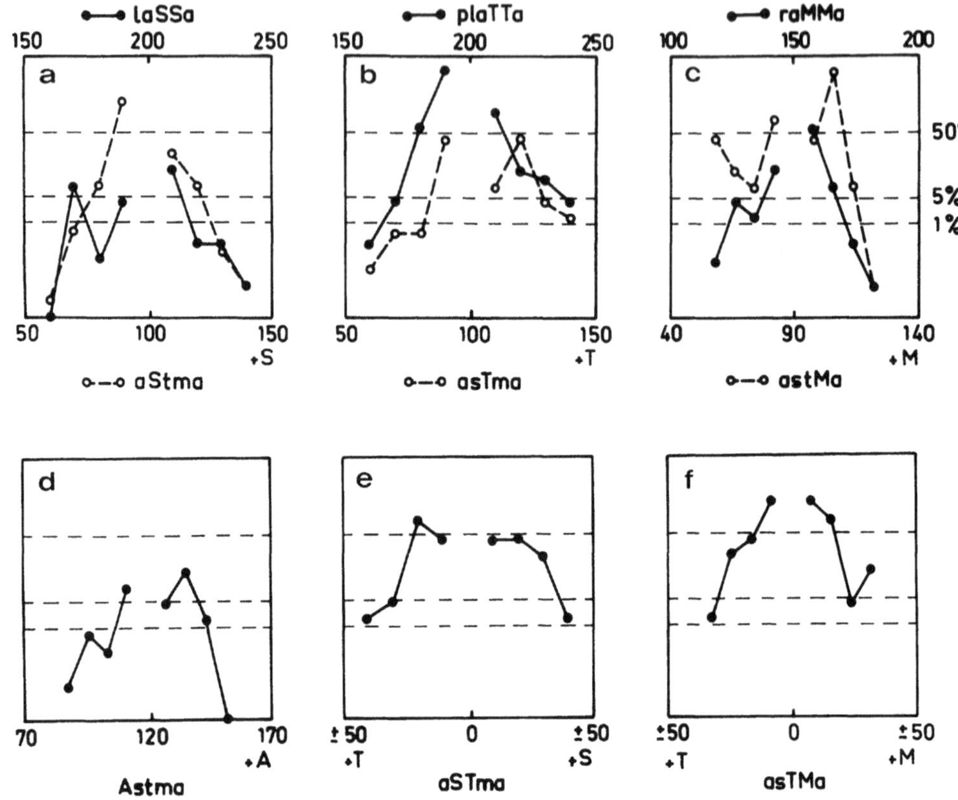

Fig. 2. Results from the AX test in experiment 2. At least one stimulus in a pair was unaltered. The horizontal axis shows the duration in msec of the manipulated segments, indicated with capitals. Note that there are two scales in Fig. "a", "b", and "c" referring to single consonants (top, solid line) and cluster consonants (bottom, dotted line). The number in the middle is the original length of the segment. In Fig. "e" and "f", where the total duration of "st" and "tm" is kept constant, the horizontal scale is the duration relative to the original length. The absolute values of these segments can be seen in Fig. "a", "b", and "c", bottom scale. The vertical axis is number of "same" responses. The 50 % line indicates mean number of same responses for **equal** pairs. Responses below the 5 % and 1 % respectively are significantly perceived different at these levels.

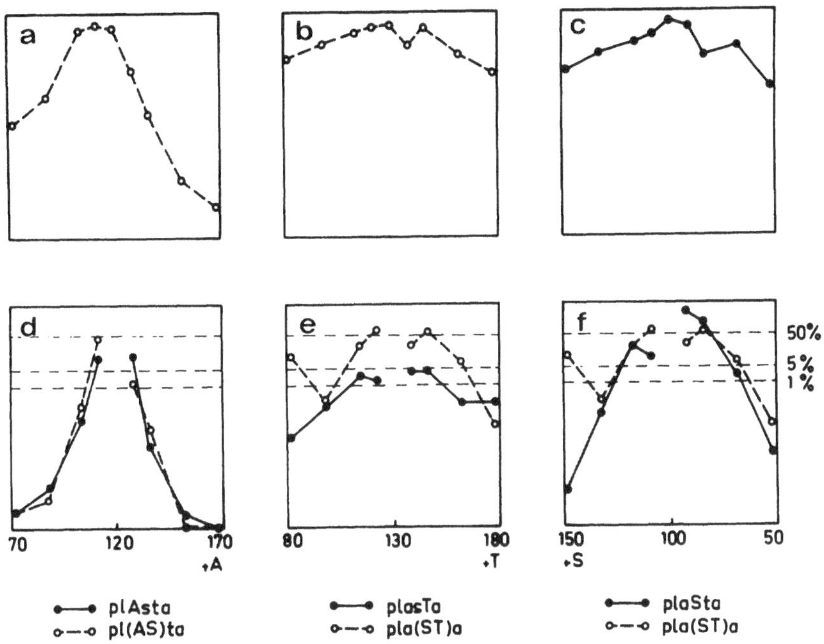

Fig. 3 "d", "e", "f". The result from the AX test in experiment 2. The display is explained under Fig. 2. The solid lines refer to manipulations of one segment only. Dotted lines correspond to stimuli where the total duration of "as" and "st" is kept constant. Note that the horizontal axis in Fig. 3 "f" is reversed for comparison. Fig. 3 "a", "b", "c" show the result from experiment 3. The vertical axis is judged acceptability. The horizontal axis and the stimuli used are the same as in Fig. 3 "d", "e", and "f", respectively.

estimate. Synthetic speech has been used previously in production experiments for verifying the perceptual importance of vowel quality (23) and word structure (19) in estimating vowel length. A very low spread (s. d. about 5 msec) was demonstrated in these experiments indicating its usefulness.

Method

In the experiments we used an expansion of our system for synthesis by explicit, ordered rules (6). By means of a joy stick external variables could be introduced in the rules. The joy stick coordinates were displayed on a screen for visual feedback.

The subject accepted the stimulus, causing a print-out, or rejected it, causing a new synthesized utterance with the present joy stick values introduced in the rules.

In this series of experiments the duration of one known segment was controlled by the subject. After each acceptance a new stimulus was produced by randomizing the duration of another segment and also randomly changing the zero point of the joy stick.

The basic durational structure of the utterance was copied from natural speech.

In a preliminary experiment PO, Fig. 4, the word "plasta" was presented in pairs where the duration of the first "a" in the second word was randomized and the "s" duration of the same word was adjusted. This situation with a fixed external reference turned out to favor the rhythmic (i. e. vowel onset) criterion with 80 % estimated compensation (i. e. negative slope of -.80 of the regression line). This is, however, a quite unnatural situation and the experiment only tells us that an almost perfect compensation could occur under certain conditions.

In the further experiments we have worked with a word embedded in a sentence without an external reference, "nu ska vi _____ båten", where the test word carries the sentence stress.

We wanted to map the perceptual dependence between segments within a word of quite common structure under a condition where the weight of different criterion should be reasonably similar to a normal listening situation. Questions that will be dealt with in connection with the individual experiments include the importance of segment type (consonant-vowel, voiced-unvoiced, same or different place of articulation) and position relative to the stressed syllable nucleus. Needless to say we will only be able to draw conclusions about perceptual demands which might (or might not) influence temporal adjustments in real speech production.

In experiment P1 presented in Fig. 4 we duplicated experiment PO, i. e. the first "a" randomized and "s" adjusted but now in sentence context and adjusted to the subjects' internal reference. As can be seen no or almost no compensation occurred. Possible asymmetries were tested by randomizing "s" and adjusting "a" (experiment P2). The picture is not entirely clear (Fig. 4). One subject has a compensation of 19 % but this figure relies quite heavily on the single s = 40 msec point which was quite an extreme stimulus. The other subject showed a slightly positive slope of the regression line. These deviations from the no compensation hypothesis are however not highly significant.

The next experiment P3 ("t" randomized, "s" adjusted) tested the compensation within a post-vocalic consonant cluster "st". Here a substantial amount of compensation was preferred, 41 % and 60 % for the two subjects (Fig. 5) (the standard deviation in compensation can be estimated at 8 % and 13 % resp.). One question to be answered is whether this ease of compensation is due to the shared place of articulation.

In experiment P4 we changed "plasta" for "plaska" randomizing "k" and adjusting "s", Fig. 5. As can be seen both subjects showed less compensation. The decrease is however much smaller than the remaining compensation. It might be argued that the postvocalic cluster, even though morphologically associated with

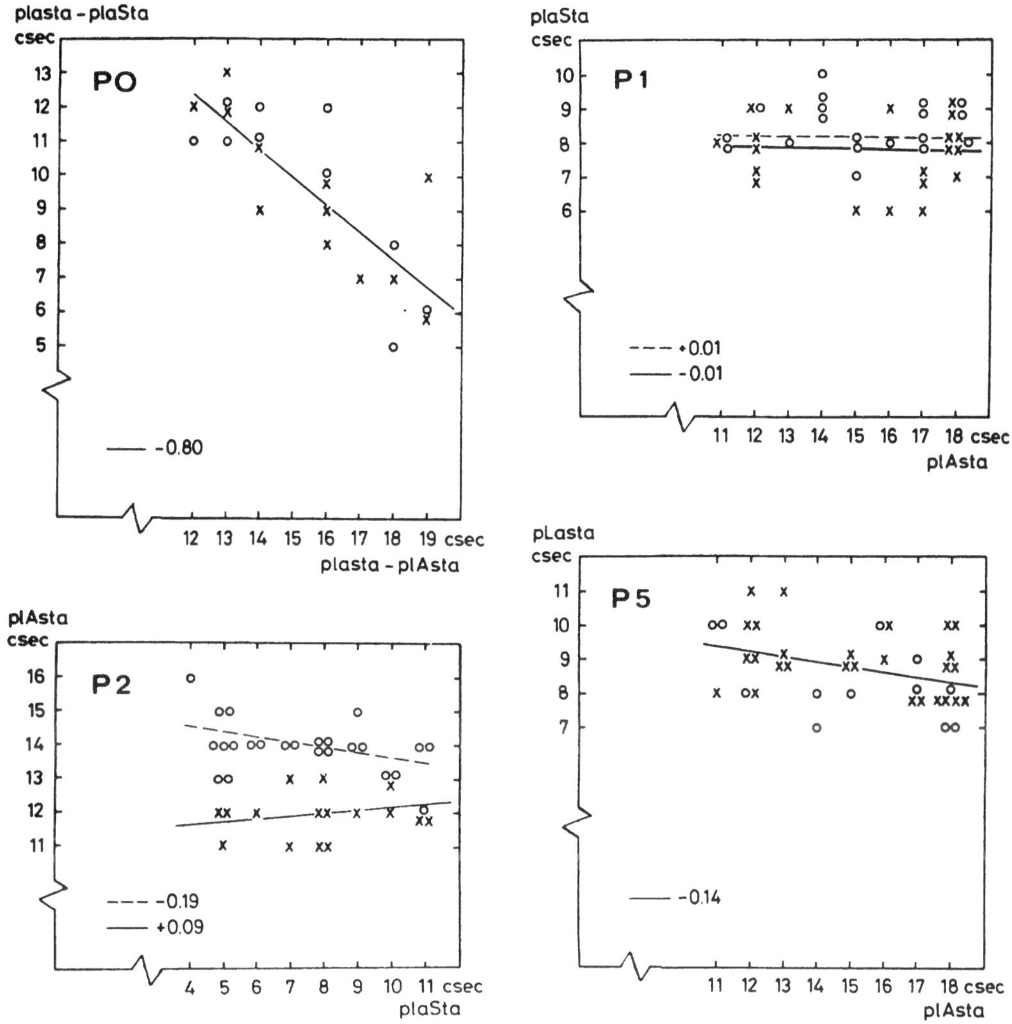

Fig. 4. Vowel-consonant interaction. Results from production experiments using synthetic speech. All durations are given in 1/100 sec (csec). The horizontal axis is the randomized duration of a segment indicated by capitals. The vertical axis is the duration of the, by the subject, adjusted segment indicated by capitals. Crosses and circles pertain to different subjects. The straight lines in the figures are regression lines the slopes of which are given in the low left corners. When two lines are drawn the data from the two subjects are separated (straight line - crosses, dotted line - circles). In experiment PO the test word is presented after a reference. In experiment P1 through P8 the test word is embedded in the sentence "nu ska vi ________ båten".

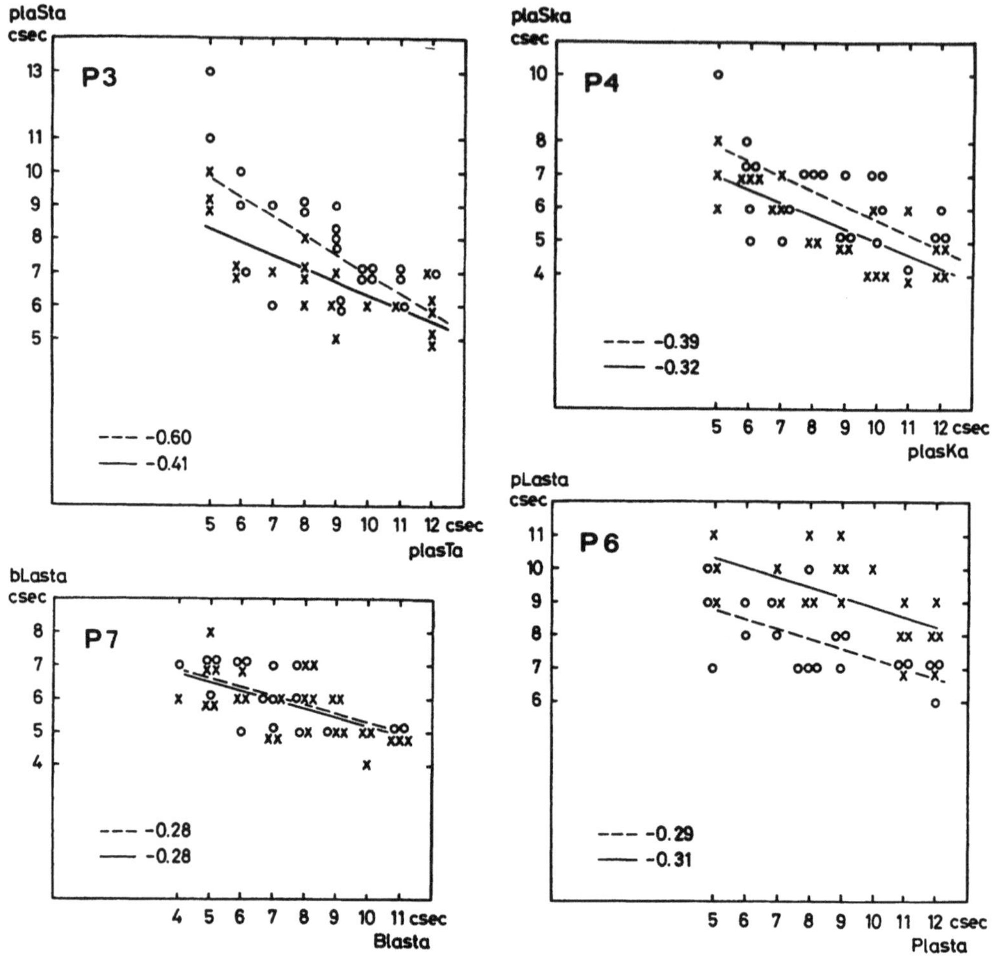

Fig. 5. Consonant-consonant interaction. Results from production experiments using synthetic speech. All durations are given in 1/100 sec (csec). The horizontal axis is the randomized duration of a segment indicated by capitals. The vertical axis is the duration of the, by the subject, adjusted segment indicated by capitals. Crosses and circles pertain to different subjects. The straight line in the figures are regression lines the slopes of which are given in the low left corners. When two lines are drawn the data from the two subjects are separated (straight line - crosses, dotted line - circles).

the preceding vowel, tends to be more related to the following vowel nucleus (i. e. open syllables as basic units (16)).

Experiment P5 (first "a" randomized, "l" adjusted, Fig. 4) tested the interaction between prevocalic consonant and the stressed vowel. In this case a slight (but not highly significant) compensation was demonstrated. In this position HUGGINS (14) could not find any indication of compensation taking this as an argument for the vowel onset hypothesis. However, it is conceivable that the voiced character of the "l" made the "la" function as a unit ("l" is partly aspirated in this position but the part subjected to variations was entirely voiced). If this was true we should find a difference in the demand for compensation in a totally voiced or unvoiced or just partially voiced consonant cluster.

For the mixed cluster in experiment P6 ("p" randomized and "l" adjusted, Fig. 5) we can see a compensation of close to 30 % for both subjects. By the way, we can see that the subjects in this case have clearly different internal references (15 msec difference in preferred "l" length). Looking at the all-voiced cluster (experiment P7, "b" randomized, "l" adjusted in "blasta", Fig. 5) we find a very similar amount of compensation 28 % indicating that the voicing is not an important determinant for temporal grouping. This figure compares nicely with the 35 % mean compensation in the final non-homorganic all-voiceless cluster in experiment P4. Comparing preferred "l" durations in experiments P6 and P7 we can note an interesting detail. The durational difference between aspirated and unaspirated "l" that has been observed in natural speech (10, 15) is nicely reproduced in our experiment supporting the view that allophonic variations of this kind have real perceptual significance.

All experiments up to this point have dealt with temporal interaction between adjacent segments. It has been argued by HUGGINS (14) that most compensation occurs across syllable boundaries. To test one such interaction we randomized the final "a" and adjusted the first "a" (experiment P8, Fig. 6). As can be seen no compensation was found, on the contrary, the vowel durations showed high positive correlation. The data points could best be described by one horizontal line corresponding to the minimum acceptable duration of the primary stressed first "a" and a regression line for durations of the secondary "a" exceeding 100 msec.

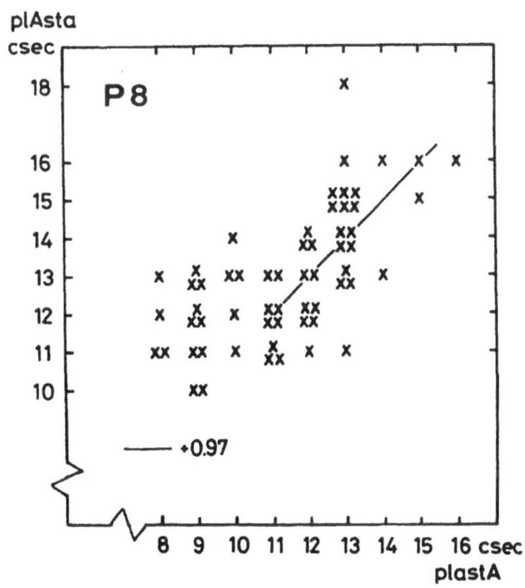

Fig. 6. FirstA adjusted when last A randomized. See Fig. 4 for further explanation.

The slope of this line is .97 pointing towards a demand on temporal balance between the two syllables. The degree of emphatic stress on the word increases with increased length of the two vowels. It should however be kept in mind that the second vowel carries secondary stress with a distinctive intonational pattern in focus position. Truly unstressed vowels may behave more like consonants and be more apt to compensation as suggested by HUGGINS (14).

The result from the production experiments P1 through P8 are summarized below in terms of mean slopes of the regression lines. Arrows point at adjusted segments.

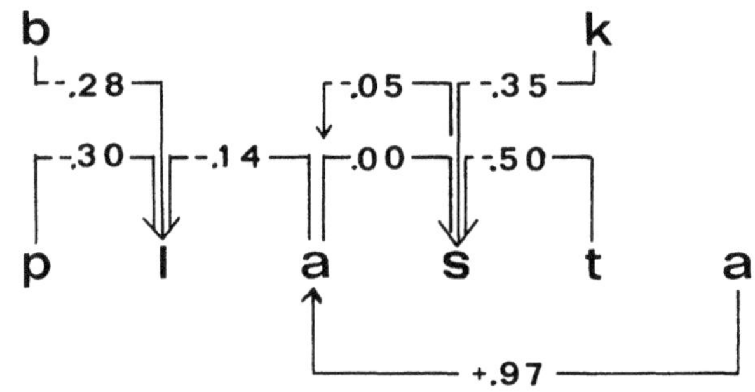

General discussion

From several studies we know that a number of factors are important in determining the durational structure of an utterance such as segment type, stress, quantity, vowel quality, degree of emphasis, position within a word or a phrase, etc. Our prime concern has, however, been with the precision and the interaction between segments within a word asked for by the perceptual system. Even if our results are preliminary in nature and in some instances lacking a high degree of statistical significance we believe that it is possible to summarize our conclusions under the headings below.

Relation to psychoacoustics

Results from tone, noise, and silence discrimination could not explain the difference in discriminability of durations in speech. The results presented by FUJISAKI et al. that all segment types were equally discriminable were not reproduced in our experiments. It should, however, be kept in mind that their results were obtained in a phoneme boundary condition.

Accuracy of segmental duration

Vowel durations showed to be most crucial in the discrimination and rating experiments, indicating that vowel length is more important than the duration between vowels or vowel onsets. In the production experiment, however, no reliable difference was found in the accuracy of adjustment of different segment types and the spread was quite great compared to results with isolated words of simple structure. This might be explained by the fact that an expanded sentence context introduces a multitude of possible criteria the weight of which is not uniquely determined.

Our results support an approach that takes vowel lengths as the basic unit of the descriptive framework.

Compensatory effects

Compensation within a consonant cluster has been shown both to decrease discriminability and increase acceptability (in the production experiments). This is in agreement with the vowel onset hypothesis, keeping vowel onset times unaltered. In homorganic clusters a higher degree of compensation might be preferred possibly depending on the less pronounced articulatory constraints. However, the demand for compensation between consonants and vowels is either very weak or non-existent irrespective of the position relative to the vowel or shared voicing cue. This points to the fact that vowel length (rather than just onset time) plays a primary role in perception. Under certain conditions where the rhythmic demands are emphasized substantial compensation occurs between vowel and consonant (experiment PO).

Perfect compensation (100 %) has, however, not been demonstrated in any case supporting our view that a number of sometimes contradictory demands have to be fulfilled simultaneously resulting in a compromise.

Perceptual balance

A tendency to keep durational ratios constant would, if it occurred between segments, counteract the tendency for compensation on the segmental level but constant ratios between syllables and longer units would in fact demand a within the unit compensation. The positive correlation between vowel lengths (experiment P8) within a word strongly indicates that the temporal balance between syllable nuclei within a word is of major importance, far more so than the normalizing effect of the sentence or the intra-syllable balance.

References

(1) ABEL, S.M.: "Discrimination of Temporal Gaps", J.Acoust.Soc.Am. 52 (1972), pp. 519-524.

(2) ABEL, S.M.: "Duration Discrimination of Noise and Tone Bursts", J.Acoust. Soc.Am. 51 (1972), pp. 1219-1223.

(3) BURGHARDT, H.: "Subjective Duration of Sinusoidal Tones", paper 20 H 1, 7th International Congress on Acoustics, Budapest 1971.

(4) CARLSON, R., GRANSTRÖM, B., LINDBLOM, B., and RAPP, K.: "Some Timing and Fundamental Frequency Characteristics of Swedish Sentences: Data, Rules, and a Perceptual Evaluation", STL-QPSR 4/1972, pp. 11-19.

(5) CARLSON, R. and GRANSTRÖM, B.: "Word Accent, Emphatic Stress, and Syntax in a Synthesis by Rule Scheme for Swedish", STL-QPSR 2-3/1973, pp. 31-36.

(6) CARLSON, R. and GRANSTRÖM, B.: "A Phonetically Oriented Programming Language for Rule Description of Speech", to be publ. in Speech Communication, Almqvist & Wiksell, Stockholm 1975, Vol. 2.

(7) CARLSON, R., ERIKSON, Y., GRANSTRÖM, B., LINDBLOM, B., and RAPP, K.: "Neutral and Emphatic Stress Patterns in Swedish", to be publ. in Speech Communication, Almqvist & Wiksell, Stockholm 1975, Vol. 2.

(8) FUJISAKI, H., NAKAMURA, K., and IMOTO, T.: "Auditory Perception of Duration of Speech and Non-Speech Stimuli", to be publ. in Auditory Analysis and Perception of Speech, Academic Press, London 1975.

(9) HAGGARD, M.P.: "Effects of Clusters on Segment Durations", pp. 1-50 in Speech Synthesis and Perception, Progress Report No. 5, 1971, Psychological Laboratory, Cambridge.

(10) HOLMGREN-RAPP, K., unpublished data.

(11) HUGGINS, A.W.F.: "The Perception of Timing in Natural Speech I: Compensation Within the Syllable", Language and Speech 11 (1968), pp. 1-11.

(12) HUGGINS, A.W.F.: "How Accurately Must a Speaker Time His Articulations?", IEEE Transac. on Audio and Electroacoustics, AU-16 (1968), pp. 112-117.

(13) HUGGINS, A.W.F.: "Just Noticeable Differences for Segment Duration in Natural Speech", J.Acoust.Soc.Am. 51 (1972), pp. 1270-1278.

(14) HUGGINS, A.W.F.: "On the Perception of Temporal Phenomena in Speech", J.Acoust.Soc.Am. 51 (1972), pp. 1279-1290.

(15) KLATT, D.H.: "Durational Characteristics of Prestressed Word-Initial Consonant Clusters in English", pp. 253-260 in QPR No. 108, 1973, Research Lab. of Electronics, Massachusetts Institute of Technology.

(16) KOZHEVNIKOV, V.A. and CHISTOVICH, L.A.: Speech: Articulation and Perception, US Dept. of Commerce Translation, JPRS 30, 543, Washington, D.C. 1965.

(17) LEHISTE, I.: "Temporal Compensation in a Quantity Language", pp. 929-937 in Proc. of the 7th International Congress of Phonetic Sciences, Montreal (ed. A. Rigault and R. Charbonneau), Mouton, The Hague 1972.

(18) LINDBLOM, B.E.F.: "Some Temporal Regularities of Spoken Swedish", to be publ. in Auditory Analysis and Perception of Speech, Academic Press, London 1975.

(19) NOOTEBOOM, S.G.: Production and Perception of Vowel Duration, Philips Res. Rep., Suppl. 5, 1972.

(20) NOOTEBOOM, S.G.: "Contextual Variation and the Perception of Phonemic Vowel Length", to be publ. in Speech Communication, Almqvist & Wiksell, Stockholm 1975, Vol. 3.

(21) OHALA, J.J.: "The Temporal Regulation of Speech", to be publ. in Auditory Analysis and Perception of Speech, Academic Press, London 1975.

(22) O'SHAUGHNESSY, D.: "Consonant Durations in Clusters", IEEE Transac. on Acoustics, Speech, and Signal Processing, ASSP-22 (1974), pp. 282-295.

(23) PETERSEN, N.R.: "The Influence of Tongue Height on the Perception of Vowel Duration in Danish", ARIPUC 8 (1974), pp. 1-10 (Univ. of Copenhagen).

(24) RAPP, K.: "A Study of Syllable Timing", STL-QPSR 1/1971, pp. 14-19.

DISCUSSION of "Perception of segmental duration"
CARLSON and GRANSTRÖM

LEHISTE:

You found a negative correlation (i.e. compensatory adjustments) between the /l/ and the /a/ in the word plasma. I presented a paper at the 1972 Bologna congress of linguists on which I reported segment durations on English words like <u>bed</u> - <u>bled</u>, <u>back</u> - <u>black</u> - <u>bank</u>. I found that timing control proceeds from the release of the first plosive to the closure for the second - the boundaries are major changes in the manner of articulation. The speaker tends to keep the duration of the whole span constant - the duration of the vowel in <u>bed</u> is comparable to the duration of the resonant - vowel sequence in <u>bled</u>.

Swedish is a language closely related to English: it might be that the same phenomenon occurs in Swedish.

CARLSON:

One of the reasons why we chose a word with a sonorant preceding the vowel was to test whether the sonorant was more closely connected to the vowel than to the other consonant in the cluster. As can be seen from our data the compensation is greater between the consonants (.30) than between /l/ and the vowel (.14). However, the sonority of the consonant may well be important, since we could see no compensation between the vowel and the following fricative.

KLATT:

You observe that there is a positive correlation between the expected durations of the two vowels in the test word. One possible explanation for this is that listeners perceive an increased or decreased speaking rate during the word. Is this a plausible explanation for positive interactions for vowels in different syllables, but negative or zero CV, VC and CC interactions?

CARLSON:

Yes, we think that this is a very probable explanation. It seems crucial that the durational balance between syllable nuclei in a word is right. That is also a possible explanation to the smaller durations j.n.d.'s in two-syllable words in your study.

FUJISAKI:

The discriminability of segmental durations reported in my study (1973) was measured in words where the duration of a particular segment discriminates between a minimal pair, and the obtained discriminability was almost equal for all types of segments, regardless of whether the crucial segment was a vowel, a nasal, a fricative, or a voiceless stop. Since the investigation by CARLSON and GRANSTRÖM do not deal with such situations, some discrepancies are naturally expected. I suggest, however, that the data should be further processed statistically in order that they may be compared more directly with existing data.

THE ROLE OF INTONATION IN SPEECH PERCEPTION

R. Collier
Belgian National Science Foundation (NFWO)
and
University of Leuven
Department of Linguistics
Blijde-Inkomststraat 21
B-3000 Leuven - Belgium
and
J. 't Hart
Institute for Perception Research
Den Dolech 2, Eindhoven 4502
The Netherlands

1. Definition of the problem

Past research of intonation at the Institute for Perception Research has been concerned with describing melodic properties of utterances in terms of perceptually relevant pitch events, without taking into consideration their possible communicative functions. The aim of this approach has been to establish the relationship between the continuously varying fundamental frequency of the speech signal and the corresponding discrete perceptual units.

In trying to reduce the wellnigh infinite variability of the acoustic phenomena to a relatively small set of perceptual categories, use has been made of an Intonator (Willems, 1966), by means of which the course of the natural fundamental frequency of an utterance can be replaced by an artificial, and hence controllable one. Starting from a faithful copy, some of the original fundamental frequency variations are left out, smoothed, or standardised, without affecting the perceptual equivalence between the original and the artificial version. The stylised perceptual equivalent of a natural fundamental frequency curve has been called a pitch <u>contour</u>. It consists of a sequence of pitch <u>movements</u> which alternate between the higher and the lower reference level of the so-called <u>declination line</u>. Several types of pitch movement have been distinguished on the basis of criteria of perceptual relevance, such as the direction of the pitch movement, its position in the syllable, and so on.

In analysing the internal composition of pitch contours it appeared that some successions of pitch movements are more coherent than others in that they build recurrent clusters. Therefore it seemed possible to distinguish a descriptive unit of a size in between that of the pitch movement and the contour. We have called this unit an intonational <u>block</u>. Its internal structure is governed by rather strict rules, whereas its external coupling to other blocks appears to be almost free in the sense that practically any other block of the same type can be substituted for it.

Once an adequate, albeit rather broad, descriptive system of the purely melodic aspects of (Dutch) intonation has been developed, it is feasible to examine its possible functions. More particularly, we would like to concentrate our efforts on establishing the role of the intonational blocks in speech perception. This interest is also stimulated by the current concern about the importance of intonational cues as mediators in speech processing (Wingfield and Klein, 1971; Svensson, 1974).

We will start from the assumption that the speaker, in producing these blocks, provides the listener with cues that help him to decode the message. This implies that both the speaker and the listener must have knowledge about "functional" rules that are operative in the implementation of the blocks. The nature of these rules can be illustrated by means of the example in Figure 1.

Figure 1 illustrates the subdivision of the contours into blocks, showing in a schematic way that for the given utterance at least nine different pitch contours are equally acceptable. This is in accordance with our earlier finding ('t Hart and Cohen, 1973) that the application of what are called optional alternative rules enables a number of different well-formed contours to be derived from one basic intonation pattern. In this example, as well as throughout this study, we will confine ourselves to only one type of basic pattern, viz. the "hat pattern" (Cohen and 't Hart, 1967). The nine alternative pitch contours of the example in Figure 1 have to be considered as "free variants": they are not uniquely related to nine different syntactic or semantic interpretations of the utterance. Nevertheless, the potential occurrence of the above mentioned melodic differences between the contours may indicate that they reflect in different ways or in different degrees certain aspects of the linguistic organisation of the utterance. It is to this aspect of the relationship between intonation and syntax that we will direct our attention in the two experiments to be described below.

$_S(_{S_1}$((wat <u>aar</u>dig)(van je))$_{S_1}$ $_{S_2}$((dat)(je)(<u>gis</u>teren)(al weer)(<u>op</u>belde))$_{S_2}$)$_S$

(how kind of you that you yesterday again called)

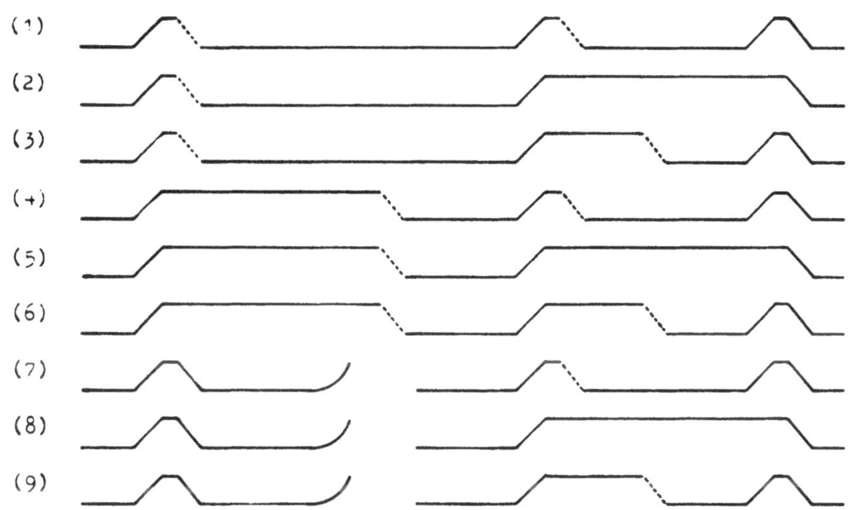

Fig. 1. The syntactic structure of an exemplary sentence and nine compatible pitch contours (with pitch accents on the underlined syllables, declination omitted). Contours 2, 5 and 8 contain two blocks, the remaining six contours contain three blocks.

Let us consider the alternative possibilities of Figure 1 in more detail. Every intonational block contains at least one prominence-lending movement (on the underlined syllables). This prominence-lending function can be implemented by either a rise, a fall or a rise-plus-fall (the syllables "aar-" in (4, 5, 6), "op-" in (2, 5, 8) and "op-" in (1, 3, 4, 6, 7, 9), respectively). An additional function of the blocks, which may help to explain the observed differences, could be that the single block in (2, 5, 8) "integrates" the constituents "gisteren" and "opbelde", contrary to what happens in the two blocks of the other contours. The melodic difference between (3), (6) and (9) on the one hand, and (1), (4) and (7) on the other is, that in (3), (6) and (9), the non-final fall, (dotted line) coincides with the end of "gisteren", whereas in (1), (4) and (7) this fall occurs, together with the rise, on the first syllable of "gisteren". The assumption that in the contours (3), (6) and (9) the delayed non-final fall (i.e. the block boundary) marks the syntactic boundary between the two adverbial adjuncts "gisteren" and "al weer" seems to be confirmed in the left-hand parts of contours (4), (5) and (6). Here, as opposed to what is the case in (1), (2) and (3), the non-final fall is also postponed in order to coincide with the syntactic boundary between the main clause (S_1) and the sub-clause (S_2). In contours (7), (8) and (9), finally, this syntactic boundary is marked by a continuation rise in the last syllable of S_1, followed by a low resumption of the pitch at the start of S_2.

If the regularities suggested by the example in Figure 1 correspond to any rules capable of generalisation, these should be recoverable as alternative strategies on the part of the speaker. At least three of these can be distinguished:

Strategy No. 1: The simplest strategy is that in which the speaker realises a rise-fall on every successive syllable that has to be marked by a pitch accent. This amounts to a decision, in respect of each syllable, as to whether it has to be accented or not. The application of such a strategy would result in a contour such as (1) in Figure 1.

Strategy No. 2: In order to realise a contour of type (2), the speaker should also know which syllables have to be accented. Moreover, he should anticipate which of these syllables is the last accentuated one in the utterance and which other is the penultimate.

Strategy No. 3: In order to produce a contour of type (6), the speaker, after having realised a prominence-lending rise, should know where the end of the current syntactic constituent* is located, so as to be able to assign the non-final fall to the first syllable of the next syntactic constituent.

* In accordance with the current terminology of generative syntax, the term "constituent" is used, to denote such words or word groups as may occur under the labels NP, VP, AdvP, etc., in a phrase marker.

A comparable strategy is needed for the realisation of a contour such as (9), in which case the continuation rise should coincide with the last syllable of the syntactic constituent.

Evidently, if the listener is aware of the existence of these strategies and of the rules for their implementation, he will find in the course of the pitch a number of clues concerning the accentual and, in certain cases, also concerning the syntactic organisation of the utterance.

In the experiments to be discussed we have attempted to verify the hypothesis that specific types of intonational blocks are used to mark the boundaries of elements in the syntactic structure of an utterance. We have limited the verification to realisations of the "hat pattern". The hypothesis is then to the effect that contours derived from the "hat pattern" delimit syntactic constituents either by the occurrence of a continuation rise on the last syllable of a constituent, of by the presence of a non-final fall on the first syllable of a following constituent. Our experimental tests thus amount to a verification of strategy No. 3, and its counter-part on the side of the listener. Strategies Number 1 and 2 will receive less attention, because they do not require the anticipation of syntactic boundaries.

2. Experiments

The hypothesis has been tested in two experiments, one in which the subjects had a listening task, and another in which they were instructed to read aloud.

For the listening experiment stimuli were produced in the following way. On a tape loop a periodic signal was recorded after it had been segmented into (13 or 15) "syllables" of 200 msec duration, separated by silent intervals of 50 msec. These syllables were fed into the Intonator in order to be provided with a controllable pitch contour, in which the position of the non-final fall was varied systematically as is shown in Fig. 2.

In the stimuli 1A to 1E prominence-lending rises are invariably located on the second and eighth syllable, and syllable thirteen always contains a prominence-lending fall. The position of the non-final fall is varied in such a way that it occurs on the second, the fourth, the fifth, the sixth and the seventh syllables in the five respective contours. Likewise, in the stimuli 2A to 2E the po-

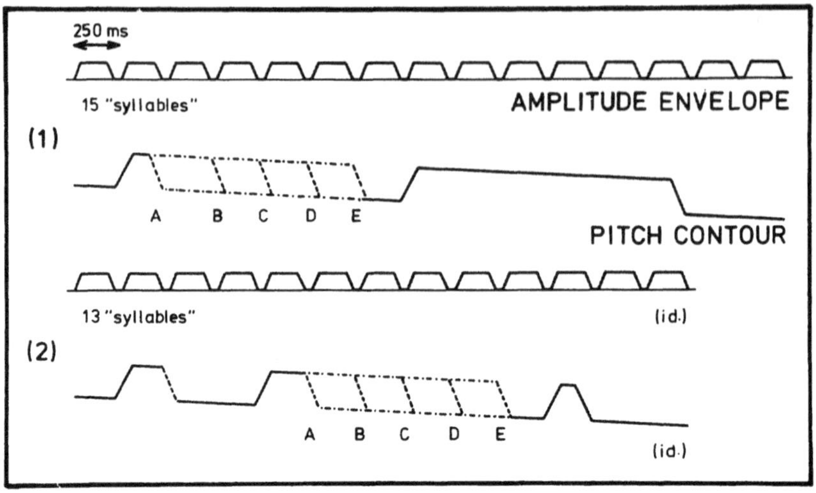

Fig. 2. Stylised representation of the stimuli used in the listening experiment.

sition of the rises is fixed on the syllables No. 2 and 5, that of the rise-fall on syllable No. 11. The non-final fall after the first rise is also fixed, viz. on the third syllable, whereas the second non-final fall is on the syllables No. 6, 7, 8, 9 or 10. The pitch variations applied are in accordance with the rules used in the synthesis of ordinary pitch contours derived from the hat pattern (e.g. the size of the rises is four semitones, that of the falls is five semitones; the durations are 100 and 75 msec, respectively; isolated non-final falls occur typically early in the syllable, etc.).

The spectral shape of the output signal can be characterised as containing a first formant at about 400 Hz, a second formant at about 800 Hz, and a low energy third formant at about 2400 Hz.

Inspired by a procedure used by Svensson (1974), we presented five listeners with the hummed pitch contours, instructing them to think of syntactically coherent but not necessarily meaningful sentences containing the same number of syllables and for which the hummed contour would constitute a suitable intonation, including the locations of the pitch accents. The ten hummed contours described in Figure 2 were played back from a tape loop and repeated as long as necessary for the subjects, to write down at least two suitable sentences for each contour. On the answer sheets the number of syllables was indicated by a corresponding number of dashes, with arrows on the accentuated syllables.

In the reading experiment we presented ten speakers with 30 typewritten sentences in which the syllables that had to be accented were underlined. These sentences were selected arbitrarily among those that had been written down by the other subjects of the listening experiment. The major systematic difference between the sentences was that one of their syntactic constituents varied in length. The subjects were phonetically naive and not familiar with any explicitly stated "rules" for Dutch intonation. They were asked to prepare their reading performance and to repeat each sentence twice. We shall now consider the analysis of the results of the two experiments.

2.1 Listening experiment

The subjects wrote down a total of 144 sentences, which, in their opinion, were compatible with the accentuation and intonation characteristics of the hummed contours. We examined these sentences in the light of our hypothesis that the position of the non-final fall is used to mark intonationally the end of a syntactically coherent whole. We therefore asked ourselves for each type of contour (1A to 1E, and 2A to 2E) whether the subjects, in thinking of compatible sentences, chose ones in which syntactic constituent boundaries and non-final falls stand in such a relationship that the non-final fall occurs on the first syllable of the following syntactic constituent. The verification of the hypothesis implies that the experimenters assign some kind of syntactic structure to the sentences proposed by the subjects. Our syntactic analysis amounted to delimiting (in the surface structure of the sentences) the syntactically coherent word groups and their hierarchical relationships*. Illustrations of such an analysis are given in Figure 1 and on examples (1), (2) and (3).

The results of the experiment divide into three categories: 103 instances that confirm the hypothesis, 16 instances that refute the hypothesis, and 25 instances that are indifferent with regard to the hypothesis.

* A random sample of sentences was submitted for syntactic analysis to an independent judge, F.G. Droste, professor of linguistics in the University of Leuven. There were no relevant discrepancies between his syntactic analysis and ours.

The 25 indifferent cases are found among the sentences that go with contours 1A and 2A. In these cases the pitch contour contains a rise immediately followed by a non-final fall. An intonational block of this type can be interpreted by the listener as the implementation of a "contrastive stress", or as the result of a speaker's strategy by which every accented syllable is marked by a rise-fall (see strategy No. 1, p. 4). Since in both cases the non-final fall is of necessity attached to the preceding rise, it cannot be delayed until the end of the current syntactic whole in order to mark its boundary. The 119 relevant cases are those in which the non-final fall does not immediately follow the rise (contours 1B to 1E, and 2B to 2E). In 103 of these cases the subjects wrote down sentences in which a major syntactic boundary was located immediately before the syllable containing the non-final fall. In other words, the number of (written) syllables between an accented syllable and the boundary of a syntactic constituent was equal to the number of (hummed) syllables between the prominence-lending rise and the non-final fall.

A few examples are given below. The brackets indicate the syntactic structure of the sentences.

(1)

(((helaas)(ben ik niet in staat))((hem)(te beluisteren)))

unfortunately am I not able him to listen to

(2)

(((of)(Peter)(nog komt))((weet)(geen mens)(totnogtoe)(met zekerheid)))

if Peter still comes knows no one as yet with certainty

(3)

(((toen)(dacht)(hij)(waarschijnlijk))((dat)(ik)(wel)(voor hem)(zou uitwijken)))

then thought he probably that I for him would pull out

Figure 3 shows for the material as a whole how the syntactic boundary shifts as a function of the increasing distance between the prominence-lending rise and the non-final fall in the hummed contours. The foreground part of Figure 3 shows the outcome of an independent check preceding the actual experiment. Here the subjects

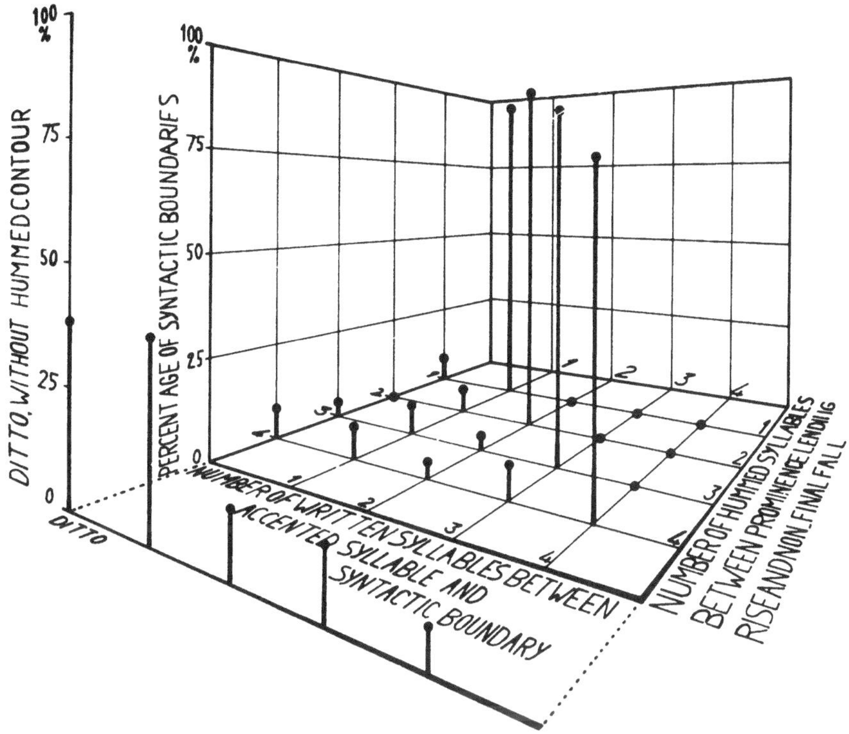

Fig. 3. Effect of hummed contour on the location of major syntactic breaks in sentences thought of in the listening experiments.

were asked to write down sentences without listening to the hummed contours, but meeting the same requirements as to the length of the utterances and the locations of the accentuated syllables.

The data of the actual listening experiment contain sixteen examples that do not confirm our hypothesis in that subjects wrote down sentences in which the non-final fall is made to occur <u>within</u> a constituent rather than at its boundary. Such counterexamples are, for example

(4)

((en <u>gis</u>teren)(op het klein <u>raad</u>huisplein)(moest ik <u>af</u>stappen))

and yesterday on the little city hall square must I get off

(5)

((Ver<u>me</u>ren)(be<u>han</u>delt)(vaak)(het onbe<u>lang</u>rijkste))

Vermeren deals often with the unimportant

(6)

((op maandag)(en ook wel op andere dagen)(zie je dat))

on mondays and also on other days see you that

(7)

((de meeste mensen)(die veel bier drinken)(gaan de bergen in))

most people who much beer drink go the mountains in

As is also shown in Figure 3, the counterexamples are always such that the syntactic boundary is placed before the non-final fall, but never beyond it. The contours that evoked these counterexamples did, of course, elicit a larger number of sentences consistent with the hypothesis: for instance, the contour of Ex. 6 is the same as that of Ex. 2.

Apart from the general tendency that the data confirm the hypothesis, a few other points deserve to be mentioned.

Upon inspection of the nature of the syntactic breaks involved, it appears that the most frequently occurring breaks are found between two clauses (32 cases), or between subject phrase and predicate phrase (26 cases). Other breaks appear between coordinate constituents of the same kind, e.g. two adverbial adjuncts, two direct objects, two subjects, etc. Thus there appears to be a tendency to mark *major* syntactic breaks.

In so far as every syntactic constituent (in the sense in which we use it here) usually receives a pitch accent and to the extent that this is implemented by a rise, there is the intonational necessity to insert a non-final fall between the successive rises. As a consequence, the non-final fall will mark the boundary between syntactic constituents of any kind or hierarchical status, provided that both constituents indeed contain a pitch accent. Interesting cases arise whenever the second of two constituents lacks a pitch accent, since then the choice exists to mark either the boundary before or the one after the second constituent. For such cases one might hypothesise that the speaker will prefer to mark the hierarchically dominant boundary. Conversely, one would expect that the listening subjects, if they think of two syntactic constituents for only one intonational block, will arrange them in such a way that the syntactically dominant boundary coincides with the non-final fall. The data contain 20 relevant cases, 18 of which confirm this specific hypothesis. A few illustrative examples are the following:

(8)

(((vanmorgen)(dacht ik al))((er)(moet)(toch wat op te vinden zijn)))

this morning thought I already there must be something to do about it

(9)

(((beloof me)(nou eerst eens))((dat jij)(er wel toe in staat zou zijn)))

promise me first that you would be able to do it

In both examples it can be seen that the boundary after the second constituent is more important than after the first, and that the former is made to coincide with the non-final fall. Other examples in support of this specific hypothesis are given under (2) and (3). One of the two counterexamples is given by the following sentence:

(10)

((toen)(Jan)(voor het eerst)((met zijn meisje)(naar bed toe)(ging)))

when John for the first time with his girl to bed went

In this case the boundary after "voor het eerst" is more important than that after "met zijn meisje"; yet the sentence is constructed in such a way that the latter constituent boundary is made to coincide with the non-final fall.

2.2 Reading experiment

The reading experiment produced 600 utterances. They were intonationally transcribed by the authors in terms of perceptually relevant pitch events. The technique of perceptual analysis and the transcription system have been explained on previous occasions ('t Hart and Cohen, 1973; 't Hart and Collier, forthcoming). In a number of cases some subjects read a sentence with an intonation pattern other than the hat pattern. In other instances they produced a hat pattern, but using either strategy No. 1 or strategy No. 2 (see p. 4). The latter strategy was used in about one quarter of the cases. Pitch contours resulting from the application of strategies No. 1 or No. 2, and contours derived from patterns other than the hat pattern are indifferent with regard to the hypothesis under verification. Such contours represent 40% of the total number of observations.

In examining the remaining 360 relevant cases we asked ourselves the following questions that are pertinent to the verification of strategy No. 3 (see p. 4): has the speaker either produced a continuation rise on the last syllable of a constituent, or a non-final fall on the first syllable of the next constituent? These two possibilities are illustrated in Figure 1, viz. as (9) and (6) respectively. 204 Utterances were produced with an intonational block ending in a non-final fall, the remaining 156 contained a continuation rise. Of the first group 192 instances supported the hypothesis, whereas in the second group 150 were in favour of it. One illustrative counterexample is given by the following sentence:

(11)

(((de oorlog)(der volkeren))((werd)(even stilgelegd)))

the war of the peoples was for a while stopped

In this case one reading subject failed to mark the major syntactic break (NP-VP), locating the non-final fall on "even".

In the great majority of the cases the location and the number of syntactic boundaries that are intonationally marked by the readers are the same as those produced by the listeners. Discrepancies between the two experiments arise when different strategies are used in the two conditions. Thus, for instance, syntactic boundaries that were not marked in the listening experiment (because the stimuli did not suggest it) are now given an intonational mark in the reading experiment (because the sentence does suggest it). An example is given in Figure 1. The sentence "wat aardig van je dat je gisteren al weer opbelde" was written down to go with a hummed contour of the type (2). Because the non-final fall in such a contour immedately follows the rise, the *major* syntactic break between S_1 and S_2 is not marked. The readers of this sentence, however, marked this boundary in 8 out of 20 cases, by producing either (4), (5), (6) or (7), (8), (9). On the other hand the fact that in the majority of the cases the readers indeed preferred to combine the rise and the non-final fall on the same word "aardig" indicates that the listener's choice of that sentence to go with the given pitch contour was not unreasonable.

Let us now return to the more specific hypothesis that a non-final fall marks the syntactically *dominant* boundary. As has been explained above, the hypothesis can only be tested when the second of two adjacent constituents lacks a pitch accent. In the listening experiment this tendency has already been confirmed. The reading experiment did not produce any counterexamples against the specific hypothesis, but it did not offer many examples in support of it either. This is due to the fact that, in spite of the instructions, many subjects produced a pitch accent on *both* of the constituents under analysis. Let us consider one example.

(12)

(((beloof me)(nou eerst eens))((dat jij)(er wel toe in staat zou zijn)))

promise me first that you would be able to do it

In 14 out of 20 cases the reading subjects added a pitch accent on the appropriate syllable of the second constituent ("eerst"). In the remaining six, possibly relevant, instances the readers had a choice between marking intonationally either the boundary after the first or that after the second constituent. In two of these six cases the readers applied strategy No. 1, which means that they did not mark any boundary at all; in the remaining four cases they marked the boundary after the second constituent, which is the syntactically *dominant* boundary.

3. Discussion and conclusion

We think that we have been able to show, in two different ways, that there is a strong tendency for intonational block boundaries and syntactic constituent boundaries to coincide. This tendency is not apparent in every sentence, which indicates that there is no obligation to always mark syntactic constituent boundaries by intonational means. This is one aspect of the freedom that exists with regard to the intonational specification of sentences, as already illustrated in Figure 1. These experiments shoud be extended into the field of spontaneous speech in order to examine how this intonational freedom is exploited there, i.e. to what extent the speaker cares to provide the listener with syntax-linked intonational cues.

For all the intonational freedom that is reflected in our data, there is nevertheless the general intonational obligation to provide pitch accents, and the more specific requirement that between two pitch accents of which the second one is not realised by a mere final fall, either a non-final fall or a continuation rise must be inserted (depending on the nature of the chosen intonational block). Our experiments have further refined this specific requirement: it must be implemented in such a way that the continuation rise coincides with the syntactic boundary, whereas the non-final fall - unless it immediately follows the prominence-lending rise - is delayed so as to occur directly following the syntactic boundary. In this way the non-final fall either delimits the most prominent syllable of the syntactic constituent or the constituent as a whole. Our experiments also showed that the syntactic boundaries that are intonationally marked are major boundaries, and that of two competing boundaries the hierarchically dominant one is selected for intonational marking.

With regard to the non-final fall, our initial strictly melodic approach to the description of intonational phenomena inevitably confined the function of this pitch movement to that of a mere "reset" to low pitch between two successive rises. Consequently, its location could not be specified in greater detail. The inclusion of functional considerations with respect to syntactic structure marking now enables us to predict its exact location in the contour.

Among the various functions of intonation, that of the implementation of linguistic stress has been generally acknowledged, and on many occasions experimentally verified. With respect to other functions it has long been supposed that intonation would be capable of marking syntactic boundaries. Our experimental results have substantiated the validity of this supposition. Indeed, they show that speech pitch may play a role in either delimiting syntactic constituent boundaries or integrating them. Whereas a speaker is obliged to mark the stresses intonationally, he seems to be free to mark the boundaries or not. Bur, in as far as he chooses to do so, he is bound to follow certain rules. To the extent that the listener shares the knowledge of these rules, he will find in the signal intonational cues for the segmentation of the surface structure of the utterance.

References

Cohen, A. and 't Hart, J. (1967). On the anatomy of intonation. Lingua 19, 177-192.

't Hart, J. and Cohen, A. (1973). Intonation by rule: a perceptual quest. Journal of Phonetics 1, 309-327.

't Hart, J. and Collier, R. (forthcoming). Integrating different levels of intonation analysis. Submitted for publication in Journal of Phonetics.

Svensson, S.G. (1974). Prosody and grammar in speech perception. Doctoral Dissertation, MILUS 2 (Monographs from the Institute of Linguistics, University of Stockholm).

Wingfield, A. and Klein, J.F. (1971). Syntactic structure and acoustic pattern in speech perception. Perception and Psychophysics 9, 23-25.

Willems, L.F. (1966). The Intonator. I.P.O. Annual Progress Report 1, 123-125.

DISCUSSION of "The role of intonation in speech perception"
COLLIER and 't HART

FUJISAKI:
To me it seems more appropriate to describe the correlation between syntax and non-final fall of pitch in terms of a statistical (probabilistic) model, rather than in terms of a deterministic hypothesis. A statistical model is not just an expression of phenomena of which we lack precise knowledge, but can be a proper expression of a relationship that involves variability or freedom.

I should also like to ask if you investigated any correlation that might exist between pause and syntactic boundary, since that seems to be more fundamental to the prosody in some other languages like Japanese.

't HART:
As long as we cannot tell what determines the speaker's decision whether or not to mark the syntactic boundary by intonational means, we agree that it seems more appropriate to speak in terms of a probabilistic model rather than in terms of a deterministic hypothesis with respect to the speaker's decision. But since in either of the alternatives, the location of the intonational marker is entirely rule governed, we prefer to maintain the deterministic way of description with respect to the location of the intonational marker.

FISCHER-JØRGENSEN:
I should like to know what you mean by "non-final fall"? Is it rather a new low start after a relatively high final pitch of the preceding non-final group? And is it possible to have such a low start in a stressed syllable?

't HART:
Contrary to what is the case with a final fall, the non-final fall does not lend prominence. Its preferred location is rather early with respect to the vowel onset of the syllable at issue to the effect that, if the prevocalic consonant (cluster) is voiceless, it is not manifest as a pitch movement. For such cases your suggestion would be fully acceptable. In voiced consonantal environments however, it will produce a clearly observable change of F_0 during the consonant

and the beginning of the vowel. We preferred to describe these acoustically different phenomena in a uniform way, by considering the non-final fall as a virtual movement in the voiceless situation. If the first syllable after a syntactic boundary is stressed, a conflict between the non-final fall and the prominence giving rise is circumvented by having the non-final fall not mark the boundary and attaching it to the previous prominence giving rise (strategy no. 1).

ANTICIPATION IN SPEECH PRODUCTION AND ITS IMPLICATIONS FOR PERCEPTION

S.G. Nooteboom and A. Cohen
Institute for Perception Research
Den Dolech 2, Eindhoven 4502
The Netherlands

1. Introduction

What we know best about speech perception is what it is not. It is most unlike the process to which one is forced in deciphering a letter in almost illegible handwriting.In the strategy adopted with this kind of emergency the analytical knowledge obtained in studying the end of the letter is used to understand the beginning and vice versa, the original letter and the gradually increasing number of filled-in interpretations or hypotheses for interpretations constantly being at one's disposal.

This situation differs from the normal perception of running speech which comes to us in an acoustically messy form analogous to that encountered in handwriting. But if we try to grasp the meaning of a stretch of speech of, say, several minutes, we cannot use analytical knowledge obtained at the end of the message for interpreting auditory stimulus properties at the beginning of the message. The auditory form of speech does not remain available so long a time, neither in general do partial conjectures of the linguistic form of the message. If we are lucky we may retain some idea of what the speaker wished to convey. Normally the intermediate auditory, phonological, morphological and syntactic interpretations are gone, as is the acoustic signal of speech.

Still there is a difference between the acoustic signal of speech and the intermediate stages of speech processing with respect to the amount of time they are at the listener's disposal. The acoustic form of speech is constantly changing, providing the listener's eardrums with nothing more than momentary perturbations of air pressure. It cannot be inspected in its physical form by the listener over any appreciable amount of time, nor can it be taken as an input more than once, as a letter may be reinspected at will by a reader. In the stages of speech processing intermediate between the acoustic signal of speech and the listener's understanding of the intention of the speaker, the situation is not quite so bad. The listener appears to have some time to inspect mental transformations of the acoustic signal and linguistic units and structures evoked by them in his mind, before he decides on what has been said.

What we are concerned with in this paper is the question of how much speech material may be present simultaneously in a listener's mind as coherent auditory structures from which interpretations may be derived. This question immediately seems to lead to considerations of short term storage capacity for auditory information. However, it is not self-evident that the results of traditional short term memory experiments are applicable to the question raised above. On the one hand, storage decay time in the perception of running speech may be strongly decreased by the constant arrival of fresh material, as in running memory span experiments. On the other hand the highly structured form of speech material, and the redundancy of language, may help to increase storage capacity considerably.

In this paper we will start from the assumption that the input speech itself provides clues as to how much a listener at least has to store in order to make efficient use of the auditory structures of speech in his decoding task. We feel that the overt acts of speech may provide many hints as to the size of stretches of speech, readied simultaneously in a speaker's mind for subsequent production, and capable of being perceived as coherent wholes in the mind of a listener.

We will borrow empirical evidence bearing on this matter from three rather divergent areas of speech study, viz. slips of the tongue, intonational patterning, and speech segment durations. Furthermore we will restrict evidence of temporal integration of speech material to cases of anticipation, that is to cases where an event coming later in the speech act exerts overt influence on a preceding event. In clear cases of anticipation it is at least plausible that the amounts of speech material over which anticipation takes place are simultaneously present in the mind of the speaker as plans or programmes for subsequent production. Moreover, particularly in anticipatory behaviour in intonation and speech segment durations, the anticipation itself may often enhance temporal integration in production in a way which makes the task of the listener easier, by providing clues on how the incoming flow of speech can be divided into coherent auditory structures suitable for further processing. By temporal integration we mean to say that the speaker imposes a set of rules on the articulatory production of syllable sequences in such a way that the actualization of each individual syllable in real time is modified by the position of that syllable within a larger structure.

2. Anticipation in slips of the tongue

Of the three areas from which we will adduce empirical evidence, slips of the tongue, intonation, and speech segment durations, the first is perhaps most difficult to relate to perception. We will not omit this type of evidence here, however, because slips of the tongue are often very revealing about chunking as a process in the use of language. After all, we use our minds for both speaking and listening, and evidence of chunking in one mode of language use may at least provide hints of possible ways of chunking in the other.

A typical slip of the tongue occurred when someone said *to colerate quite a lot* instead of *to tolerate quite a lot*. We may assume that the speaker mistakenly replaced the first phoneme of *tolerate* by the first phoneme of *quite*. This implies that at least the words *tolerate* and *quite* were simultaneously present in the speaker's mind, which amounts to a sequence of four syllables. A speaker may, as in the above slip, mistakenly select a phoneme yet to come, which is an example of anticipation. He might also select a phoneme already pronounced, as in *prodruct* instead of *product*, which is called perseveration, or he may transpose two phonemes, as in *to shut him court* for *to cut him short*. As in both anticipations and transpositions an event yet to come affects an earlier event, we will concentrate on those two classes of errors of speech, and will not consider perseverations.

It has been observed many times (cf. Fromkin, ed., 1973) that in slips of the tongue where single phonemes or phoneme clusters are shifted, these phonemes belong to different syllables. When phonemes are part of the same syllable they do not usually take each other's places. This appears to indicate that at the mental level of programming where phonemic errors occur, the phonemes themselves, although psychologically real shiftable properties of syllables, do not stand in serial order as the letters of a compositor's rule. What we call the "order of phonemes" seems to differ in nature from the order of syllables.

We regard phonemic speech errors, then, as interactions between different syllables, which are simultaneously present in a mental programme for speech production. The interactions come about in the substitution, omission, or addition of a phoneme or phoneme combination belonging to one syllable, under the influence of the phonemic form of another syllable.

The distance between these two interacting syllables as observed in the overt speech act, may be taken to be a measure of the

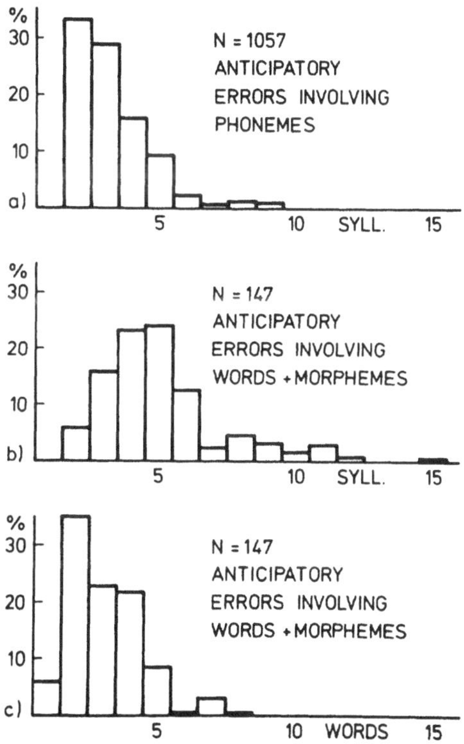

Fig. 1. Frequency distributions of material spans in anticipations and transpositions in German and Dutch errors of speech. a. Errors involving phonemes and phoneme clusters, material spans in syllables. b. Errors involving morphemes and words, material spans in syllables. c. Ditto, material spans in words.

amount of speech material which at least was simultaneously present in the production programme. Fig. 1a presents a frequency distribution of such distances as observed in two collections of errors of speech, viz. 265 anticipations and transpositions in German, to be found in Meringer (1908), and 792 anticipations and transpositions in Dutch collected by Cohen (1966) and Nooteboom (1969). Obviously the form of this distribution does not tell us what a frequency distribution of lengths of speech programmes would look like: When two adjacent syllables interact there may be much more in the programme. It seems reasonable to conclude, however, that in general the length of production programmes does not exceed the "magical number" of seven syllables (Miller, 1956). The speaker does not seem to look much further ahead, at least not as far as the phonemic make-up of syllables is concerned.

We predict from this that temporal integration of stretches of speech, affecting the execution of individual syllables, will generally not comprise much more than seven syllables. This sets an order of magnitude to the amount of speech material over which a listener might expect auditory aspects of syllable productions to be conditioned by speech material coming later in the utterance.

In errors of speech not only single phonemes or meaningless combinations of phonemes may take each other's places, but also meaningful units such as morphemes and words. Looking for an indication of whether errors involving meaningful units do or do not stem from the same production programme as that from which phonemic errors originate, we made a frequency distribution of the distances between these interacting units, again expressing the distances in syllables. The errors were taken from the same collections as the phonemic ones, 107 from Meringer and 40 from the Dutch collection. The result is presented in fig. 1b. The form of this distribution differs clearly from that of the phonemic errors. The differences in the location of the peaks of the distributions may be related to the fact that the meaningful units themselves often comprised more than one syllable. This does not explain, however, why the cases comprising more than seven syllables are relatively more frequent than with the phonemic errors. It seems as if errors involving meaningful units may occur over greater distances than those involving meaningless units.

Let us assume that the mental programme where slips involving words and morphemes occur is not identical to the one where phonemic slips take place. We may conceive of a more abstract programme, not of a sequence of syllables, but of a sequence of words, readied for production. These words consist of one or more morphemes. A slip in this programme may occur when a morpheme or combination of morphemes is substituted, omitted or added to the word under the influence of another word in the programme. Within this conception we should rather express the distance between interacting units in words than in syllables. This is done in fig. 1c. We notice that the shape of the distribution becomes much more similar to the one for phonemic speech errors. Notably the length of the distances appears to be limited to about seven words.

The suggestion that there may be a mental programme of meaningful entities which is logically prior to the programme of syllables is also supported by errors in which the actual phonemic form of the units is affected by their new position within the syntactic structures. This may be exemplified by the following slip, cited from Meringer: bis er's bei Dir abholt instead of bis Du's bei ihm abholt. For a more elaborate treatment of hierarchical processing in speech production as apparent from such errors of speech see Fromkin (1971).

The greater length in syllables of the programme for meaningful entities than of that involving syllable sequences suggests that a listener has to expect morphological and syntactic conditioning between words over greater stretches of speech than conditioning of syllable production aspects. We would expect, though, that such morphological and syntactic conditioning, in spontaneous speech, at least in so far as it is anticipatory, will rarely occur over more than seven words.

These observations on errors of speech make it plausible that the size of amounts of speech material over which anticipation may take place depends on the level of linguistic programming, and that correspondingly the size of effective decision units in speech perception may vary with the level of processing. We will now focus our attention on temporal integration of syllable strings brought about by prosodic patterning.

3. Anticipation in intonational patterning

Although actual F_0 contours in real speech often look rather chaotic owing to seemingly unsystematic fluctuations, it has been shown that the pitch contours of well formed utterances in Dutch speech can be replaced by much simpler stylized contours giving the same melodic impression. Such contours are said to be "perceptually equivalent" to the original ones ('t Hart and Cohen 1973). This possibility of replacing pitch contours of real speech by simple stylized approximations without changing the melodic impression makes it feasible to describe intonational patterning in terms of sequences of simple perceptually relevant pitch movements, omitting all those capricious details which are irrelevant to the perception of speech melody. About 10 types of such pitch movements appear to be necessary for the description of Dutch intonation. These movements differ in direction, extent of movement, slope, and, above all, in their timing with respect to the syllables on which they occur (Collier 1972, 't Hart and Cohen 1973).

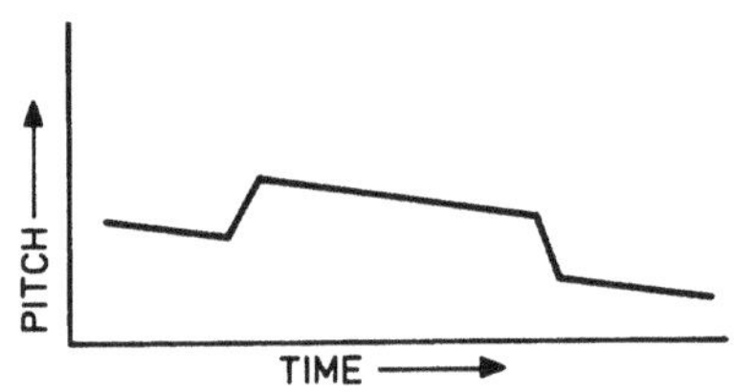

Fig. 2. Stylized version of a "hat pattern" pitch contour, consisting of an accentuating rise plus an accentuating fall, superimposed on a gradually lowering declination line.

Some of these perceptually relevant pitch movements are realizations of pitch accents, others are not.

We will not attempt here to give an extensive description of Dutch intonation. This has been done in the publications mentioned above (see also Cohen and 't Hart 1967, 't Hart and Collier forthcoming, and Collier and 't Hart, this issue). For our present purposes we will limit the discussion to contours of the type presented in fig. 2. This is a stylized realization of the so-called "hat pattern", the most frequent intonation pattern in Dutch. It typically contains an accentuating pitch rise followed by an accentuating pitch fall, both being superimposed on a gradually lowering line of pitch.

A speaker using this contour is deliberately accentuating the lexically stressed syllables of two words, or, where the rise and the fall occur on the same syllable, the lexically stressed syllable of one word, of the phrase under production. Which words cán and which words cannót receive a pitch accent in a particular phrase, depends on the semantic role of the words in that phrase.

Many phrases contain only one pitch-accentuated word, which is then usually spoken with a rise plus a fall on its stressed syllable. If a speaker produces a pitch rise which is not immediately followed by a pitch fall, he must be sure that within the same phrase he will meet another word which can be accentuated, in that case by a pitch fall, in order to avoid producing a phrase which is either not well formed or not conveying what he wants to say. Errors of this kind are often observed in reading aloud, notably by news readers on radio and television. Typically, the speaker produces a rise not long before the end of a sentence, keeping the pitch high after the rise, just as in fig. 2, and then, because each pattern of this type has to include an accentuating fall towards the end, he is forced to produce a pitch accent on a word which should not have been accentuated. The following example may give an impression of such an intonational mistake:

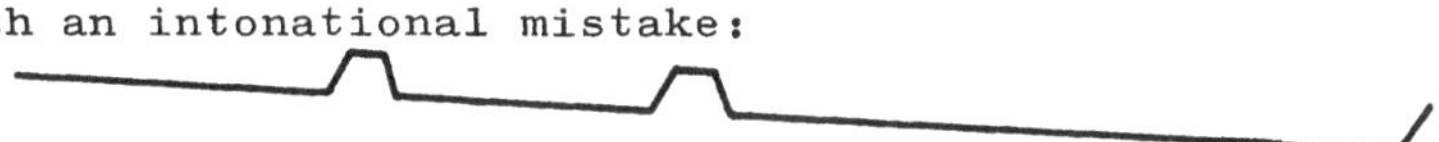

In een gesprek dat Kissinger voerde met Sadat
(In a conversation that Kissinger had with Sadat)

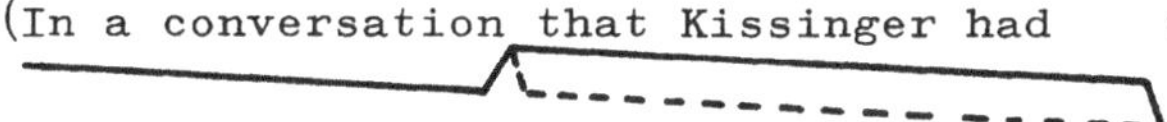

werd hij het eens met de president
(he reached agreement with the president)

The pitch accent on presidént, which word in Dutch has lexical stress on the final syllable, wrongly suggests that president does not refer to Sadat, but rather to a person not yet mentioned. The dashed line gives the correct pitch contour. Such errors in pitch contours, due to a lack of anticipation of the number of words to be accentuated before the end of a sentence, occur quite often in reading aloud . From time to time newsreaders are heard not to make a final fall at all at the end of a sentence, keeping their pitch high to the very end. Noticing their mistake they sometimes add:

Punt (= full stop).

In an extensive survey of pitch contours in spontaneous Dutch such intonational errors are seldom if ever observed. Apparently in spontaneous speech, speakers in generating contours of this type, are adequately planning ahead, and only keep the declination line high if they know that a suitable word will come for reali-izing a pitch accent through the final fall in pitch.

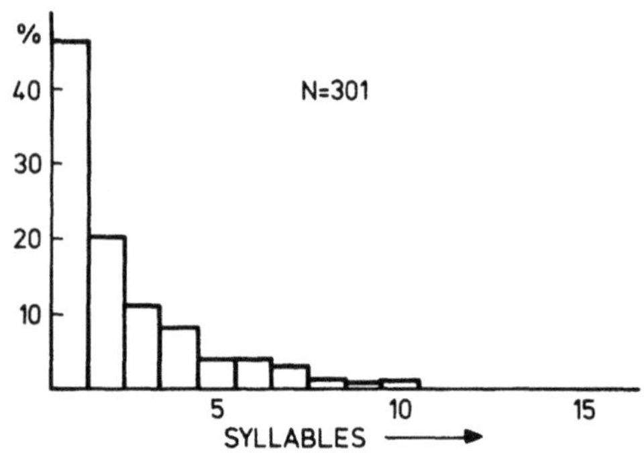

Fig. 3. Frequency distribution of material spans covered by rise-fall pitch contours of the type shown in fig. 2. Rises and falls are included in the spans as first and last syllables respectively. Data from Collier (1972).

These observations suggest that the amount of speech material comprising a rise and a subsequent fall has in some way to be present in the speaker's mind while he is programming a pitch contour of this type. In fig. 3 we show a frequency distribution of lengths of syllable sequences over which such intonational anticipations take place. The distribution was derived from data in Collier (1972), concerning an extensive corpus of Dutch pitch contours in both spontaneous speech and quasi-spontaneous stage speech. We excluded all monosyllabic utterances, and included utterances with only one pitch accent realized by a rise and a fall on the same syllable.

It should be noted again that such a frequency distribution is not one of lengths of mental programmes of speech production. Such programmes will probably often be considerably longer than the distance between a rise and a fall. The distribution may give an indication, however, of how far speakers at least are planning ahead in realizing pitch contours. This planning appears to extend over syllable sequences similar in length to those involved in errors of speech.

It is interesting to speculate that pitch contours of the type described represent important aspects of rule-governed temporal integration of speech acts, giving clues to the listener on how far he has to stretch his span of attention in order to inspect coherent auditory structures of speech, which might be effective decision units in processing the auditory form of speech.

We have restricted our discussion of intonational anticipation to one single, very simple type of pitch contour in Dutch. For a much more extensive discussion of possible perceptual functions of pitch contours, and an experimental evaluation, we refer to the contribution by Collier and 't Hart to this symposium.

4. Anticipation in speech segment durations

From the beginning of this century (Cf. Roudet 1910, 237), many investigators have drawn attention to the fact that speech segment durations are systematically affected by their position in an utterance. One of the most important effects appears to be a shortening of speech segment durations as a function of the number of syllables yet to be produced in word or phrase (Lindblom and Rapp, 1973, Nooteboom 1973).

An example of such anticipatory compensation is presented in fig. 4. These data were obtained by De Rooij of our Institute. The duration of a vowel segment /e:/, combined with the immediately following occlusion interval, in realizations of the phrase initial word /pe:t/, is plotted as a function of the number of syllables in the phrase. The phrases were made to increase in length by adding monosyllabic words to the end. Thus the number of syllables is identical to the number of words. The data are averaged over three realizations of each of two subjects. Measurements were made from high quality oscillograms.

We see that in the overt speech acts timing of articulatory movement appears to be adjusted to the amount of speech material remaining to be produced.

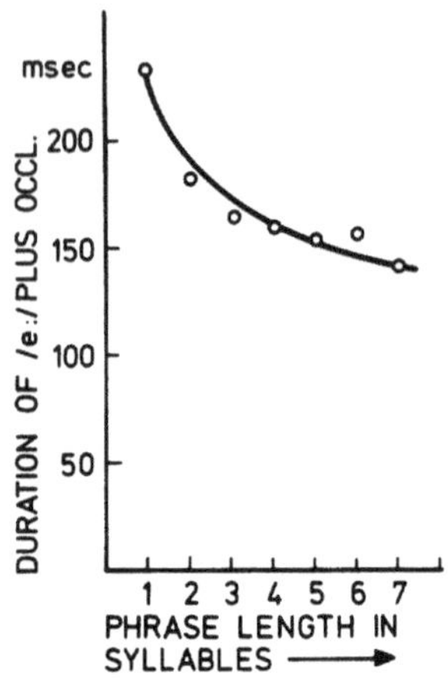

Fig. 4. Anticipatory shortening of the durations of a speech segment consisting of the vowel /e:/ plus the immediately following occlusion time in utterance initial realizations of the word /pe:t/, as a function of the number of monosyllabic words in the utterance. Averaged over 3 realizations of each of two subjects. Measurements from high quality oscillograms.

It is difficult to give a fair estimate of how far this kind of anticipatory compensation extends. From fig. 4 we would say that anticipation occurs over at least 4 or 5 syllables.

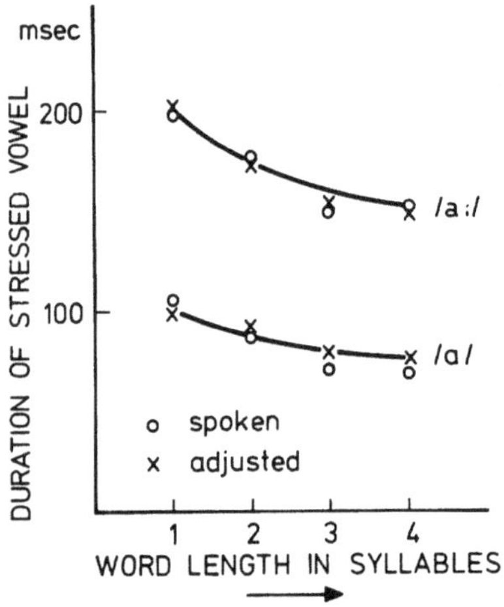

Fig. 5. Anticipatory shortening in spoken and perceptually adjusted durations of /a:/ and /ɑ/ sounds in word initial syllables as a function of word length in syllables. Spoken durations measured from spectrograms and averaged over 2 realizations of each of 3 subjects. Adjusted durations averaged over 10 realizations of each of the same 3 subjects.

Some years ago we demonstrated that the auditory effects of such temporal regularities belong to the implicit knowledge language users have about the way speech should sound in their language (Nooteboom 1973). This was done by asking Dutch subjects to adjust, by turning a knob, vowel durations in different positions in synthesized words, such that the words were made to sound as natural as possible. An example concerning stressed vowels in initial syllables of words with increasing word length is shown in fig. 5. The adjusted durations of both long /a:/ and short /ɑ/ nicely follow the durations of the same vowels in actually spoken versions of the same words.

Recently we attempted to find out whether listeners can actually use such knowledge in speech perception. We reasoned that if both long and short vowels are shortened as a function of the number of syllables remaining to be produced, as is apparent from fig. 5, then the cue function which vowel duration has for the perception of phonemic vowel length category should also be affected by word length. More precisely, one would expect that in monosyllables the switch-over point from short vowel responses to long vowel responses on the stimulus dimension of vowel duration will be highest, and will decrease with the number of syllables added. Such a finding would be evidence that the "interpretation" of auditory cues may be affected by a number of syllables coming later in the utterance, and thus would suggest auditory decision units which are considerably greater than the syllable, and might parallel the amounts of speech which are temporally integrated in production.

We will briefly describe part of our perceptual measurements designed to answer this question. A more extensive description is given in Nooteboom (1974).

We obtained a vowel sound which was completely homogeneous in spectral composition, intensity and periodicity, by repeating one glottal period of a digitally stored spoken /a:/ sound. The duration of this homogeneous vowel was varied from 60 tot 130 msec. in steps of 5 msec. by means of an electronic gate. Rise and decay time were set at 10 msec. to avoid auditory clicks. The resulting vowel stimuli were made audible in four context conditions, viz. /Vt/, /Vtə/, /VtəXə/, /VtəXə̇rə/, in which /V/ stands for the stimulus vowel sound. In each context condition we measured short vowel/long vowel response curves in a forced choice listening experiment with ten subjects, all native speakers of Dutch. The results are presented in fig. 6.

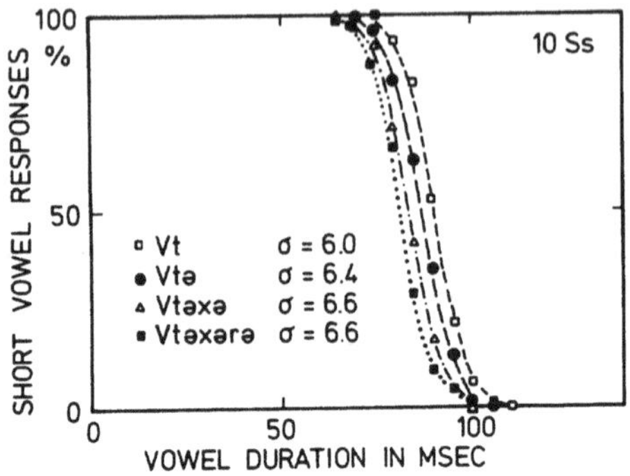

Fig. 6. Percentage of short vowel responses as a function of vowel duration in a forced choice listening experiment, in 4 context conditions. The choice was between Dutch /a:/ and /ɑ/. Response curves of individual subjects were averaged after they were shifted in the horizontal dimension so as to make their estimated 50% points agree with the mean of the 50% points.

These data show a small but systematic effect of word length on phonemic categorization of vowel duration. This supports the idea that auditory properties of speech can be interpreted within auditory structures, which are considerably longer than a syllable. Admittedly our forced choice listening experiment with nonsense words took us a long way from the normal spontaneous perception of running speech. Also the disambiguation of phonemic vowel length category by temporal patterns, used here as a tool for perceptual measurement, may itself be a marginal function of such patterns, the real value of which may be that they provide auditory coherence to syllable sequences. It is comforting, though, that the effect was predicted from anticipatory compensation in speech production, which has been found by so many investigators in so many languages that we may feel reasonably sure that it represents a universal tendency of human speech.

5. Discussion

The examples of anticipatory behaviour in speech production discussed in this paper focus attention on a level of organization of speech which goes beyond the syllable but does not necessarily extend to complete sentences. This particular level of organization seems to stem from a short term storage in which actual articulatory behaviour is planned and temporally integrated. The main tenor of our paper is that such temporal integration in speech production results in temporally coherent auditory structures which play a central role in the perception of speech. We will now attempt to speculate somewhat further on the role of these coherent auditory structures and to relate our ideas to other findings in the literature.

Massaro (1972, 1974), on the basis of an extensive survey of the literature and of his own backward recognition masking experiments, concludes that the first stage of speech recognition must operate on acoustic segments that are of less than 250 msec. duration, that is of roughly syllable length. He therefore suggests that syllables are the most likely candidates for being the perceptual units of speech, arguing that the perceptual unit employed in speech perception could not be larger than the temporal life of a preperceptual image of the speech stimulus. What remains, after the time delay of 250 msec. or less, he calls synthesized information. If we accept the assumption that backward recognition masking provides evidence for a kind of passive store of acoustic information, our usage of the term auditory information is certainly not limited to that store, but rather refers to what Massaro calls synthesized information. We would still hold, however, that this auditory information can be, although is not necessarily, preperceptual or precategorical, in the sense that it is not yet completely determined in terms of to-be-recognized linguistic units such as phonemes, morphemes and words. If preperceptual auditory information did not survive longer than about 250 msec., we could hardly expect the interpretation of some auditory cues to be dependent on auditory context extending over some syllables that arrive later at the auditory system, as suggested by our experiment on phonemic categorization of vowel duration. We think of auditory information as precategorical images of the acoustic stimulus, which may have undergone some initial processing, for instance in terms of pitch, auditory durations, spectral quality and intensity, but is not yet interpreted in terms of linguistic categories. We suggest that auditory information in this sense may have a temporal life corresponding to a number of syllables.

It does not seem likely, though, that all perceptual decisions are delayed until a coherent auditory structure of a number of syllables is stored. There is, for example, increasing evidence that within the perceptual system there are feature detectors which most likely operate without much delay (Stevens 1973, Eimas and Corbit 1973, Cooper 1974a, b, Cooper and Blumstein 1974).

However, listening to gated-out portions of running speech of, say, a few hundreds of msec, one will soon notice that such speech fragments more often than not are unintelligible. Interestingly enough, results of Pickett and Pollack (1963) and Pollack and Pickett (1963) show that the intelligibility of such excerpts of fluent and conversational speech on the average only reaches 100% if the duration of the excerpts increases to about 1.5 sec. In a later study Pollack and Pickett (1964) showed that it is the auditory rather than the structural context which is needed by the listeners. This also encourages the view that auditory information, existing alongside the derived linguistic information, may have a temporal life of about 1.5 sec, corresponding to roughly seven syllables.

The idea that auditory information may still be available after the corresponding linguistic responses are activated is basic to the theoretical concept of a Precategorical Acoustic Storage (PAS) of the kind put forward by Morton (1969, 1970) and Crowder and Morton (1969). As the temporal life of PAS is estimated to be some 1.5 sec (Crowder 1969), it is tempting to relate our coherent auditory structures to PAS. The information in PAS, however, is typically strongly interfered with by the arrival of fresh stimulus information, and does generally not contain more than 2 stimulus items. It seems difficult to apply such findings to the perception of running speech. Apparently we are not yet in a position to explain the results of traditional short term memory experiments and the little we know of short term storage in the perception of running speech, by one and the same theoretical concept.

We would like to suggest that the amount of, perhaps partially decoded, auditory information stored for further perceptual processing is not so much a function of a fixed mental time window, passively scanning the acoustic input, but is rather controlled by those properties of the auditory stimulus signal which are responsible for what we, rather vaguely, call auditory coherence. We prefer to think of chunking rather than of scanning.

Our assumption that auditory "chunking" is controlled by auditory attributes presupposes some initial processing of the auditory stimulus. More precisely, we have to assume that those stimulus properties which constitute auditory coherence are continuously being analyzed before any division of the auditory flow into coherent auditory structures. This brings to mind the concept of "input channel" in the literature on selective attention. In so far as an input channel is defined with respect to stimulus attributes, such as spatial location or voice identity, it presupposes initial processing prior to the processes of selective attention (Cf. Swets and Kristofferson 1970).

This analogy between auditory coherence and input channel does not seem to be accidental. In pilot experiments in our Institute with speech assembled from prerecorded words spoken in isolation by a single speaker, and made audible without acoustic pauses, some listeners reported hearing more than one speaker at the time, speaking from different directions (Slis 1969). This rather dramatically demonstrates that the prosodic aspects of speech, providing auditory coherence, may have a particular relevance to the definition of an input channel in speech perception. This is in line with the results of an extensive study on "temporal coherence in tone sequences" by Van Noorden (1975). Van Noorden showed that it is possible to define rather precisely the stimulus conditions under which a tone sequence can, and under which it cannot be heard as a single melodic pattern.

If it is correct that the prosodic properties of speech, pitch contours, temporal organisation, and overall intensity contribute to the auditory coherence, then we have to assume that these prosodic properties have a dual function in speech perception. They also contain important, language-specific cues to the perception of linguistic aspects of the message, such as stress, grammatical categories, and linguistic boundaries. The relevance of pitch and duration for the perception of accent and stress is too well known to enlarge upon. Blesser (1969) and Svensson (1974) showed that if the prosodic structure of speech alone is communicated to the listener, information on grammatical categories is partly preserved, or at least the number of possible response categories is decreased. Wingfield and Klein (1971) showed that listeners locate the moment of switching input from one ear to the other more precisely if the switch occurs on a major syntactic boundary than if it occurs within a syntactic clause. If, however, prosodic and syntactic structures conflict, it appears that prosodic boundaries may be at least as important as syntactic ones in bringing about this effect. Collier and 't Hart, in their contribution to this symposium, show that pitch contours aid in the perception of syntactic boundaries.

What emerges from this discussion is that the coherent auditory structures of speech, resulting from rule-governed patterning in speech production, and extending over a few words or a number of syllables, are of primary importance in speech perception. They appear to have a dual function, on the one hand preserving perceptual continuity, enabling the listener to interpret the auditory information at all in terms of coherent patterns of speech, and on the other hand mediating the perception of grammatical categories, syntactic boundaries, and possibly also the semantic structures of messages.

We think that future research efforts could be fruitfully directed towards these dual functions of prosodic aspects of speech. Attempts should be made to establish what the initial processing of speech stimuli consists of, which particular extracted auditory attributes are responsible for auditory coherence, and how language-specific prosodic patterns are derived from the seemingly chaotic signals of speech. Not less important, we should study the short term storage capacity for melodic and rhythmic aspects of speech, how this relates to short term storage of words and word groups, and how the latter may be enhanced by the first.

The suggestions made in this paper are in line with the general idea underlying models of the analysis-by-synthesis kind. Such models, however, are often too specifically concerned with events on the segmental level, as is the case with the motor theory of speech perception (Liberman, Cooper, Shankweiler, Studdert-Kennedy 1967), or too loosely linguistically oriented, not impinging on what enters the perceptual system in an auditory form, as appears to be the case in several Artifical Intelligence approaches (Schank and Colby 1975). We feel that the link between production and perception of running speech is at least partly established through speech prosody. If one is to accept the idea of analysis-by-synthesis, then the study of speech prosody, both in production and in perception, may help us in finding out what, actually, a listener is synthesizing on the auditory level of speech perception.

References

Blesser, B.A. (1969). Perception of spectrally rotated speech. Unpublished dissertation, M.I.T.

Cohen, A. (1966). Errors of speech and their implication for understanding the strategy of language users. Zeitschrift für Phonetik, Sprachwissenschaft und Kommunikationsforschung 21, 177-181. Also in: Fromkin (ed.) 1973.

Cohen, A. and Hart, J. 't (1967). On the anatomy of intonation. Lingua 19, 177-192.

Collier, R. (1972). From pitch to intonation. Unpublished dissertation, University of Louvain.

Cooper, W.E. (1974) Contingent feature analysis in speech perception. Perception & Psychophysics 16, 201-204.

Cooper, W.E. (1974). Perceptuomotor adaptation to a speech feature. Perception & Psychophysics 16, 229-234.

Cooper, W.E. and Blumstein S.E. (1974). A "labial" feature analyzer in speech perception. Perception & Psychophysics 16, 591-600.

Crowder, R.G. (1969). Improved recall for digits with delayed recall cues. Journal of Experimental Psychology 82, 258-262.

Crowder, R.G. and Morton, J. (1969). Precategorical acoustic storage. Perception & Psychophysics 5, 365-373.

Eimas, D.E. and Corbit, J.D. (1973). Selective adaption of linguistic feature detectors. Cognitive psychology 4, 99-109.

Fromkin, V.A. (1971). The non-anomalous nature of anomalous utterances. Language 47, 27-52. Also in: Fromkin (ed.) 1973.

Fromkin, V.A., ed. (1973). Speech errors as linguistic evidence. Mouton, The Hague.

Hart, J. 't and Cohen, A. (1973). Intonation by rule: a perceptual quest. The Journal of Phonetics 1, 309-327.

Hart, J. 't and Collier, R. (forthcoming). Integrating different levels of intonation analysis. To be submitted to the Journal of Phonetics.

Liberman, A.M., Cooper, F.S., Shankweiler, D.F. and Studdert-Kennedy, M. (1967). Perception of the speech code. Psychological Review 74, 431-461.

Lindblom, B. and Rapp. K. (1973). Some temporal regularities of spoken Swedish. PILUS (Papers from the Institute of Linguistics, University of Stockholm) no. 21.

Massaro, D.W. (1972). Preperceptual images, processing time, and perceptual units. Psychological Review 79, 124-145.

Massaro, D.W. (1974). Perceptual units in speech recognition. Journal of Experimental Psychology 102, 199-208.

Morton, J. (1969). The interaction of word information in word recognition. Psychological Review 76, 165-178.

Morton, J. (1970). A functional model of memory. In: Models of human memory, Norman (ed.), Academic Press, New York.

Nooteboom, S.G. (1969). The tongue slips into patterns. In: Nomen, Leyden studies in linguistics and phonetics. Mouton, The Haque-Paris. Also in: Fromkin (ed.) 1973.

Nooteboom, S.G. (1973). The perceptual reality of some prosodic durations. The Journal of Phonetics 1, 25-45.

Nooteboom, S.G. (1974). Some context effects on phonemic categorization of vowel duration, IPO Annual Progress Report 9, 47-54.

Pickett, J.M. and Pollack, I. (1963). The intelligibility of excerpts from fluent speech: effects of rate of utterance and duration of excerpt. Language and Speech 6, 151-165.

Pollack, I. and Pickett, J.M. (1963). The intelligibility of excerpts from conversation. Language and Speech 6, 165-171.

Pollack, I. and Pickett, J.M. (1964). The intelligibility of excerpts from fluent speech: auditory vs. structural context. Journal of Verbal Learning and Verbal Behavior 3, 79-84.

Roudet, L. (1910). Eléments de phonétique générale. H. Welter, Paris.
Slis, I.H. (1969). Synthese van zinnen met behulp van losse gesproken woorden. (Synthesis of sentences from words spoken in isolation.) I.P.O. Internal Report no. 131.
Stevens, K.N. (1973). The potential role of property detectors in the perception of consonants. M.I.T. Research Laboratory of Electronics, Quarterly Progress Report 110, 155-168.
Svensson, S.G. (1974). Prosody and grammar in speech perception. MILUS (Monographs from the Institute of Linguistics, University of Stockholm) 2.
Swets, J.A. and Kristofferson, A.E. (1970). Attention. Annual Review of Psychology 21, 339-366.
Noorden, van L.P.A.S. (1975). Temporal coherence in tone sequences. Doctor's thesis, Eindhoven.
Wingfield, A. and Klein, J.F. (1971). Syntactic structure and acoustic pattern in speech perception. Perception & Psychophysics 9,23-25.

DISCUSSION of "Anticipation in speech production and its implications for perception"
NOOTEBOOM and COHEN

LEHISTE:

I would like to comment about two points.

It seems to me you are distinguishing between speech programming and linguistic programming and claiming that the mental program for speech production does not exceed the "magical number seven" units - syllables or words. I wish you could clarify this difference. There are languages like German that seem to require preplanning that looks much farther ahead toward the end of the sentence, implying a mental program that encompasses the whole sentence.

(Example: "Er hat gestern Abend nach Arbeitsschluss von allen seinen langjährigen Arbeitsgenossen und zu diesem Anlass eingeladenen Freunden festlich Abschied genommen".)

My second point concerns the matter of anticipatory compensation. You describe a situation in which the segmental sounds of the first syllable in an utterance become progressively shorter, the more syllables are added on to the end. You conclude that the shortening depends on anticipating the number of syllables that remain to be produced in the utterance. I have done a study in which test words were placed in frames of different lengths. Two such frames were "Say the word --- again" and "Sometimes it's useful to say the word --- again". The number of syllables remaining to be produced is the same in both frames, but the test words were much shorter in the longer frame. It seems that what matters is the duration of the whole utterance.

NOOTEBOOM:

I would really doubt whether in the German example you gave all individual words are selected and ordered by the mental program when the speaker produces the beginning of the sentence. He may select the words "Abschied genommen" first, and afterwards fill in the preceding words. With respect to your second point, one may find both an effect of preceding number of syllables and an effect of following number of syllables. In our own data the latter is generally stronger than the former, which makes it inconvenient to collapse the two in a single factor "utterance length".

HAGGARD:

The disambiguation of phonemic length by the following context does indeed offer the possibility of measuring the duration of indecision and/or the duration of the necessary echoic memory.

However, until the data of Figure 6 on polysyllables are supplemented with studies varying the length of the various segments, we cannot rule out that most of the effect might be due simply to the one or two following syllables.

KLATT:

Your analysis of speech errors endeavors to estimate the maximum length of interaction between units of a sentence as a function of tye type of unit. For example, you present data suggesting that phonemic exchange and intrusion errors are limited to about a 7-syllable interval. In this example, a 7-syllable limit is statistically proved, but the explanation for this limit may not involve number of syllables as the critical variable. Analysis of speech errors by Fromkin (1971) and by Shaltuck (in a recent M.I.T. Thesis) suggest constraints on this type of error that depend on other linguistic variables such as stress, position within the word or morpheme syntactic classes of the pair of words. and whether the words are in the same syntactic phrase.

BERLIN:

Contextual anticipation or even backward disambiguation should not be considered unique to spoken language. I'm indebted to Dr. Henry Truby for this example: In response to "MOCK" - most English speakers have already coded a word meaning "mimic". However, if I add BA to the end we now have the word <u>MOCKBA</u> which Russian readers in our group recognize as the city <u>Moscow</u>. The final two characters disambiguated the first four.

In music, similar anticipation and backward disambiguation also occur in the playing of rapid sequences either before or after a "rest" or silence. Most performers punctuate the last note before a silence or accentuate the first note after a silence (or long note) without instructions to accentuate. Listening to a segment of music I suspect listeners can tell if a silence is <u>coming</u> or has recently passed by the presence or absence of the accented note. Similar backward disambiguations are seen in activities such as tennis and chess (notorious for its vindictive disambiguations sometimes eight to twelve moves after the gambit is offered) among many other examples.

THE INTONATION-SYNTAX INTERACTION: PROSODIC FEATURES IN PERCEPTUAL PROCESSING OF SENTENCES

Arthur Wingfield

Department of Psychology
Brandeis University
Waltham, Massachusetts 02154 U.S.A.

Speech is an intrinsically temporal event in which the auditory input is constantly changing over time. No single instant of the speech signal carries sufficient information to be fully discriminating, and recognition on the phonemic level, for example, is heavily dependent on the context of neighboring linguistic elements (Cole & Scott, 1974). For this reason, logic alone would demand that speech must be processed in "units" of some useful and manageable size. Arguments for the phoneme as one such unit have been offered (Liberman, et al., 1967), as have arguments for units closer to syllable size (Massaro, 1972).

Running parallel with this controversy has been the simultaneous realization that perceptual processing on the sentential level must be no less dynamic, and no less susceptible to the logic of the "units" argument. The time-honored line, "I was going to take a train to New York, but decided it would be too heavy," makes the point. The sentence is sequentially processed on a phonemic level, perceptually segmented into syntactic units of a size bounded by the capacity of immediate memory, and each unit given preliminary analysis for meaning. A final stage follows with a higher-order analysis of the combined product of these sub-span analyses. Hence, "I was going to take a train to New York" is given meaning, "but I decided it would be too heavy" is given meaning, and the two units are then combined to give the joke its humor.

As reasonable as this picture of levels of processing may appear, it contains a minor paradox. If we agree that analysis for meaning follows correct perceptual segmentation, what determines the initial segmentation? The answer that segmentation is based on meaning seems somehow unsatisfactory if we also argue that meaning analysis is based on units that have already been correctly grouped.

The purpose of this paper is to examine the stimulus variables which determine the formation of perceptual units in the understanding of connected speech, and to discuss the implications of the "units" hypothesis for active speech perception. The level of analysis with which we are primarily concerned is that of supra-morphemic segmentation on units the size of linguistic clauses.

Syntactic Segmentation

There is considerable evidence that surface clause boundaries within sentences play a critical role in perceptual organization by marking the boundaries of the perceptual units in which the sentences are analysed (Fodor, Bever & Garrett, 1974, pg. 329). Much of this evidence rests on a paradigm introduced by Ladefoged and Broadbent (1960), where clicks were inserted into recordings of either sentences or strings of random digits. Listeners had great difficulty in reporting exactly where the clicks occurred, which would be surprising if perception were based on a sequential, sound by sound analysis. Attributing the inaccuracy to listeners' tendency to shift the subjective location of the clicks to the boundaries of perceptual units, they concluded that decoding speech information "(must) involve operating on units which are...larger than the duration of a single speech sound" (pg. 169). Because the magnitude of localization errors were larger for sentences, Ladefoged and Broadbent further related the size of such processing units to speech content.

Fodor and Bever (1965) later used the same paradigm to define these units in terms of major linguistic constituents. Clicks placed at major syntactic boundaries of sentences were located correctly more often than those placed within constituents, and the direction of errors showed a strong tendency to "migrate" toward these boundaries. Especially important to our present interests, the same essential results were obtained even when a splicing technique was used to remove normal intonation pattern of the sentences (Garrett, Fodor & Bever, 1966).

It is certainly the case that perceptual segmentation based on clausal boundaries can be demonstrated in the absence of intonation cues which might normally accompany such boundaries (Lieberman, 1967, pg. 125; Abrams & Bever, 1969). On the other hand, the relationship between syntax and intonation pattern is an intimate one in both the perception and production of connected speech. The likelihood that prosodic features do play a key role in perceptual organization thus requires further examination.

Segmentation by Source or Structure

One way to view the role of intonation pattern in fluent speech is that it adds, in the Information Theoretic sense, redundancy to the speech signal. That is, it adds perceptual information to the sentence by reducing uncertainty at or around clausal boundaries. As such, it can be eliminated (e.g., by speaking in a monotone, or reading printed words) with lexical content alone allowing correct syntactic resolution. On the other hand, under circumstances when structure is less easy to determine, it can be of measureable help. Such circumstances might obtain under poor listening conditions, in cases of hearing impairment, or most importantly, in ordinary ongoing speech which is only rarely of coherent syntactic form (cf., Goldman-Eisler, 1968).

In an earlier study (Wingfield & Klein, 1971), we hoped to provide evidence for this view by exploiting a method analogous to the "click" studies cited. We changed the methodology in two respects. The first change was based on the suggestion that rather than eliminating intonational marking, the way to reveal its effects might be to put it in direct conflict with objective sentence structure. How listeners resolved this conflict would be of primary interest.

Pairs of sentences were constructed which shared a common word sequence, but where the major syntactic boundary (MSB) occurred in different positions within the common sequence. An example was the sentence pair, (a) "To avoid any attempts to influence voting, machines were installed," and (b) "Due to our new mayor's influence, voting machines were installed." Using the procedure of Garrett, et al. (1966), each sentence was separately recorded, and then the identical word sequences of each sentence (underlined) were cross-spliced into the linguistic frame of the other sentence. For example, sentence (a), with a lexically defined MSB between voting and machines, was now heard in the intonation pattern of sentence (b) where the MSB occurred between influence and voting. The reverse would be true for the spliced version of sentence (b). Splicing was carefully done so as to be unnoticeable in playback. The sentences in anomalous intonation were as clear and as "naturally fluent" as the originals. Unlike Garrett, et al., who used this procedure to control for intonational marking, the sentence types used here were specifically chosen for their clear and unambiguous intonational marking when normally spoken. The spliced versions would therefore not subdue intonation, but would highlight the conflict.

The second change to the earlier methodology was to use source localization rather than extraneous clicks, to overcome some criticism of that technique (cf., Ladefoged, 1967). Forty sentences were presented over dichotic earphones. A sentence would begin in one ear, and then be switched without warning to the other ear. The listeners' task was to repeat the full sentence and to indicate at which point the source switched between ears. Half of the sentences were in normal intonation, and half were in anomalous conflicting intonation.

For the sentences heard in normal intonation, results were exactly as might be predicted from the earlier click studies. Source localization was significantly more accurate when the switch occurred at the MSB (93% correct), than when it occurred within a constituent either before (62% correct), or after (59% correct) the MSB. Of the errors, 44% "migrated" to the MSB. Clearly, the integration of a single message from two sources is more influenced by segmentation imposed by the listener, than by segmentation in terms of source. Where this segmentation occurred, however, showed a striking difference for the sentences heard in anomalous intonation.

In the case of the anomalous intonation sentences, there was no significant difference in accuracy of localizing switches which occurred at the lexically defined MSB (59% correct), and when they occurred either before (53% correct) or after (53% correct) that point. It was when the switch occurred at the point marked by the intonation pattern associated with the MSB from the spliced version, that localization was most accurate. Localization was significantly more accurate when the switch occurred at the intonationally marked MSB (77% correct), than when it occurred either before (57% correct) or after (44% correct) that point.

It would be a mistake to conclude, however, that intonational features alone determine perceptual segmentation. First, localization accuracy attributed to intonation alone in the anomalous sentences (77%) was not as great as that of the normal sentences where the two potential cues of syntax and intonation coincide (93%). Second, mislocated switches in the anomalous sentences tended to "migrate" with equal frequency to either the lexically defined, or prosodically defined MSB (30% in both cases). Again, these values are less than the 44% migration to the MSB in the normal sentences where lexical and intonational markings were consonant. We find evidence for perceptual segmentation based on both objective syntactic structure and intonational marking, but their effects are not simply additive. The two seem to interact in a way consistent with our hypothesis. Syntactic clausal boundaries determine perceptual boundaries, with normally accompanying prosodic cues aiding in the resolution of this syntactic structure.

This conclusion implies that prosodic features such as pauses do not, in ordinary speech, simply "chunk" sentences as an auditory analog of the Von Restorff effect. It also offers a possible solution to the earlier mentioned segmentation paradox. Prosodic features may operate to influence an initial structural hypothesis as a prelude to a self-corrective, active perceptual model of the sort proposed by Halle and Stevens' (1964) analysis-by-synthesis formulation. The process suggested is an ongoing one, in which preliminary perceptual (structural) hypotheses are formed, and then modified as the processing continues. Prosodic features such as intonation do not directly determine perceptual segmentation, so much as add information about clausal boundaries which do.

Intonation and Intelligibility

One outcome of the first experiment was a slight recall advantage for the sentences heard in normal intonation over those in anomalous intonation. This was curious, since the recorded segments themselves were identical in both conditions. This would seem to suggest an effect of perceptual segmentation on recognition accuracy for the entire sentence; the final level in the processing stages previously described.

If intonation supplies redundancy to the recognition process, its effects should be magnified if the same sentences were presented under conditions of increased information load. Presumably, the more difficult the perceptual task, the greater would be the observed effects of prosodic marking on recognition. One way to test this hypothesis would be to use artificially accelerated, or time-compressed speech, since it can increase input load up to and beyond tolerable limits, while leaving the relative timing and prosodic pattern within the sentences intact (Wingfield, 1975).

The previously described normal and anomalous intonation sentences were compressed through the so-called "sampling method" on an electro-mechanical compressor of the Fairbanks type. This method periodically deletes small (20 msec) tape segments at regular intervals, and then abuts the remaining segments in time. When played back at normal speed, the result is speech reproduced in less than normal time, but without the distortion in pitch or quality that would, for example, accompany tape recorder playback at faster than normal speed. Degree of compression is controlled by the frequency with which the tape segments are deleted. Although dependent on message complexity, good intelligibility can be demonstrated for speech reproduced in as little as 50% of normal time (Foulke, 1971).

Fig. 1 shows mean intelligibility scores for normal and anomalous intonation sentences heard at normal rate (207 words per minute), and when compressed to 80, 70, 60, 50, and 40% of ordinary playing time (corresponding to speech rates of 259, 296, 345, 414, and 518 words per minute). The sentences, in a balanced design, were presented binaurally over earphones to 10 listeners who were instructed simply to listen to each sentence, and to report as accurately as possible what was heard.

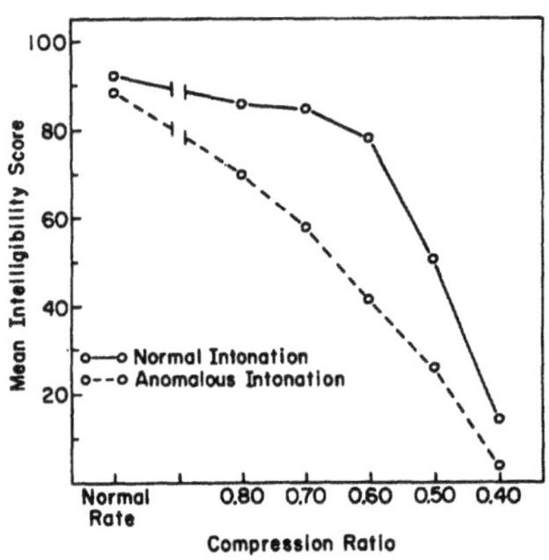

Fig. 1. Intelligibility of sentences heard in normal and in anomalous intonation as a function of degree of time compression of the speech signal (Wingfield, 1975).

As one would expect, intelligibility declines with increasing speech rate [$F(5,45) = 76.36$, $p<0.001$], and sentences heard in normal intonation better withstand the effects of compression than those in anomalous intonation [$F(1,9) = 21.71$, $p<0.01$]. Of greater interest, however, is the pattern of intelligibility decline. Sentences heard in normal intonation show generally good intelligibility up to 60% compression, losing only about 15% intelligibility between that point and normal speech rate. Only beyond 60% compression is there a marked decline. Sentences heard in anomalous intonation show a simpler pattern of results: an essentially linear decline with increasing compression. The hypothesis that normally accompanying intonation pattern adds redundancy to the speech signal is thus confirmed. The question of how this redundancy operates is revealed by an error analysis of the listeners' perceptual responses.

Fig. 2 shows a classification of error responses to normal and to anomalous intonation sentences for each speech rate tested. The upper curve shows the proportion of responses to normal intonation sentences which, while not identical to the test sentences, were nevertheless fully grammatical, and had their MSB at the same point. This represents the majority of errors so long as intelligibility remains high (normal rate through 60% compression), ceasing to be the majority only beyond this point when non-grammatical utterances or sentence fragments begin to characterize the errors.

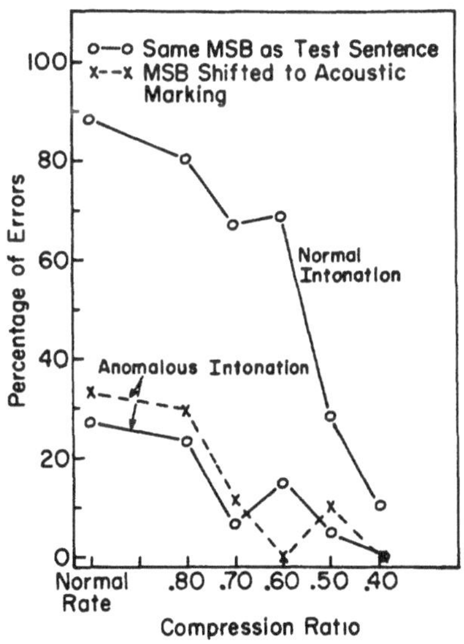

Fig. 2. Percentage of grammatically correct error responses and structure for normal and anomalous intonation sentences as a function of degree of time compression.

Three examples typify these errors. The first shows simple omissions and substitutions, the second shows migration of the word trouble to keep the erroneous response syntactically consistent, and the third shows how a misinterpretation of the word made for its homophone maid generated a shift of the to an earlier position, and a change of grammatical ending to keep the response syntactically consistent with the test sentence.

(1) Test sentence: "Except for one minor delay, traffic near the tunnel moved."
Response: "Except for momentary delay, traffic in the tunnel moved."

(2) Test sentence: "Despite the stoppage unions tend to cause, trouble was avoided."
Response: "Despite the trouble that students tend to cause, violence was avoided."

(3) Test sentence: "After pleading, we made her cook the dinner that night."
Response: "After pleading, the maid cooked dinner that night."

The error pattern for sentences heard in anomalous intonation is shown by the two lower curves in Fig. 2. Again, the majority of errors at normal speech rate and at 80% compression remain fully grammatical sentences. In this case, however, where the objective structure and acoustic marking conflict, error responses are distributed between the two possible resolutions. In some cases the MSB of the response conforms with that of the test sentence, while a slightly greater proportion show a restructuring to produce a sentence with the MSB shifted to the point marked acoustically by intonation. This pattern of results obtains until 60% compression, where overall intelligibility scores are extremely low, and non-grammatical fragments become common.

Two examples of resolution to shift the MSB to correspond with intonational markings are shown below. The first is a test sentence with the MSB between cities and to trust, spliced so as to be heard in the intonation pattern of a sentence with the MSB between trust and a stranger. In the response, to trust was heard as with trucks, and a was reported as the. The result is a grammatical sentence, but of a "new" structure, with the MSB shifted to conform with the intonational marking. The second example is taken from a test sentence with the MSB between influence and voting, but heard in the intonation pattern of a sentence with the MSB between voting and machines. No words are deleted in this response. Rather, there is a perceptual restructuring to resolve the conflict in favor of the intonational marking through addition of the words way of, and an ing ending to influence.

(1) Test sentence: "In larger cities, to TRUST A STRANGER MUST BE dangerous."
Spliced version: "In a position of TRUST, A STRANGER MUST BE supervised."
Response: "In larger cities with trucks, the stranger must be dangerous."

(2) Test sentence: "Due to our new mayor's INFLUENCE, VOTING MACHINES WERE INSTALLED."
Spliced version: "To avoid any attempts to INFLUENCE VOTING, MACHINES WERE INSTALLED."

Response: "Due to our new mayor's <u>way of</u> influenc<u>ing</u> voting, machines were installed."

Several general conclusions can be drawn. First, neither in the case of normal nor anomalous intonation sentences were the errors representative of a word by word decoding of the speech stream. Listeners seem to gain an overall sense of meaning and structure, followed by a cognitive reconstruction to keep the response consistent with that understanding. Second, while compression may collapse stress and timing pattern, the overall recognition superiority of normal intonation sentences maintains throughout the rates tested, and continues to function in generally the same way as at normal speech rates. Even when the words themselves become unrecognizable, prosodic features continue to be transmitted, supporting the notion that prosodic features may well be the most resistant parts of the speech waveform to any naturally occurring form of distortion (cf., Huggins, 1972). Finally, error analysis, which gives a better indication of perceptual operations than correct responses, shows perceptual structuring of sentences to be determined as much by prosodic features as by formal structure. In normal intonation where the two coincide, resolution is aided by this acoustic marking, which results in the high recognition scores. The final stage of analysis in the active perception of the sentences is dependent on, and is guided by, the successful completion of this syntactic resolution.

<u>What Determines the Size of the Perceptual Unit</u>?

We have taken the position that there is no single "unit" of speech perception, so much as a hierarchy of processing units, each with its own natural size (syllables, morphemes, clauses, etc.). The segmentation paradox with which the discussion opened, results from a view of the temporal course of speech perception as a strict sequential one, in which each stage of analysis follows completion of the preceding one. We have tried to show that depth, rather than sequence of processing, better describes the dynamics of the perceptual operation over the course of time. Later stages importantly influence the perceptual output of the earlier ones. Thus, for example, phonemic judgments completed at one level, are modified, or even rejected, as analysis continues to clausal and sentential levels. Demand characteristics of experiments which terminate processing at the level of phonemes, syllables or words, for example, thus yield rules valid only for those levels. Later levels of processing include review of all evidence, and may perhaps reverse earlier decisions in the light of further information.

There is an upper limit on the size of processing units, bounded by the capacity and duration of the memory store in which they are presumably held for analysis. For example, analysis at the syllabic level has been identified with the existence of an auditory memory image lasting perhaps 100-250 msec, but certainly under one second (Crowder & Morton, 1969; Massaro, 1972; Huggins, 1974).

On the other hand, analysis at the clausal level implies the characteristics and duration of storage typically associated with immediate, or short-term memory (STM) (cf., Peterson & Peterson, 1959; Waugh & Norman, 1965; Broadbent, 1971, pp. 324-399). Reaction times to clicks are longer near clausal boundaries, suggesting that the processing load on the storage system is particularly high at these points (Abrams & Bever, 1969). Capacity estimates of STM and typical sentential clause length are close, although estimates of the former have been based primarily on unidimensional stimuli (Miller, 1956). Simultaneous retention of prosodic pattern to aid analysis is also required. Similarly, while the upper limits on STM capacity may be fixed, its effective size will vary with coding strategies imposed on the stimuli in store (Wingfield & Branca, 1970; Wingfield & Byrnes, 1972). As length and number of clauses in a speech sample increase, the issues of STM capacity, processing time, and coding routines become increasingly critical.

Because of these considerations, the upper limits on segment size as determined by STM capacity and by speech content require clarification if we are to fully understand perceptual processing of continuous speech. Compressed speech may be particularly useful in this regard, in view of the suggestion that deteriorated intelligibility results primarily from elimination of available processing time, rather than compression of the acoustic signal itself (Aaronson, Markowitz & Shapiro, 1971). One way to approach this question is to examine intelligibility of compressed speech samples with pauses of various durations inserted at various critical points within speech passages. In this way, one might shed further light on the effects of temporal and content parameters in the temporary storage and perceptual processing of speech. Work along these lines is now in progress.

Conclusions

The research described in this paper has attempted to make several main points. (1) Speech is processed not on a sequential, segment by segment basis, but in terms of a hierarchy of units, the one examined here being of clausal length. (2) Perceptual segmentation is an intermediate stage of processing, dependent on an interaction of syntactic structure and prosodic cues, with the latter aiding directly in the resolution of the former. (3) Even when word intelligibility becomes diminished through time compression, prododic pattern resists disruption, and continues to influence syntactic resolution. (4) Finally, the factors thought to constrain unit size in the perceptual processing of connected speech were considered, along with a methodological approach which may allow further clarification of these issues to be forthcoming.

Acknowledgments

I thank Jan F. Miller for aid in preparing the stimulus tapes used in these experiments, Dr. Emerson Foulke, Center for Rate Controlled Recordings, University of Louisville, for assistance in time-compression of the speech materials, and John L. Wheale for helpful discussion and criticism during the preparation of this paper. Financial support through PHS Grant NS-09767 from the National Institutes of Health and Grant BMS 75-22200 from the National Science Foundation is gratefully acknowledged.

References

Aaronson, D., Markowitz, N., & Shapiro, H. Perception and immediate recall of normal and "compressed" auditory sequences. Perception and Psychophysics, 1971, 9, 338-344.

Abrams, K., & Bever, T. G. Syntactic structure modifies attention during speech perception and recognition. Quarterly Journal of Experimental Psychology, 1969, 21, 280-290.

Broadbent, D. E. Decision and Stress. New York: Academic Press, 1971.

Cole, R. A., & Scott, B. Toward a theory of speech perception. Psychological Review, 1974, 81, 348-374.

Crowder, R. G., & Morton, J. Precategorical acoustic storage. Perception and Psychophysics, 1969, 5, 365-373.

Fodor, J. A., & Bever, T. G. The physchological reality of linguistic segments. Journal of Verbal Learning and Verbal Behavior, 1965, 4, 414-420.

Fodor, J. A., Bever, T. G., & Garrett, M. F. The Psychology of Language. New York: McGraw-Hill, 1974.

Foulke, E. The perception of time compressed speech. In D. Horton & J. Jenkins (Eds.), The Perception of Language. Columbus, Ohio: Merrill, 1971.

Garrett, M. F., Bever, T. G., & Fodor, J. A. The active use of grammar in speech perception. Perception and Psychophysics, 1966, 1, 30-32.

Goldman-Eisler, F. Psycholinguistics: Experiments in Spontaneous Speech. New York: Academic Press, 1968.

Halle, M., & Stevens, K. N. Speech recognition: A model and a program for research. In J. A. Fodor & J. J. Katz (Eds.), The Structure of Language. Englewood Cliffs, N. J.: Prentice-Hall, 1964.

Huggins, A. W. F. On the perception of temporal phenomena in speech. Journal of the Acoustical Society of America, 1972, 51, 1279-1290.

Huggins, A. W. F. On perceptual integration of dichotically alternated pulse trains. Journal of the Acoustical Society of America, 1974, 56, 939-943.

Ladefoged, P. Three Areas of Experimental Phonetics. London: Oxford University Press, 1967.

Ladefoged, P., & Broadbent, D. E. Perception of sequence in auditory events. Quarterly Journal of Experimental Psychology, 1960, 13, 162-170.

Liberman, A. M., Cooper, F. S., Shankweiler, D. P., & Studdert-Kennedy, M. Perception of the Speech Code. Psychological Review, 1967, 74, 431-461.

Liberman, P. Intonation, Perception, and Language. Cambridge: MIT Press, 1967.

Massaro, D. W. Preperceptual images, processing time, and perceptual units in auditory perception. Psychological Review, 1972, 79, 124-145.

Peterson, L. R., & Peterson, M. J. Short-term retention of individual verbal items. Journal of Experimental Psychology, 1959, 58, 193-198.

Waugh, N. C., & Norman, D. A. Primary memory. Psychological Review, 1965, 72, 89-104.

Wingfield, A. Acoustic redundancy and the perception of time-compressed speech. Journal of Speech and Hearing Research, 1975, 18, 139-147.

Wingfield, A., & Branca, A. A. Strategy in high-speed memory search. Journal of Experimental Psychology, 1970, 83, 63-67.

Wingfield, A., & Byrnes, D. L. Decay of information in short-term memory. Science, 1972, 176, 690-692.

Wingfield, A., & Klein, J. F. Syntactic structure and acoustic pattern in speech perception. Perception and Psychophysics, 1971, 9, 23-25.

DISCUSSION of "The intonation syntax interaction: Prosodic features in perceptual processing of sentences"
WINGFIELD

MASSARO:
You state that intonation only adds redundancy to the information already given by syntactic structure. I do not see how it is possible to determine whether intonation or syntactic structure is primary in disambiguating the sentence. More importantly, it is necessary to determine how these two sources of information are combined in sentence comprehension: your experiments, as presently designed, cannot do this. I would suggest conveying both the amount of syntactic structure and intonation information in a factorial experiment to evaluate how these two sources of information are combined.
Finally, I think that it is necessary to determine the stage(s) of processing contributing to the results in these experiments.

LEVELT:
I have two problems with your paper, a major and a minor one. The major problem is related to what Dr. MASSARO brought up a moment ago so I can be short about it. The point is that I have found no evidence in your paper that the phenomena that you report are perceptual in the sense that they arise during the intake of information. Rather they could very well be the result of response bias arising during the reconstructive phase when the subject generates his judgment with respect to what he heard some time ago. In this respect the problem is equivalent to what critics like Ladefoged, and Reber & Anderson brought forward with respect to the click-studies. At the end of your paper you remark that it might be important to the subject "to keep the response consistent with that understanding". If this would explain all of your results, I would feel unhappy by calling your effect perceptual.
The second point relates to your statement that the effects of syntactic structure and intonation are not simply additive. I see no evidence in your data that this is so. It should appear from, for instance, there being interaction effects in an analysis of variance. What, then, is your motivation for that statement?

WINGFIELD:

Perhaps in these and some other experiments we would be on safer ground in referring to "perceptual reports", rather than "perception". Our primary concern in this paper is the relationship between acoustic cues and syntactic structure, and it is often difficult to say exactly at what stage in the total perceptual process these factors operate. The present results using time-compressed speech suggest the effects of structuring occur at a relatively high level of processing.

In respect to switch localization, Jan Klein Miller and I did make an attempt to reduce memory bias by giving listeners printed copies of the sentences as they were heard. There was no need to recall the sentences in this case, but merely to mark the location of the switch as it was heard. (This study is reported in the Wingfield and Klein (1971) paper referred to in the text). While the results showed a relative increase in accuracy of switch localization, the main findings as reported here were still seen under these conditions.

My statement on additivity is intended as a comment on the role of acoustic features in sentence segmentation as I see it. I suggest that the two factors should not be treated independently: acoustic pattern can act as a critical marker to aid determination of major syntactic boundaries which, in turn, determine segmentation for analysis.

PISONI:

The question of what perception is and the relationship of perception and memory are obviously quite relevant topics to this meeting. If I understand MASSARO correctly he is suggesting that listeners perceive speech from the outside in. Obviously such an approach assumes that the relevant syntactic information for perceptual analysis is marked in the physical signal and all the listener has to do is passively respond to these physical attributes to construct a semantic representation.

BERTELSON:

I have two comments.

1. There seems to be a serious difficulty about using click migration, or shift point migration in your case, to get information regarding the size of processing units. The idea, first but forward by Fodor and Bever, that the system somehow tries to avoid interruption of perceptual units by the click, predicts complete

migration into the boundary between two units. Now, what is generally observed, and it seems to be the case in your experiments as well, is partial migration towards the boundary. This of course does not help preserving the integrity of constituents, whether grammatical or prosodic. Is there another explanation for the occurrence of partial migration?

2. Like LEVELT, I feel the idea that click localization is determined at a rather late stage of perceptual interpretation operations is still very much alive. It seems however that the term "response bias" produces some difficulty when applied in this context. The idea is that there is an a priori tendency to report the click as occurring in one particular position - e.g. in the boundary. This is clearly demonstrated by the silent click manipulation: migration towards the boundary occurs even in the absence of sensory evidence regarding the click. But there is no implication that processing the speech material is not necessary: obviously, the sentence must be constructed before the location of the boundary is established.

WINGFIELD:

I do agree that the results of the so-called "click" studies must be treated with caution. It might be fair to conclude the issue by considering the early "click" studies as a heuristic serving to point in fruitful directions for further enquiry, rather than offering a definitive test of hypothesis. As such, however, they have been of value.

LEHISTE:

It seems to me that you are equating prosody and intonation, and claiming that syntactic structure is (as a rule) signalled by intonation. I should like to refer briefly to a study of mine (with J. Olive and L. Streeter) in which we took syntactically ambiguous sentences, monotonized them after analysis-resynthesis (using a method developed by J. Olive) and manipulated the time dimension. Appropriate changes in timing enabled the listeners to disambiguate most of the ambiguous sentences. A similar manipulation of the fundamental frequency (starting from a temporally neutral form of the utterance) was very much less successful in providing syntactic cues.

WINGFIELD:

I welcome your comment and may say that I do not at all mean to deny the importance of temporal patterning or hesitation pause as potential acoustic markers for correct segmentation. We are both aware of evidence which suggests that this may be so, and I am glad you raised the point.

RHYTHMIC EXPECTANCY IN CONTINUOUS SPEECH PERCEPTION

J. G. Martin, Department of Psychology
University of Maryland, College Park 20742

The model concerns moment-to-moment listener expectancies based upon the dynamics of patterned speech movement. For convenience here we distinguish between (a) temporally short-lived (segmental) cues including coarticulatory effects and (b) all other cues (call them suprasegmentals) in continuous speech perception. Suprasegmental cues enable the listener to expect or anticipate the rough outlines of speech not yet heard. These suprasegmental redundancies include timing and hence generate listener expectancies in real time. At some given moment after prior context has been heard, suprasegmental processing load is lighter, with correspondingly greater efficiency in segmental processing. That is, segmental and suprasegmental processing can be said to interact.

Two experimental paradigms are reported together with data. The general methodological approach is to produce small changes in the acoustic stimulus by means of tape-splicing or other experimental intervention, and then to observe listener performance in tasks requiring continuous speech processing. In one paradigm, the task requires reaction time to phoneme targets in sentences in which targets are temporally displaced by means of pre-target interventions. In the other paradigm, the task requires immediate recall of spoken digit strings whose "articulatory integrity" has been violated by means of tape splicing.

Experimental results point to an interaction between the effects of segmental and suprasegmental cues as they are encoded into continuous speech.

1. Introduction

The experimental approach and results described in this paper flow from a model for perception and production discussed in a paper several years ago (Martin, 1972), wherein it was suggested that the notion of rhythmic structure might be fruitfully applied to speech. Before discussing experiments, however, it may be useful here to mention some reasons why we have found the particular notion of rhythm a more helpful heuristic than the more general notion of prosody. It may also be useful to clarify some aspects of the theoretical model to which we were led. The notion of rhythm has many attractive implications; for example, no other notion more richly expresses the simultaneity and shared purpose of the speaker's and listener's activities. Unfortunately, however, certain implications of rhythm, e.g., temporal regularity, can be misleading in the context of speech research, as will be discussed later. But other implications are important *in principle*. To consider these implications is to begin to think about the dynamic structure of speech, and thence to entertain certain possibilities for a theoretical model, and experimental approaches for a test of the model, that otherwise might escape notice. Some of these implications may be mentioned very briefly.

As everyone knows, elementary communication theory notions regard the input signal as a sequence of "states" ("elements," "feature arrays," etc.) with conditional probabilities between states represented in terms of redundancy, or predictability. In principle, to call a sequence rhythmic is to add timing to the aspects of the sequence which can be redundant. This is perhaps one of the more obvious albeit misleading implications, particularily if one is thinking of musical rhythms in which the timing is regular. But rhythm also implies (a) that redundancy (timing or other) is not limited to adjacent states, and (b) that redundancy is not limited to past and present states but also includes future states. The significance of (a) for perception through time is that it entails <u>pattern</u> redundancy so that, for example, ambiguities in the signal at time T_N may be disambiguated not only at time T_{N-1} (as in a sort of Markov-chain fashion) but also at other times. The significance of (b) is that to the extent that future aspects of the sequence are redundant, i.e., <u>predictable</u>, they are to that extent known. Hence one might say that at a given instant in time the "percept" often consists in the combination of information from not only past and present but also <u>future</u> states. The advantages in principle of a schema like this for explaining rapid, automatic auditory perception are fairly obvious. For one thing, the greater the amount and extent of temporal pattern redundancy in the input signal, the lesser would seem to be the need for postulating an extended acoustic ("echoic") memory store, and this would be especially the case when some of the redundancy concerns aspects of the signal not yet heard.

The principles of temporal patterning and temporal predictability are assumed to apply to auditory temporal pattern perception in general (Sturges and Martin, 1974). In the particular case of speech, and communication models thereof, the patterning principle raises empirical questions which should be postponed until after a consideration of the predictability principle and its empirical demonstration. With respect to this principle, the listener has been described as in part a sort of dynamic feed-forward device who, given cues from early elements in an utterance, can in effect predict future vocal-tract movements of the speaker (Martin, 1972). There is abundant evidence suggesting that the listener is sensitive to the acoustic effects of constraints on production (e.g., Cooper, Delattre, Liberman, Borst and Gerstman, 1952; Dorman, Raphael, Liberman and Repp, 1975; Liberman, Cooper, Shankweiler and Studdert-Kennedy, 1967). Most of this evidence comes from research with single syllables and is limited to models for phoneme recognition. But there is of course no reason in principle for such a limitation. The general principles of temporal patterning and redundancy mentioned above suggest a communication model of larger scope. Such a model would say, for one thing, that more than phonemes and syllables are encoded in the course of producing a continuous utterance, that the speech encoding process and its decoding consequences are richer and have a longer temporal reach. In discussing the model here it is convenient to make a general distinction between events having a shorter or longer temporal life. Then one may think of the pre-programmed speech act as divided into two kinds of encoding movements (or activities); those which are temporally short-lived and those whose consequences extend throughout much of the duration of the act. It is possible to imagine that the effects of these two kinds of movements interact. The acoustic record of these movements (speech) contains information which, again, can be conveniently divided into two sorts; "segmental" cues to specific phonemes including coarticulatory effects, and "suprasegmental" cues to everything else, the latter of which includes the signalling of events over a longer temporal range. These latter cues might permit the listener to expect or anticipate the rough outlines of speech not yet heard. If suprasegmental cue redundancies include timing, which is most likely since speech is produced by patterned movement, accordingly such cues might generate listener expectancies concerning later events in real time. If one were to discuss such a model in generalized analysis-by-synthesis terms, one might say that what is synthesized involves not only what has been heard but what is yet to be heard. To the extent that it is appropriate to speak of segmental and suprasegmental information as separable or separately processed, one might also say that there is reduced suprasegmental processing load after some prior speech context has been heard and that, based on limited capacity notions, there is then correspondingly greater efficiency in the processing of segmental cues. That is, the two presumed processes might be said to interact. These considerations of course are intended to apply to the phonetic, that is, segmental and suprasegmental structure of an utterance and not to "higher" levels. However, certain aspects of suprasegmental structure and surface syntax appear to be correlated (e.g.,Blesser, 1969; Svensson, 1974), and so it may even be that the main outlines of syntactic structure often are communicated relatively early in an utterance. In any case, a communication model based on timing redundancy in movement patterns and correlated listener expectancies would appear to have potentially great advantages over models without these characteristics, and such a model warrants sustained research effort.

Given the considerations just discussed, the initial task in our research program appeared to be to demonstrate that temporally long-ranged suprasegmental cues exist in continuous speech and are used in perception. Since it is not known what these cues are, how they interact, how they sound, or what they look like, the best course of action seemed to be first to manipulate the acoustic input signal in some way and then to observe perceptual consequences (if any) when the listener's task required continuous processing. Two experimental paradigms using this approach together with selected illustrative experiments and their results are described below.

In these paradigms, changes are produced in the acoustic stimulus by means of some experimental intervention such as tape splicing, so that the result presumably could not be spoken by a human talker. If the intervention has effects on performance in the listener's assigned task, the result is attributed to listener expectancies which are violated or disconfirmed by the consequences of the intervention. Often we expect poorer performance, since experimental interventions degrade the acoustic signal by violating the natural flow, or more generally, the "articulatory integrity" of the utterance. However, the use of experimental interventions might raise the suspicion that they produce non-speech effects, that is, that they disrupt or distract the listener in ways unrelated to the effects on speech processing and produce poorer performance for that reason. Such non-speech disruption effects would not be easy to specify, given the present state of perceptual research, nor do they seem to particularly relevant unless they can account for all of the poorer performance. Nevertheless, comparison or control conditions were included in both experimental paradigms to counter that possible objection, by showing that intervention effects do not necessarily interfere with but can actually facilitate performance, depending upon how the intervention is related to the listener's experimental task.

2. Reaction time to temporally-displaced phoneme targets in sentences

In this paradigm, the approach was to temporally displace stretches of the running acoustic speech stimulus by means of tape-splicing or other experimental interventions so that listener expectancies presumably were violated, and then to observe reaction time (RT) to a phoneme target (Foss, 1969) which occurred later in the sentence and was therefore temporally displaced (Meltzer, Mills, Imhoff, Zohar and Martin, 1975). Although the finding of slower RT to temporally-displaced targets than to targets not temporally displaced would be consistent with a timing-expectancy hypothesis of the sort discussed earlier, there remains the question of whether other intervention effects besides speech timing might be responsible for negative effects on RT. While on *a priori* grounds the opposite result seems just as likely, i.e., cue effects when pre-target interventions are used, it is nevertheless useful to show that the combined net speech and non-speech intervention effects are not necessarily negative. One goal of Experiments I and II below was to show that interventions under certain conditions can produce faster rather than slower RT as a special case, and hence that interventions by themselves probably cannot provide a plausible account for slower RT in other cases where all relevant conditions have been met, as in Experiment III.

Experiment I. The target was /b/ and there were four experimental conditions each of which were constructed by tape-splicing copies of the same sentence. The cut for the splice was made at the identical place on each copy, namely, 50 msec prior to the target, so that the 50 msec of speech immediately preceding the target was always intact although the manipulations preceding the cut were different. The conditions differed primarily in terms of whether or not the target was displaced and whether or not speech information had been discarded by the splicing, and secondarily in terms of whether or not a patch of masking white noise had been added at the locus of the splice. The reason for using the noise mask rather than blank tape is the interesting fact that it is difficult to tell by ordinary listening if the speech "behind" the mask has been changed at all.

The sentences were spoken naturally; no musical "beats" or similar other constraints were imposed. An example of a sentence used was: He laughed and laughed till his belly wiggled like jelly. Table I below illustrates the sequences making up the different experimental conditions.

Table I

Experimental Sequences

NORMAL:	x a b B
EARLY:	x b B
LATE:	x * * b B
ON-TIME:	x * b B
LATE-INTACT:	x a * b B

Consider first the Normal sequence (not used in this particular experiment) and imagine that it represents a sentence on audiotape. Capital B is the target location, pre-target interval b is 50 msec long, interval a is 100 msec long, and x represents the remaining prior context of the sentence. In the first experimental condition (EARLY), interval a was spliced out of the sentence and the ends abutted. Hence the target was displaced 100 msec "early" relative to prior running context, but (presumably) a part of the cost of the displacement was the speech information in the discard interval. In the LATE condition, interval a was spliced out but 200 msec of white noise (indicated by **) was spliced in. Hence the target was displaced 100-msec "late" and again speech information in the discard interval was lost. In the ON-TIME condition, interval *a* was spliced out and an equivalent interval of noise was spliced in. Hence this condition contained an "on time" target, although it shared a noisy interval with the LATE condition and discard interval with both the EARLY and LATE temporal displacement conditions. In the LATE-INTACT condition, no speech was discarded although 100 msec of noise was spliced in. Hence the target was displaced "late" although in this condition the sentence remained "intact" in the sense that no linear speech information was lost.

On the timing hypothesis, which is that disrupted timing "derails" normal on-line perceptual processing, the straight-forward prediction is relatively slower RT to EARLY and LATE targets compared to ON-TIME targets. On the same hypothesis one of course might predict slower RT to LATE-INTACT targets also. However, only in the LATE-INTACT sequence was no pre-target speech interval discarded. If coarticulatory information about the target reached beyond 50 msec prior to the target and into this interval in the original sentence (interval *a*), perhaps the listener could use this information to anticipate the target across the noise displacement interval. Since such information would be received even earlier relative to the target than in the case of the original sentence, RT as fast or even faster to LATE-INTACT targets might be expected than to ON-TIME targets. It is important to note that this condition is a special case of intervention effects; the coarticulatory hypothesis applies only when the intervention is in close proximity to the target and predicts faster RT because target information is provided *prior* to the intervention. We would expect, however, that other response measures besides target RT, e.g., sentence memory, might suffer because of intervention effects in this condition just as in the other displacement conditions.

The results are shown in Table 2. Compared to ON-TIME targets, RT was reliably slower (Newman-Kuels tests) to EARLY and LATE targets but reliably faster to LATE-INTACT targets, as expected. The slower RT results were interpreted in terms of disrupted ongoing perceptual processing as a result of disrupted timing expectancies, whereas the faster RT to INTACT targets, on the other hand, was taken to show that coarticulatory cues in interval *a* provided target information in advance, so that the listener could more than overcome the effects of disrupted timing.

Table 2

RT in csec to Phoneme Targets in Experiment I

Condition:	EARLY	LATE	ON-TIME	LATE-INTACT
RT:	78.1	76.5	70.0	65.8

The next experiment was done to check on some alternative explanations for the latter result.

Experiment II. Possibly the advantage of LATE-INTACT targets over ON-TIME targets in Experiment I was not advance information but simply "complete" information, however temporally deployed, since this was the only condition in which no speech was discarded. Accordingly, LATE-INTACT sequences like those in Experiment I were compared to copies of the original which were not spliced at all (NORMAL sequence in Table I). A further experimental comparison involved the assumption that coarticulatory effects in speech are temporally restricted and so their effects on target anticipation should be related to target proximity. Accordingly, proximity of advance information to target was varied through the use of three LATE-INTACT conditions which differed only in the time intervening between the displacement interval and target. These can be imagined by looking at the LATE-INTACT sequence in Table I, wherein interval b was either 50, 75, or 100 msec in duration so that interval a correspondingly varied in target proximity in the NORMAL sequence prior to splicing and hence presumably contained less coarticulatory target information the farther away it was from the target. The farther away from the target, then, the less should be their facilitative effects on target RT. The fourth experimental condition contained the NORMAL (uncut) version of the sentences. With no white noise in the NORMAL sequence, there was no point in using white noise in the LATE-INTACT sequences either, so noise was reduced to background level in those sequences. The effect often sounded like a slight hesitation.

The results are shown in Table 3, wherein LATE-INTACT is abbreviated to LI. RT was reliably faster (Newman-Kuels tests) to LI-50 targets than to either NORMAL or LI-100 targets. As can be seen, the closer the cut to the target the faster was the RT, as would be expected on coarticulatory notions. Furthermore, RT was faster to targets preceded by experimental interventions than to targets in sentences with no interventions at all. These results appear to rule out the possibility that faster RT to LATE-INTACT targets in Experiment I was based on completeness of linear speech information rather than advance information, since speech information was complete in all conditions of Experiment II. The results of both experiments appear to show that net disruptive effects on RT are not a necessary

Table 3

RT in csec to Phoneme Targets in Experiment II

Condition:	NORMAL	LI-50	LI-75	LI-100
RT:	62.1	56.2	57.6	60.4

consequence of interventions in the running speech signal, particularly since sentences with interventions produced faster target RT than did sentences containing no interventions at all. These results make unlikely an interpretation of intervention effects solely in terms of non-speech disruption effects.

Experiment III. Coarticulatory cue effects in Experiment II diminished inversely with proximity to target, so that RT to displaced (LATE-INTACT) targets and non-displaced (NORMAL) targets was nearly equal when the interval between intervention and target had been increased to 100 msec. Those results were assumed to be due to the cancelling of (positive) coarticulatory effects and (negative) temporal displacement effects, an interpretation which implies that experimental interventions earlier in the sentence would produce net negative displacement effects on RT. Accordingly, interventions in Experiment III preceded the target by a substantially longer time interval, namely, two syllables. Segmental cues to target phoneme recognition then were presumably left intact, so that any intervention effects applied to the suprasegmental structure of the signal. Since any speech information preceding the intervention now was presumably far in advance of any possible target-related coarticulatory effects, slower RT was predicted to temporally-displaced targets than to the same targets in the normal (non-displaced) version of the sentence.

All targets were the phoneme /g/ carried by a nonsense word in an otherwise normal sentence. All sentences had approximately the same structure as in the example:

The farmer could not sell more milk,
so he had to/produce GOLrem instead.

The slash indicates the location of the experimental intervention, which was always followed by a stop consonant.

The experimental intervention consisted of adding 200 msec of silence to the silence normally preceding a stop consonant so that in effect the target was displaced 200-msec "late" relative to running speech context preceding the intervention. Intervention effects on target processing then necessarily spanned a time interval equal to the duration of two syllables. It is important to emphasize that this experimental intervention, which added time to the pre-stop-consonant silent interval, nevertheless left the sound sequence completely intact. As in the case of Experiment II, this intervention often produced the impression of a slight hesitation also.

The results were overall slower RT to temporally- displaced targets, as expected. Two additional experiments were also done, differing only in the duration of the temporal displacement. In one, 100 msec rather than 200 msec of silence was added; in the other, 67 msec was "subtracted" so that the target was "early" rather than "late." The results of all experiments were essentially the same; slower RT to displaced targets.

Discussion of target experiments. Slower RT to temporally-displaced targets is in line with the view that on-going speech perception is based upon timing expectancies generated by suprasegmental cues to the continuous phonetic structure. An alternative account solely in terms of disruptive or other non-speech intervention effects appears unlikely in view of the RT effects in Experiment II, in which intervention effects produced faster rather than slower RT, and in view of the fact that lengthening pre-stop-consonant silence left the sound sequence intact in Experiment III. Concerning the latter, it is of course possible (or even likely) that extending pre-stop-consonant silence may do violence to the cues to the stop consonants themselves plus neighboring phonemes, but it seems unlikely that the negative effects would extend into processing two syllables later, nor does there appear to be any theory of speech perception (besides the present one) which would predict such an effect. The results of these and other experiments using the RT paradigm lead to the conclusion that speech is organized with timing redundancy over relatively long stretches of an utterance. The resulting cues are compelling and generate expectancies which are actively used in the course of continuous perception. Further experiments appear to show that these cues are provided very early in the utterance, e.g., RT slow-downs occur when the pre-target intervention follows the first syllable of the sentence. But what processes are involved? What happens when a target is displaced? When 200 msec of silence is artificially introduced into a running sentence, the speech signal in effect stops all at once, albeit briefly. But, one might say, by then the listener cannot stop all at once; the "momentum" in the listening process carries him on, so that he must stop and then start over.

3. Immediate memory for digit strings

In this paradigm, the purpose of the experimental intervention was to alter not necessarily (or not only) timing but more generally the articulatory integrity of the acoustic stimulus, by splicing together stretches of speech which were spoken in independent utterances. The listener's task, immediate recall of a string of spoken random digits, is a demanding one with "overload" characteristics, and hence small changes in the acoustic stimulus might produce differential effects whereas one would not expect memory for materials such as sentences to be affected. The task is normally used in studies of memory, but in our usage we think of this paradigm as perceptual since it tests effects of variations on the physical stimulus.

The experiment described here (Mills and Martin, 1974) has since been replicated and extended in a variety of ways (Mills and Martin, 1975). It concerns what is known as the stimulus prefix effect (Dallett, 1964), namely, you speak (and tape-record) the prefix _zero_ followed by a string of seven digits, you tell the listener to recall the digits without the _zero_, and you find that the _zero_ nevertheless depresses recall of the seven digits. The prefix is always there, it is always _zero_, the subject is told to ignore it, but apparently he cannot. In later work using this experimental paradigm, Neisser, Hoening and Goldstein (1969) simply recorded the prefix in a different voice from the voice speaking the recall digits and found that the prefix effect nearly disappeared. Our interpretation of this result was that when the same speaker pronounces the prefix and the recall digits, this constitutes one organized articulatory stimulus containing not seven but eight digits. The listener responds to it as such, which means that he or she will (or rather must) perceptually process an intact eight-digit string and then somehow suppress the prefix before reporting the seven digits. Obviously with two speakers there cannot be a single intact articulatory stimulus, so the listener should find it easier to separate the prefix from the digit string, as Neisser _et al_ showed. Our experiment was done to provide further support for our interpretation, namely, that part of their prefix effect was phonetic and could not be explained solely in terms of "higher" levels of processing which might invoke gestalt notions such as "grouping" or "similarity." In the experiment there were two prefix conditions and a no-prefix condition which are relevant here. Following are examples for convenient reference:

INTACT:	0 3 4 2 1 8 6 5
SPLICED:	0/3 4 2 1 8 6 5
NO-PREFIX:	3 4 2 1 8 6 5

In one prefix condition, a speaker pronounced the prefix plus recall digits as usual, at a 2/sec rate (INTACT). Hence the prefix plus recall digits constitutes one intact articulatory unit. In the other prefix condition (SPLICED), the _same_ speaker pronounced the prefix and the recall digits, but the prefix was spliced into the tape prior to the digit string; it had previously been cut out of another utterance spoken by the (same) speaker. Hence there could not be any articulatory continuity between the prefix and following recall digits. The seven recall digits _themselves_ were an intact articulatory unit, in fact a tape copy of the seven-digit NO-PREFIX string, although now they were preceded by a spliced-on prefix. A number of controls were used to insure that the transition across the splice was smooth, for example, that pitch and loudness were the same, or nearly so; the significance (or lack thereof) of these controls is discussed below. The timing across the splice was controlled by the use of clicks on the second channel. As the clicks were recorded they were also monitored by the speaker and used to time his pronunciation of the digits. Since he monitored through headphones the clicks were not audible on the experimental tape. In some of these strings, you could not even tell that they had been spliced. In sum, the result of these interventions was that the SPLICED strings started with a prefix, but the following seven digits were by themselves an intact articulatory unit, so that presumably the experimental intervention now favored the listener in performing his task, a part of which was to ignore the prefix. The INTACT version was a prefix plus seven recall digits, but in this case it was one intact articulatory unit containing eight digits, so that in the absence of the experimental intervention the listener presumably was hampered in his experimental task.

The results were that SPLICED strings were recalled more easily than INTACT strings, though not as easily as the NO-PREFIX strings. These results appear to support the articulatory unit hypothesis since they show that a part of the prefix effect can be accounted for by phonetic continuity in the stimulus.

Discussion of recall experiments. It is important to make clear what the real theoretical issue is in these experiments. As mentioned earlier, care was taken to insure a smooth transition across the splice. Nevertheless, cues signalling tape-splicing discontinuity probably were present and detectable by the appropriate psychophysics in many if not all strings. Furthermore, the effect of switching speakers (Neisser _et al_) was greater than the effect of only switching utterances (our experiment). These considerations might suggest that although there are differences of degree between these two methods, nevertheless in both cases there are cues signalling discontinuity and that the prefix can be separated perceptually for that reason.

To say that, however, is to explain very little and may even obscure the fact that the real problem is to explain why the prefix effect is greatest when the prefix-plus-digits string is articulatorily intact. What is it about the INTACT string which makes the prefix more difficult to ignore? An explanation is required in terms of cues produced in the acoustic stimulus as the prefix is spoken and in terms of the immediate effects of these cues on the listener's expectancies. It is as if the mere act of speaking _zero_ engages the listener's perceptual process and sets it into motion, and that this process prepares him to expect certain acoustic attributes (and not others) at certain times (and not others). These expectancies may not even be limited to the first one or two digits, at least on the evidence from serial position curves (Mills and Martin, 1975), i.e., the poorer recall of INTACT strings was not limited to the digits immediately following the prefix but was actually greatest in the middle of the string, the usual locus of difficult recall. We are trying to identify the acoustic cues that might be responsible for the fact that the SPLICED prefix is easier to separate than the INTACT prefix. A far harder question is that of the processes involved. For instance, having just heard a SPLICED prefix, which formerly was utterance-initial, and now hearing the first recall digit, which was also utterance initial, does some process recycle or start over, in effect erasing the alien prefix?

4. General Discussion

All of the experiments reported above involve intervention effects and all seem to point to the same general conclusions, but here we wish to single out Experiment III of the target RT experiments for discussion since the acoustic results of the experimental intervention therein can be exactly stated, and hence the implications of the experiment can be seen most clearly. Recall that in that experiment, the silence preceding a stop consonant was experimentally lengthened by 200 msec; otherwise the sound sequence was not interrupted. Presumably all segmental cues to target recognition were intact since the target followed two complete syllables later, so any experimental intervention effects would seem to apply to the suprasegmental structure of the sentences. The fact that target RT nevertheless suffered appears to demonstrate temporal predictability as postulated earlier, but it also appears to mean that there is an interaction in the effects of segmental and suprasegmental cues as they are encoded in continuous speech. Shields, McHugh and Martin (1974) found similar interaction effects in an earlier experiment. They recorded RT to targets carried by stressed and unstressed syllables in running sentence context and also to the same targets when they were excised out of sentence context and placed in a nonsense sequence. The RT differences between stressed and unstressed targets nearly disappeared in the nonsense context condition, which seems to show that part of the normal RT advantage of stressed over unstressed targets in sentence context is based on their relation to the remainder of the sentence, i.e., to its suprasegmental structure.

Consideration of segmental-suprasegmental cue interactions raises further questions about the structure of continuous speech. One question not fully answered yet is that of how much segmental (allophonic) variation in continuous speech may be accounted for by interaction with suprasegmental structure as distinct from immediate context. Another question concerns the putative "blurring" or "disappearance" of segmental cues in rapid spontaneous speech. Perceptual success in that case is often attributed to the listener's knowledge of the syntax of the language, etc., but the question remains open as to the extent to which the cue structure in such cases is phonetically adequate but simply more highly encoded. Questions like these concern complex cue interactions and do not even arise except in the context of work with continuous speech. Surely there are many interesting facts yet to be learned by further study of the continuous speech signal (see Kuhn, 1975, for a recent demonstration of this point).

The experimental results above and others have been taken to show that continuous speech encoding provides "long-range" timing and other cues used in efficient moment-to-moment perception. What then is the next step? Experiments are in progress to determine the temporal reach and other aspects of such cues, but a more important eventual goal is to identify these cues and the way in which they are patterned. However, in the absence of an adequate theory of auditory temporal pattern perception (continuous speech perception being the special case of present interest), the best research strategies are far from obvious. It is not clear that identification, discrimination and other methods found so useful in speech perception work up till now will also be useful in the case of continuous speech. As in visual perception, there are difficulties in the use of perceptual reports as a clue to the relations between stimulus and percept. Further difficulties can be predicted if one prejudges perceptual effects to be expected on the basis of measureable aspects of the physical stimulus. Such difficulties can be seen in the confused status of certain issues in linguistic research having a perceptual basis. One example, closely tied to the notion of rhythm, is the issue of "isochronous stress." As is well known, perceptual impressions of isochrony have been reported for certain languages, particularily English (e.g., Abercrombie, 1967). Equally well known is that measurements of interstress intervals on the time line have shown that the corresponding physical isochrony is usually not there (e.g., Allen, 1973; 1975; Bolinger, 1965). One might conclude from these facts that isochrony is "only" perceptual. But that is a hasty and possibly misleading conclusion; perhaps there is more to be learned. It may be useful (a) to ask why the listener reports perceptual isochrony, and also (b) to ask if it is reasonable to require the listener to make judgments about single physical dimensions in speech. A possible answer to the first question is that the reported isochrony is based on probabilities, e.g., interstress intervals in the physical speech are close to equal on the average and so the listener interprets what he hears in terms of his past experience (his knowledge), i.e., the explanation lies in the listener rather than in the stimulus. Such an explanation is often quite respectable in perceptual research, particularily when the stimulus is ambiguous. A more stimulus-oriented approach (cf. Garner, 1970; Gibson, 1966), on the other hand, and one that might provide a possible answer to both questions, would be to take the listener's report of what he heard more seriously and look to see what kinds of stimuli might be responsible for producing the reported impression. For example, it may be that speech cues along several physical dimensions interact (Martin, 1972) in such a way that stress is temporally predictable and is reported as isochronous for that reason, or, what may amount to the same thing, that the perceptual result of cue interactions is that interstress intervals are heard as equal. Similar interactions were reported by Woodrow long ago (1909). Working with non-speech rhythmic patterns, he showed that as relative physical intensity differences increased between alternating loud and soft tones, the differences in physical inter-tone time intervals preceding and following the loud tone could be increased so that they differed by 50% or more and were still reported as equal. Woodrow's results indicate, for one thing, that there is not necessarily a one-to-one correspondence between single dimensions of stimulus and percept when cue dimensions are allowed to interact. Furthermore, they remind us that temporal predictability is not equivalent to strict regularity. The ticking of a clock must be (almost) strictly regular to be heard as regular (and predictable) because time is the only relevant physical cue dimension in that rhythmic pattern. If the ticking is irregular, it is easily heard as such.

Seen in this light, the fact that the listener cannot isolate single dimensions of the acoustic stimulus as they have been encoded into continuous speech is no more surprising than that he cannot isolate first and second formants. Similar remarks hold for the well-known problems in work on the perception of linguistic stress and also for other perception work with continuous speech. In visual pattern perception it is commonplace that stimulus and percept are complexly related and that "elements" ("features," "dimensions") in context have different properties than in isolation and different properties in different contexts. Unfortunately, there is little reason to suppose that auditory temporal pattern perception will turn out to be much less complicated.

Articulatory and acoustic phonetics are advanced endeavors whereas "perceptual phonetics" lags far behind. The reason is plain; perception is part of a discipline with methodological problems severe enough that they have only begun to be faced within the last hundred years. Nevertheless, the obvious (but far from trivial) fact is that the communicative significance of events in the speech chain ultimately will be determined not on acoustic or physiological grounds, but on perceptual grounds. Studies focussing on continuous speech as inputs, and real-time factors in processing, must precede any adequate account of the perceptual process.

The logical place to begin is by studies to determine the contribution of the acoustic speech stimulus. We need to know a great deal more than we know now about the kind of input data the (perceptual) program is prepared to accept before we will begin to make useful guesses (or theories) about the way the program works.

5. Acknowledgments

This work was supported by grants from the National Institute of Mental Health, United States Public Health Service (MH 16726) and the U. S. Army Research Institute for the Behavioral and Social Sciences, and by a grant from the Bio-Medical Sciences Support Committee, University of Maryland, to the Center for Language and Cognition.

6. References

Abercrombie, D. *Elements of general phonetics*. Chicago: Aldine, 1967.

Allen, G. D. Rhythm. Section contributed to Netsell, R. Speech physiology. In F. Minifie, T. Hixon, and F. Williams (Eds.), *Normal aspects of speech, hearing and language*. Englewood Cliffs: Prentice-Hall, 1973.

Allen, G. D. Speech rhythm: Its relation to performance universals and articulatory timing. *Journal of Phonetics*, 1975 (in press).

Blesser, B. A. *Perception of spectrally rotated speech*. Ph.D. Thesis, Department of Electrical Engineering, Massachusetts Institute of Technology, 1969.

Bolinger, D. L. Pitch accent and sentence rhythm. In I. Abe and T. Kanekiyo (Eds.), *Forms of English*. Cambridge: Harvard Press, 1965.

Cooper, F. S., Delattre, P., Liberman, A. M., Borst, J. M., and Gerstman, L. J. Some experiments on the perception of synthetic speech sounds. *Journal of the Acoustical Society of America*, *24*, 597-606, 1952.

Dallett, K. M. Effects of a supplementary prefix on recall. *Journal of Experimental Psychology*, *67*, 296-298, 1964.

Dorman, M. F., Raphael, L. J., Liberman, A. M. and Repp, B. *Some masking-like phenomena in speech perception*. Paper presented at Acoustical Society of America, Austin, Texas, April, 1975.

Foss, D. J. Decision processes during sentence comprehension: Effects of lexical item difficulty and position upon decision times. *Journal of Verbal Learning and Verbal Behavior*, *8*, 457-462, 1969.

Garner, W. R. The stimulus in information processing. *American Psychologist*, *25*, 350-358, 1970.

Gibson, J. J. *The senses considered as perceptual systems*. Boston: Houghton Mifflin, 1966.

Kuhn, G. M. On the front cavity resonance, and its possible role in speech perception. *Journal of the Acoustic Society of America*, 1975 (in press).

Liberman, A. M., Cooper, F. S., Shankweiler, D. P., and Studdert-Kennedy, M. Perception of the speech code. *Psychological Review*, *74*, 431-461, 1967.

Martin, J. G. Rhythmic (hierarchical) versus serial structure in speech and other behavior. *Psychological Review*, *79*, 487-509, 1972.

Meltzer, R. H., Martin, J. G., Mills, C. B., Imhoff, D., and Zohar, D. Reaction time to temporally-displaced phoneme targets in continuous speech. (submitted for editorial review) 1975.

Mills, C. B., and Martin, J. G. Articulatory organization in the stimulus prefix effect. *Perception and Psychophysics*, *16*, 309-314, 1974.

Mills, C. B., and Martin, J. G. *Effects of articulatory organization of speech on digit recall*. Paper presented at American Psychological Association Annual Meeting, Chicago, August 1975.

Neisser, U., Hoenig, Yvette, J., and Goldstein, Evelyn. Perceptual organization in the prefix affect. *Journal of Verbal Learning and Verbal Behavior*, *8*, 424-429, 1969.

Shields, J. L., McHugh, A., and Martin, J. G. Reaction time to phoneme targets as a function of rhythmic cues in continuous speech. *Journal of Experimental Psychology*, *102*, 250-255, 1974.

Sturges, P. T., and Martin, J. G. Rhythmic structure in auditory pattern perception and immediate memory. *Journal of Experimental Psychology*, *102*, 377-383, 1974.

Svensson, S. -G. *Prosody and grammar in speech perception*. Doctoral Dissertation, MILUS 2 (Monographs from the Institute of Linguistics, University of Stockholm),1974 .

Woodrow, H. A quantitative study of rhythm. *Archives of Psychology*, *18*, No. 1, 1909.

DISCUSSION of "Experiments with Long-Range cues in continuous speech perception and immediate memory" MARTIN

BERLIN:

You should consider that the reason you get such uniform results, regardless of the locus of the "cut", or whether you use natural or synthetic speech. is that you are not sampling coarticulatory cues but simply <u>reaction time</u> to a warning noise burst.

MARTIN:

1. Results are quite different between natural and synthetic speech.
2. The noise bursts do not account for the differences between experimental conditions in any consistent way, and in fact the main conclusions hold up when noise is not used.

BERTELSON:

In your paper you pay little attention to the possible effects of expectancies which can be induced in the subjects by your experimental manipulations themselves. In the first experiment for instance, the target is in three out of four cases preceded by white noise. Unavoidably, the noise must come to be used as a warning signal. That kind of effect must presumably depend on how often the different types of trials occur in a given experimental run, but you do not try to manipulate that kind of variable. To me it seems as if you were assuming that the subject goes on expecting a normal sentence.

ON THE DYNAMIC USE OF PROSODY IN SPEECH PERCEPTION[1]

C. J. DARWIN
Laboratory of Experimental Psychology
University of Sussex
Brighton BN1 9QG, England

Abstract

Two roles that prosodic variables might play in the perception of speech are reviewed and two experiments described. Prosody is seen on the one hand as helping to direct attention to a particular speaker and to the potentially most informative parts of his speech. It is also seen as modifying the hypotheses a listener might entertain about a sentence as he listens to it. The use of models of perception taken from natural language understanding programs is suggested as a way to begin modelling this second role of prosody.

There is little doubt that prosodic variables make a significant contribution to the intelligibility of speech. We lack, however, a suitable framework for modelling, in perception, the relation between prosodic variables and segmental information. Part of the reason for this is that many very different types of information can be conveyed by prosodic variables, and it would be a brave soul who denied any of them a role in intelligibility. The segmental distinctions carried by normally prosodic dimensions, such as change in pitch as a weak cue to voicing (Fujimura, 1961; Haggard, Ambler & Callow, 1970) and lexical distinctions carried by stress (e.g., see Fry, 1970), are not particularly difficult to integrate into a scheme for the perception of lexical items from segmental cues. But these aspects of prosody exclude perhaps the bulk of the contribution that prosody makes to the intelligibility of speech. Sentences made up of concatenated words which have been spoken in isolation are less intelligible than those spoken fluently (Stowe & Hampton, 1961; Abrams & Bever, 1969), a finding which is almost paradoxical when one considers (Huggins, 1972) that words excised from fluent speech are less intelligible than the same words spoken in isolation (Lieberman, 1963). The resolution of this paradox can be achieved in part by knowing that coarticulation is no respecter of word boundaries, but it is likely, if not compelling, that the prosodic information present in fluent speech helps perception

[1] This paper was written while the author was on leave of absence at Haskins Laboratories and the University of Connecticut. The first experiment reported here was carried at in Sussex, the second in Connecticut. The work was supported by grants from the British S.R.C. and the American N.I.C.H.D.

of the fluent utterance. When coarticulatory artifacts are made less likely, prosodic perturbations less drastic than word-splicing still influence intelligibility. When sentences which share a common five-word portion, but have major syntactic boundaries at different places within this common string are cross-spliced so that an intonation contour inappropriate to the sentence's syntactic structure is heard, the intelligibility of these mosaic sentences is lower than those with normal intonation (Wingfield & Klein, 1971; Wingfield, 1975 and this volume). This technique of cross-splicing introduces two perturbations into the prosody; first it produces an abrupt change in both pitch and rhythm at the splice points and second it gives an inappropriate placement of the intonational cues to the syntactic boundary. While both of these changes may contribute to the reduced intelligibility of the mosaic sentences, the latter certainly has a specific effect since subjects' transcriptions of these sentences include ones with the syntactic boundary at the point suggested by the intonation contour. This result can be compared with the outcome of other experiments which have shown that subjects can extract considerable information about the stress-pattern of speech even when there is virtually no segmental information present, either by virtue of its being hummed (Svensson, 1974) or spectrally rotated (Blesser, 1969). If indeed information which is a potentially useful indicant of syntactic structure can be extracted independently of segmental information, do we need to postulate any <u>interaction</u> between segmental and prosodic information while the sentence is being perceived? Perhaps Wingfield and Klein's subjects simply reinterpreted the sentences they heard in the light of an independently perceived intonation contour. Not very dynamic, but certainly easy to model!

Life would be more interesting and speech perhaps easier to perceive if prosody played a more dynamic role. This paper aims to review the evidence that bears on a potentially dynamic role of prosody in the perception of speech and to present two new fragments of data which contribute to this discussion. The review will be based on two putative roles of prosodic variables, the role of rhythm and pitch contour in both allowing temporal prediction and giving continuity to an attended channel, and the role of prosodic variables in delimiting higher syntactic structures.

Temporal patterning, prediction and attention

Reaction time techniques provide a useful way of studying the <u>on-going</u> process of sentence perception. The most extensively used technique, the phoneme-monitoring task, was introduced by Foss and Lynch

(1969). Here subjects, while listening to a sentence which they subsequently must recall, have to press a key whenever they hear a word beginning with a particular phoneme (usually /b/). Changes in this reaction time have been found to depend on a number of syntactic and semantic variables. Reaction time is longer when the target word occurs in a self-embedded than in a right-branching sentence (Foss & Lynch, 1969), in sentences with deleted rather than intact relative pronouns (Hakes & Cairns, 1970; Hakes & Foss, 1970), in sentences with reduced rather than intact complements (Hakes, 1972) and after high rather than low frequency words (Foss, 1969; at least for adjectives of different frequencies, Cairns & Foss, 1971). These differences have been interpreted as reflecting changes in the processing load imposed by the sentence (cf. Aaronson, 1968) and should not be thought of as reflecting directly the extraction of segmental information. The conscious access to the phonemic level that this task requires is made after (or at least is influenced by) decisions about what the word containing the target is. Reaction times are shorter if the target starts a word rather than a nonword (Rubin, Turvey & van Gelder, in prep.; see also Foss & Swinney, 1973; Treisman & Tuxworth, 1974, for disucssion on related points).

The phoneme monitoring task has been used in two contexts which relate to the use of prosodic information. An experiment by Cutler and Foss (1973) provides a convenient starting point. Following an earlier finding that reaction time to an initial stop consonant was faster when it appeared at the beginning of a content than of a function word, they showed that this difference was attributable to differences in the stress with which function and content words are normally spoken. When stress and lexical category were varied independently the faster reaction time accompanied the stressed word. Although unstressed function words gave slower reaction times than unstressed content words, they felt this could be due to different degrees of stress.

This experiment alone can support a number of interpretations: (i) stressed words may be articulated more precisely so that there is more segmental information available, (ii) stressed words may be processed more efficiently on account of their intrinsic stress, (iii) stressed words may be processed more efficiently through the preceding stress pattern suggesting that they will be stressed. This last hypothesis is compatible with recent comments on the use of rhythm by Martin (1972).

The first two interpretations attribute the faster reaction time for stressed words to factors intrinsic to the word itself, and can now be dismissed with some confidence. Shields, McHugh and Martin (1974) measured reaction time to nonsense disyllables beginning with a target phoneme. The disyllable was pronounced with the stress either on the first or the second syllable. When the word occurred as part of a fluent sentence, subjects were faster to a target in a stressed syllable than in an unstressed one, provided that the target did not occur too close to the end of the sentence. However, when the same target words were spliced out and presented in isolated list form, the differences between stressed and unstressed syllables disappeared. The curious lack of difference when the targets occur at the end of the sentence may be simply a floor effect, since reaction times decreased throughout the sentence for both types of syllables. This study is complemented by a recent experiment by Cutler (cited in Cutler, 1975) who spliced target words into a sentence context, whose prosody suggested that the target word would or would not be stressed. She found that reaction time was faster when subjects expected a stressed word than when they did not. The stress difference thus seems to be independent of the intrinsic stress of a word and depends rather on the preceding prosodic pattern allowing the subject to anticipate the forthcoming stressed word.

If some form of anticipation is the cause of the stressed syllable advantage then we might expect two further effects; some time should be necessary for subjects to get the rhythm of the utterance on which prosodic predictions might be based, and local disturbance of the rhythm should upset the stressed syllable advantage. There is evidence for both. Aaronson (1968) asked her subjects to monitor a list of digits, spoken at a constant rate, for the occurrence of a target digit. There was a decrease of about 100 ms in reaction time to the target over the first three serial positions. This occurred whether subjects subsequently had to recall the list of digits or not. Reaction times remained steady in subsequent serial positions for subjects who had only to monitor the lists, but they increased after about the third item until the end of the list for subjects who had to recall the lists as well as monitor. We will return to this latter finding below; for the present it is enough to note that the initial decrease in reaction time may well reflect subjects accommodating to the rhythm of the list. Cutler (1975) showed that local temporal disturbances influence phoneme-monitoring reaction times. She found that inserting a quarter-second period of silence before a pair of monosyllabic words embedded in a fluently spoken sentence influenced the reaction time to a phoneme target at the beginning of either of the words. The reaction time was slowed to the second word, which was initially spoken with stress, and quickened to the unstressed first word. Cutler failed to find any advantage for the stressed word over the unstressed when the silent interval was absent; this may possibly be attributable to the targets' position in the sentence (cf. Shields et al., 1974).

The picture emerging from these studies then is of subjects' attention being allocated preferentially towards the stressed syllables in a sentence (Shields et al., 1974). This is made easier by the timeing of speech (at least in a stress-timed language) being determined by stressed syllables. Allen (1972), for example, showed that subjects can tap with less variability to stressed syllables than to unstressed (even when the sentence in which they were embedded was heard repeatedly), and Huggins (1972) finds subjects to be more tolerant of timing distortions which preserve stressed syllable separations. As Cutler and Foss (1973) point out, allowing processing to be directed towards the stressed parts of the sentence allows the focus of the speaker's sentence to control the listener's perception. However, we might note in passing that it is not yet clear how much the results of these experiments depend on using a response measure which is temporal. The task of targetting for a phoneme might cause an undue advantage to those portions of the speech stream which are temporally predictable. It would strengthen the case if other measures, such as perhaps the detectability of feature substitutions, showed a similar increase in stressed syllables. Some support comes from an experiment by Dooling (1974). He found that subjects' immediate recall of sentences heard in noise was improved if they had previously been exposed to sentences with a similar stress pattern. His results also showed that repetition of stress pattern was much more effective at improving performance than was repetition of syntactic structure, when these two variables were independently manipulated.

Although Martin's theorizing emphasizes the predictive nature of rhythm, others have shown that pitch contour can play a similar role. The phenomenon of "primary auditory stream segregation" (Bregman & Campbell, 1971) illustrates this, and again, like Martin's work, embraces both speech and music. A random sequence of six notes (3 high, 3 low), when played rapidly, will perceptually segment into a high and a low tune, despite lack of any greater rhythmic cohesion within than between tunes. The analogy with speech perhaps lies both in the use of frequency continuity of formants to help in their tracking (cf. Dorman, Cutting & Raphael, 1975) and in the use of continuity of pitch to help in attending to one voice against competing sounds. This last point has been pursued in an experiment that Davina Simmonds carried out with me in 1973. While this experiment was motivated by the Bregman and Campbell findings, it does not in fact distinguish between rhythm and pitch, but looks at the influence of prosodic variables in general on the ability to attend to a particular speech source.

Experiment 1

The technique of shadowing, although complex for the subject, provides a valuable way of studying the process of selective attention to a particular speaker. Treisman (1960) asked subjects to shadow a passage of continuous speech lead to one ear and to ignore a similar passage lead to the other. At some point during the passage the two channels were switched so that the passage that had been shadowed continued in the other ear. She found that subjects occasionally gave as part of their shadowing response words which had in fact occurred immediately after the switch on the ear they had been instructed to ignore, and that these intrusions were more common the more redundant the passages. In as much as this effect has any sensory basis, we can ask whether the tendency for subjects to shadow words from the unattended ear is being determined by some semantic/syntactic priming as Treisman maintained, or rather whether the continuity in rhythm and pitch across the ears at the switch point is sufficient to cause a momentary change in the ear from which the attended auditory input is drawn.

Our experiment involved basically a repetition of Treisman's but with independent manipulation of prosodic and semantic factors. Pairs of passages of about 50 words each were selected from short stories by H. E. Bates. From each pair of passages four recordings were made by the same female speaker. Two were of the original passages, and two were made by reading the first part of one passage followed smoothly by the second part of the other. The switch point between the passages was later than half-way through each passage and was always prior to a word beginning with a stop consonant (to facilitate subsequent splicing) but was otherwise placed at random (although never at a major clause or sentence boundary). From these four original recordings, four different dichotic conditions were made: a normal condition in which the two original passages were paired together (aligned to give simultaneity of the stop closures); a semantic change condition made by pairing together the other two original recordings; an intonation change condition made by switching the latter pair of passages after the stop closure; and a condition in which both semantics and intonation changed, made by switching the two original passages after the stop closure. The first and last of these conditions repeat Treisman's conditions, the other two vary independently semantic and intonational continuity on the attended ear. All four conditions were made up with the same number of splicings and re-recordings to prevent the preparation of the tapes selectively introducing artifacts into the experiment.

Each of fourteen subjects was instructed to shadow the passage on one ear with as little lag as possible between hearing the speech and saying it. They were told to try not to chunk the speech into phrases before speaking it, and were given five practice trials on similar dichotic passages which did not have any switches. They then took eight different dichotic passages with each of the four experimental conditions appearing twice in a counter-balanced order.

Two types of errors are distinguished in the results. A particular trial is classed as having an omission error if the subject misses at least two words over the break point, and it is classed as an intrusion error if the subject shadowed any words from the unattended ear over the break point. Table 1 shows the distribution of errors over the four different types of break.

Table 1

Percentage of trials on which omission or intrusion errors occurred for different types of discontinuity and for subjects who either shadowed continuously or 'chunked' their responses.

		Type of discontinuity on shadowed ear			
		No break	Intonation	Semantic	Both
Chunked (5 Ss)	Intrusions	0	20.0	0	0
	Omissions	10.0	40.0	90.0	50.0
Continuous (9 Ss)	Intrusions	0	77.8	5.6	55.6
	Omissions	0	11.1	55.6	22.2
TOTAL (14 Ss)	Intrusions	0	57.1	3.6	35.7
	Omissions	3.6	21.4	67.8	32.1

In the results for all the subjects combined, there is a significant change in the number of intrusion and omission errors across the three experimental conditions ($p<.01$ and $<.05$, respectively). This variation is due to there being significantly more intrusion errors in both the conditions with intonation changed than when only the semantics changed ($p<.05$), and through there being significantly more omission errors in the condition with the semantics alone changed than in the other two experimental conditions ($p<.01$). A closer look at the subjects responsible for the various types of errors showed, in addition, that very few intrusion errors were made by the five subjects, who, despite their instructions, chunked their shadowing responses. Omission errors, by contrast, were distributed more evenly between the subjects, but tended to be made more by the subjects who chunked their responses. In other words, for subjects who shadow continuously, an abrupt change of the intonation contour between the ears causes intrusion errors to occur, while an abrupt change in the semantic content of the message, but without any switch in the intonation contour gives omission errors. Subjects who chunk their shadowing responses likewise produce a large number of omission errors on semantic-switch trials, but show a different pattern from the subjects who shadow continuously on the intonation-switch trials, giving more omission than intrusion errors.

Clearly the subjects who chunk their responses are much more sensitive to the semantic constraints on the message they are shadowing, in that they show omission errors when there is a semantic discontinuity on the attended ear. But nevertheless they do not show any tendency for intrusions to occur from the other ear when only the semantic information switches ears. Although the subjects who shadow continuously are disrupted less by the semantic discontinuity than their chunking colleagues, they too show no tendency for intrusions to occur from the semantically appropriate unattended ear. Rather than assuming that intrusion errors are caused by some semantic priming of a contextually likely word it seems, at least in this experiment, that they are occurring only when there is a continuity of intonation across the two ears. For a brief time after a switch in intonation has occurred the continuity of the intonation contour overrides the ear of entry as a criterion for selecting the speech that needs to be attended. If subjects are shadowing continuously, this leads to intrusion errors from the ostensibly unattended ear. But if the subject is chunking his response he has time between hearing the potentially intrusive words and making his response to omit the words from the inappropriate ear. This, however, causes omission errors, particularly when there is no semantic continuity on the attended ear to help him retrieve the words that occurred there immediately after the break.

This finding, that prosody helps a listener to attend to a particular speaker, complements the work reviewed earlier on the predictive use of prosody in anticipating stress. Both types of experiment emphasize the role prosody plays in controlling which parts of the speech stream are attended to, whether the selection is made temporally or spatially.

Prosody and the structures of speech

As well as allowing anticipation of the speech stream, prosodic factors doubtless also play a role in segmenting speech into higher order structures. One rather simple way in which prosody delimits complex structures in speech is through pauses. Goldman-Eisler (1968) claims that even at its most fluent over two-thirds of speech consists of utterances of six words or less. Pauses virtually never appear within a word, and their length is determined by the type of syntactic juncture that they occur in. They tend to be longer at the end of sentences than at the beginning of subordinate clauses, and before a subordinate clause their length is determined by the type of clause (Goldman-Eisler, 1972). Pauses have been used as experimental variables in studying memory for lists of items that do not naturally fall into larger structures, and we might get some insight into their use in natural language, by looking for a moment at their effect in unnatural situations.

A brief period of silence after an item in the middle of a list of digits provides an opportunity for rehearsal of the preceding group of digits (Ryan, 1969; Kahneman, Onuska & Wolman, 1968; Kahneman & Wright, 1971), as well as allowing the additional information from an auditory memory to be utilized (Crowder & Morton, 1969). Merely indicating by instructions or interposed tones in a regular list how that list should be grouped gives substantially poorer overall performance and a reduced scalloping of the serial recall curve (Ryan, 1969). This is perhaps because less capacity can be allocated to rehearsal when there are competing perceptual demands. As well as delimiting groups of items for rehearsal, pauses also serve to delimit larger structures for coding in long-term memory. The cumulative learning that occurs when supra-span strings of items are repeated occasionally throughout a short-term memory experiment (Hebb, 1961) is reduced when these strings are grouped differently by pauses (Bower & Winzenz, 1969), but this reduction only seems to occur when the items that compose the string are amenable to some higher level recoding (Laughery & Spector, 1972).

Coming closer to real speech, reading a list of nonsense syllables in a natural intonation can increase the number of syllables remembered immediately after, provided the syllables contain additional bound morpheme clues to the intended syntactic structure (O'Connell, Turner & Onuska, 1968). Neither intonation, nor the presence of bound morphemes alone is sufficient to increase the number of syllables recalled. Incidentally, both these additional cues also seem to be necessary to procure a right-ear advantage (Zurif & Sait, 1970; Zurif & Mendelsohn, 1972; Zurif, 1974).

It is tempting to extrapolate from these experiments to a view of clause or sentence perception that delays syntactic processing of the clausal unit until a clause boundary has been reached, so that the end of the clause is, like the pause in a string of digits, a time of great mental effort (cf. Kahneman, Onuska & Wolman, 1968; Wright & Kahneman, 1971), but there is some evidence against this. As we remarked earlier, Aaronson (1968) found that reaction time to a target digit in a string that the subject had to remember increased towards the end of the list, reflecting the increased memory load. By contrast, there is very little evidence that a similar increase in reaction time occurs for a phoneme target in a sentence. Foss & Lynch (1969) did find such an increase, but all relevant subsequent studies have found a decrease in reaction time throughout the sentence (Foss, 1969; Hakes & Foss, 1970; Shields, McHugh & Martin, 1974). Similarly, reaction time to a click played at the end of a clause <u>decreases</u> with increasing length of the clause, while reaction time to a click presented between clauses shows no consistent increase with length of the first constituent (Abrams & Bever, 1969). Nor is reaction time to a click consistently faster at the beginning of the second clause than in the clause break (Abrams & Bever, 1969; but see Bever & Hurtig, 1975). It is clear then that the constituent words of a clause are not being held in memory like so many digits. Something much more dynamic is happening. But what? Perhaps the best type of model to use here would be one based on Winograd's (1972) integrated language understanding system. In this program both syntactic and semantic information is used, as each successive word in a sentence is encountered, to construct procedures which can subsequently be used to take action on the sentence. The end of a clause does not in this scheme imply any particularly energetic syntactic activity on the part of the computer, provided the syntactic organization that it had presumed during perception of the preceding word string is compatible with the string's ending. Incompatibility requires back-tracking, but the program's use of semantic constraints for eliminating alternative syntactic strucutres as it goes along reduces the likelihood of this being necessary. If these sort of syntactic and semantic constraints are also being used dynamically in natural perception, then we might suppose that prosody is used in a similarly dynamic and interactive way to guide the search towards an appropriate syntactic organization of the sentence.

Some of the results of an experiment run recently by John Capitman in collaboration with myself and Susan Brady bear a little on these issues. Our experiment was designed primarily to be a replication of Wingfield and Klein's interesting finding that inappropriate intonation contours impair immediate recall of a sentence.

Experiment 2

The sentences we used in this experiment were similar to those used by Wingfield and Klein (1971; see Wingfield, this volume). Each sentence that the subjects heard was one member of a group of four sentences derived from pairs of sentences sharing a common string of words. Two of the sentences in the group of four were the originals with appropriate intonation, the other two were generated by cross-splicing the common string of words between these two sentences. This string always started with a stop consonant to facilitate silent splicing, and continued to the end of the sentence to reduce the rhythmic discontinuity that splicing can introduce. The sentences were between 10 and 14 words long and the common string was between 6 and 10 words long. Unlike the Wingfield material, the major syntactic boundary in each sentence was both preceded and followed by at least two words of shared material, this insured that the intonationally suggested boundary in the mosaic sentences generated by cross-splicing was also preceded and followed by at least two words of shared material. As in the Wingfield and Klein experiments the sentences were presented monaurally and the ear of presentation was switched within the sentence. This switch was always made during a stop closure to prevent extraneous clicks. The switch point always occurred within the shared material and was never at the intonationally suggested boundary, but was otherwise unconstrained. The recordings were made on the Haskins Laboratories PCM facility.

Each of 24 subjects heard one sentence of the four generated from each of the eleven pairs in an order that counterbalanced conditions (normal versus cross-spliced). Subjects were instructed to write down each sentence as soon as they had heard it. In order to make the task more difficult they were told to listen for the occurrence of stop consonants and to circle these in their answer. Their performance on this task was ignored.

On average 55.5% of the normally intonated (unspliced) sentences were recalled correctly while only 44.5% of the cross-spliced sentences were recalled correctly. In examining the errors that subjects made, three types of error are distinguished:

(i) Omission errors: Each word omitted, whether or not this omission changed any other aspect of the sentence, was scored as an omission error.

(ii) Lexical errors: Each changed word in a string that did not affect the primary grammatical relations (the truth conditions of the sentence) was scored as a lexical error. Changes in tense, mood and aspect were also counted as lexical errors.

(iii) Syntactic errors: Each change in any segment of a string that resulted in a change in the grammatical relations or the truth conditions of the sentence was scored as a syntactic error (except for errors of tense, mood and aspect). No attempt was made to distinguish those changes that seemed to conform to the intonation pattern from those that did not. One other error type was defined which depended on the position of the error but not on its type:

(iv) IB2 errors: [This score applied only to the cross-spliced sentences.] Each error of any type occurring within one word on either side of the intonationally suggested boundary was scored as an IB2 error.

Summing together all the first three types of errors (omission, lexical, syntactic), Capitman found significantly more errors on the cross-spliced than on the normally intonated sentences. This difference is significant by a Wilcoxon test both across subjects ($p<.01$) and across sentences ($p<.02$). Moreover this increase in errors is due to an increase in the omission ($p<.02$ by subject; $p<.05$ by sentence) and syntactic errors ($p<.01$ by both subject and sentence), but not to an increase in lexical errors. An exception is seen in three of the sentence pairs which show a decrease in syntactic errors in the crossed intonation condition, though still an increase in omission errors. All of these sentences have an "If...then...." construction, which may impose greater syntactic or semantic constraints on the subjects' responses, causing them to give up on their response rather than change the truth conditions of the sentence.

Although these general results confirm Wingfield and Klein's findings on intelligibility with a different set of sentences, they do not get us much nearer to the question of how the inappropriate intonation contour influences recall. Inspection of the IB2 errors does bear on this question. When listening to the crossed intonation sentences, subjects make more errors within one word on either side of the intonationally suggested boundary when this boundary precedes the major syntactic boundary than when it follows it ($p<.025$ by subject; $p<.05$ by sentence). This is not simply due to subjects making more errors earlier in the sentence since the reverse tends to be the case. This asymmetry in the disruptive effects of an inappropriate intonational cue to a clause boundary is what one might expect on the notion that intonational information is being used dynamically to revise syntactic hypotheses as the subject listens to the sentence. If prosody were being used solely to restrict alternatives after initial perception of the sentence, it is difficult to see why the relative positions of the syntactic and intonational boundaries should matter. On a more dynamic view it is possible that the inappropriate intonational boundary leads to back-tracking throughout the previous incomplete clause; this clause will be much longer, and so the effects of back-tracking will be more disruptive when the incomplete clause is the first one of the sentence. Hence having the intonational boundary before the syntactic boundary will be more disruptive than having it after it. Looking at errors only in the region of the intonation boundary appears to be more sensitive a measure than looking at errors over the whole sentence since there did not appear to be any significant difference in errors over the sentence as a whole as a function of these relative locations. This perhaps suggests that back-tracking affects perception of the part of the sentence that is being actively perceived when it occurs. But we cannot entirely rule out the possibility that this effect is being caused in part by the rhythmic discontinuity that cross-splicing introduces immediately after the breakpoint. Although these conclusions are extremely speculative and the evidence on which they are based is not strong, we feel that the questions raised by this sort of approach are extremely interesting and hope to pursue them in subsequent experiments.

In summary, this paper has tried to draw attention to some of the possible dynamic roles that prosody might play in the perception of speech: the rhythmic and melodic aspects of speech may allow the listener to predict when potentially important speech material will arrive (and perhaps allow him to allocate his processing capacity accordingly); they may also allow him to attend selectively to one voice among many more readily. In addition, prosody undoubtedly plays a role in delimiting higher-order structures in speech and the suggestion is made here that this is done by dynamically modifying the hypotheses the listener entertains while listening to a sentence. While we neither pretend that this exhausts the uses of prosody, nor believe that the study of isolated sentences is a particularly insightful way to study a variable which depends so much on intersentence context (see Smith & Goodenough, 1971, for an interesting example of intonation interacting with context in a perceptual task), we do hope that the issues raised here will help to stimulate work on an unduly neglected area of speech perception.

REFERENCES

Aaronson, D. Temporal course of perception in an immediate recall task. Journal of Experimental Psychology, 1968, 76, 129-140.

Abrams, K., & Bever, T. G. Syntactic structure modifies attention during speech perception and recognition. Quarterly Journal of Experimental Psychology, 1969, 21, 280-290.

Allen, G. D. The location of rhythmic stress beats in English: An experimental study, I & II. Language and Speech, 1972, 15, 72-100, 179-195.

Bever, T. G., & Hurtig, R. R. Detection of a nonlinguistic stimulus is poorest at the end of a clause. Journal of Psycholinguistic Research, 1975, 4, 1-7.

Blesser, B. Perception of spectrally rotated speech. Ph.D. dissertation, Massachusetts Institute of Technology, Cambridge, Mass., 1969.

Bower, G. H., & Winzenz, D. J. Group structure, coding and memory for digit series. Journal of Experimental Psychology, 1969, 80, no. 2, part 2.

Bregman, A. S., & Campbell, J. Primary auditory stream segregation and perception of order in rapid sequences of tones. Journal of Experimental Psychology, 1971, 89, 244-249.

Cairns, H. S., & Foss, D. J. Falsification of the hypothesis that word frequency is a unified variable in sentence processing. Journal of Verbal Learning and Verbal Behavior, 1971, 10, 41-43.

Crowder, R. G., & Morton, J. Pre-categorical acoustic storage (PAS). Perception and Psychophysics, 1969, 5, 365-373.

Cutler, A. Rhythmic factors in the determination of perceived stress. Paper presented to the 89th meeting of the Acoustical Society of America, Austin, Texas, April 1975.

Cutler, A., & Foss, D. J. The importance of lexical item stress for lexical access. Paper presented at the 44th annual meeting of the Midwestern Psychological Association, Chicago, May 10-12, 1973.

Dooling, D. J. Rhythm and syntax in sentence perception. Journal of Verbal Learning and Verbal Behavior, 1974, 13, 255-264.

Dorman, M. F., Cutting, J., & Raphael, L. J. Perception of temporal order in vowel sequences with and without formant transitions. Journal of Experimental Psychology: Human Perception and Performance, 1975, 2, 121-129.

Foss, D. J. Decision processes during sentence comprehension: Effects of lexical item difficulty and position upon decision times. Journal of Verbal Learning and Verbal Behavior, 1969, 8, 457-462.

Foss, D. J., & Lynch, R. H. Decision processes during sentence comprehension: Effects of surface structure on decision times. Perception and Psychophysics, 1969, 5, 145-148.

Foss, D. J., & Swinney, D. A. On the psychological reality of the phoneme: Perception, identification and consciousness. Journal of Verbal Learning and Verbal Behavior, 1973, 12, 246-257.

Fry, D. B. Prosodic phenomena. In B. Malmberg (Ed.), Manual of Phonetics. Amsterdam: North-Holland, 1970.

Fujimura, O. Some synthesis experiments on stop consonants in the initial position. Quarterly Progress Report, Research Laboratory of Electronics, M.I.T., 1961, 61, 153-162.

Goldman-Eisler, F. Psycholinguistics: Experiments in Spontaneous Speech. London: Academic, 1968.

Goldman-Eisler, F. Pauses, clauses, sentences. Language and Speech, 1972, 15, 114-121.

Haggard, M. P., Ambler, S., & Callow, M. Pitch as a voicing cue. Journal of the Acoustical Society of America, 1970, 47, 613-617.

Hakes, D. T. Effects of reducing complement constrictions on sentence comprehension. Journal of Verbal Learning and Verbal Behavior, 1972, 11, 278-286.

Hakes, D. T., & Cairns, H. S. Sentence comprehension and relative pronouns. Perception and Psychophysics, 1970, 8, 5-8.

Hakes, D. T., & Foss, D. J. Decision processes during sentence comprehension: Effects of surface structure reconsidered. Perception and Psychophysics, 1970, 8, 413-416.

Hebb, D. O. Distinctive features of learning in the higher animal. In J. F. Delafresneye (Ed.), Brain Mechanisms and Learning. Oxford: Blackwell, 1961.

Huggins, A. W. F. On the perception of temporal phenomena in speech. Journal of the Acoustical Society of America, 1972, 51, 1279-1290.

Kahneman, D., Onuska, L., & Wolman, R. E. Effects of grouping on the pupillary response in a short-term memory task. Quarterly Journal of Experimental Psychology, 1968, 20, 309-311.

Kahneman, D., & Wright, P.. Changes of pupil size and rehearsal strategies in a short-term memory task. Quarterly Journal of Experimental Psychology, 1971, 23, 187-196.

Laughery, K. R., & Spector, A. The roles of recoding and rhythm in memory organization. Journal of Experimental Psychology, 1972, 94, 41-48.

Lieberman, P. Some effects of semantic and grammatical context on the production and perception of speech. Language and Speech, 1963, 6, 172-187.

Martin, J. G. Rhythmic hierarchical versus serial structure in speech and other behavior. Psychological Review, 1972, 79, 487-509.

O'Connell, D. C., Turner, E. A., & Onuska, L. A. Intonation, grammatical structure and contextual association in immediate recall. Journal of Verbal Learning and Verbal Behavior, 1968, 7, 110-116.

Ryan, J. Grouping a short-term memory: Different means and patterns of grouping. Quarterly Journal of Experimental Psychology, 1969, 21, 137-147.

Shields, J. L., McHugh, A., & Martin, J. G. Reaction time to phoneme targets as a function of rhythmic cues in continuous speech. Journal of Experimental Psychology, 1974, 102, 250-255.

Smith, F., & Goodenough, C. Effects of context, intonation and voice on the reaction time to sentences. Language and Speech, 1971, 14, 241-250.

Stowe, A. N., & Hampton, D. B. Speech synthesis with pre-recorded syllables and words. Journal of the Acoustical Society of America, 1961, 33, 810-811.

Svennson, S.-G. Prosody and grammar in speech perception. MILUS no. 2, Institute of Linguistics, University of Stockholm, 1974.

Treisman, A. M. Contextual cues in selective listening. Quarterly Journal of Experimental Psychology, 1960, 12, 242-248.

Treisman, A. M., & Tuxworth, J. Immediate and delayed recall of sentences after perceptual processing at different levels. Journal of Verbal Learning and Verbal Behavior, 1974, 13, 38-44.

Wingfield, A. Acoustic redundancy and the perception of time-compressed speech. Journal of Speech and Hearing Research, 1975, 18, 139-147.

Wingfield, A., & Klein, J. F. Syntactic structure and acoustic pattern in speech perception. Perception and Psychophysics, 1971, 9, 23-25.

Winograd, T. Understanding Natural Language. New York: Academic Press, 1972.
Wright, P., & Kahneman, D. Evidence for alternative strategies of sentence retention. Quarterly Journal of Experimental Psychology, 1971, 23, 197-213.
Zurif, E. B. Auditory lateralization: Prosodic and syntactic factors. Brain and Language, 1974, 1, 391-404.
Zurif, E. B., & Mendelsohn, M. Hemispheric specialization for the perception of speech sounds: The influence of intonation and structure. Perception and Psychophysics, 1972, 11, 329-332.
Zurif, E. B., & Sait, P. E. The role of syntax in dichotic listening. Neuropsychologia, 1970, 8, 239-244.

DISCUSSION of: "On the dynamic use of prosody in speech perception"
DARWIN

BERTELSON:

Since the author is absent, we may perhaps for once be complimentary and say this is a nice and an important paper.

As Mark Haggard noted in his presentation the first experiment imposes a complete vision of the interpretation of the post-shift intrusion phenomenon discovered by Anne Treisman and which has played a great role in discussions on the nature of selective attention in the last ten years. On the other hand, DARWIN's data are much more intrusion than those of Treisman, where subjects only produced an infinitesimal proportion of intrusion above 6% in the best case.

THE PHONETIC STRUCTURE OF PARAGRAPHS

Ilse Lehiste
Department of Linguistics
Ohio State University

The present paper describes a pilot study designed to clarify some issues that need to be taken into account in planning a larger-scale investigation of the phonetic structure of paragraphs. The investigation is expected to make a contribution to our knowledge about discourse-level phonetics and phonology, and thus deepen our insights about the general process of communication by spoken language. It is clear that people communicate by isolated sentences only in exceptional circumstances. A message is often long and complicated enough to require that it be cast in paragraph form. (The planned study will be limited to paragraph level, but there exist units of still higher levels, e.g. anecdotes, sermons, or narrative reports.) Sentences within a paragraph may be formally linked by the use of definite and indefinite articles, deictic adverbs and pronouns, sequence of tenses etc. It is hypothesized also that paragraphs possess a suprasegmental structure that indicates the beginning and end of paragraphs and characterizes the body of the paragraph. For example, it may be that the intonation contour applied to a sentence produced in isolation (constituting a one-sentence paragraph) will differ from the intonation contour applied to the same sentence in the beginning, middle and end of a paragraph. In other words, it is possible that a paragraph is characterized by an overall intonation structure to which the intonation contours of its constituent sentences are subordinated. Similar relationships are expected to prevail with regard to timing and stress, i.e. the manner in which prominence is realized within a sentence and within a paragraph. Underlying this kind of reasoning is a basic belief that connected speech is not amorphous, but has a linguistically determined internal structure which is reflected in the phonetic realization of connected discourse.

A corollary hypothesis is that since speakers and listeners share the same code, listeners are capable of deciding whether a sentence has been produced in isolation or as part of a larger structure.

Materials. To test these hypotheses, a set of materials was selected consisting of a three-sentence paragraph. The sentences were such that each could be used in isolation, and each of the six possible orderings of the three sentences within the paragraph appeared equally possible. The sentences had neutral emotional content, and were of the kind that one might easily hear within the setting of a linguistics department. They were taken from an article by K. L. Pike. Their original order was as follows: "Language is a composite of form and meaning. If a person tries to study meaning without reference to the formal structure of language, he may end up with no structuring at all. Meaning does not occur in isolation, but only in relationship to forms."

An example of a different order might be the sequence 231: "If a person tries to study meaning without reference to the formal structure of language, he may end up with no structuring at all. Meaning does not occur in isolation, but only in relationship to forms. Language is a composite of form and meaning."

Method. A set of nine 5 x 8 cards was prepared on which the three sentences appeared in isolation as well as in the six possible orderings. The cards were placed in random order. A graduate student of linguistics (a male native speaker of American English) recorded the material on the cards on tape in an anechoic chamber using high quality recording equipment. At the time of the recording, the speaker did not know the purpose of the experiment, although he was aware that I had been studying suprasegmental structures for some time. (The materials were actually recorded twice, but only the first recording was used in the subsequent listening tests.)

Broad-band and narrow-band spectrograms were made of the sentences on a Voiceprint spectrograph. The recordings were also processed through a Frøkjaer-Jensen pitch meter and intensity meter, and displayed graphically by means of a Mingograf. A listening test was prepared, in which each sentence was isolated from its context and presented twice, for a total of 42 sentence presentations. The test was given to 30 listeners (native speakers of American English), who were asked to judge whether the sentence had occurred in isolation or in initial, medial, or final position within a paragraph. To avoid possible bias, the order of the columns on the test instrument was systematically varied.

Results. The results of the listening test are presented in Table 1. The table also contains chi-square values. It appears that the distribution of listener responses is significantly different from chance distribution in 13 out of 21 instances, or in approximately 62% of the cases. A difference between the sentences becomes evident from these responses: the first sentence shows the greatest amount of listener agreement, followed by the third sentence. The second sentence has the greatest number of random responses. There seems to be a connection between this finding and the length of the sentences: the average duration of the first sentence was 1975 msec, the duration of the third was 3727 msec, and the duration of the second 6006 msec.

The results also indicate that while agreeing among themselves, the listeners may assign the sentence to a position that is different from the position in which the speaker actually produced the sentence. There were three instances in which the listeners agreed significantly with the speaker, and six instances in which the listeners agreed among themselves, but their interpretation of the production differed from the speaker's intention. (When the speaker served as listener, he was correct in 14 out of 21 instances. His responses are not included in the overall averages.) Two comparably high scores occurred in four instances. In one of these, both were wrong (i.e. different from the actual production); in three cases, one of the two high scores was correct. Totally random responses occurred in eight instances. (I should insert here that if the content of the sentences would have introduced a semantic bias, there would probably not have been any random responses.)

I concluded from these results that the listeners have certain expectations regarding the suprasegmental shape of the sentence as it might appear within various positions in the paragraph or in isolation. The next step in the study was to establish the possible causes of listener agreement: to find out what the productions have in common that the listeners classify as belonging in a certain position.

Table 2 presents a partial set of measurements. For each occurrence of a sentence, the table contains its overall duration, the duration of the first and last word (or phrase) of the sentence, and the fundamental frequency contour on those parts of the sentence, i.e. the initial and terminal intonation contours. A study of these data makes it possible to draw some tentative conclusions. I will restrict my comments to a discussion of the first sentence; it seems rather clear that the length and the complex nature of the second and third sentences have interfered either with the signals of paragraph structure employed by the speaker or with the ability of the listeners to focus on them.

A first observation is that the sentence is longer in isolation than when it is part of the paragraph. (This holds true for the other two sentences also.) Length of the sentence seems to cause listeners to identify it as having been produced in isolation. Table 3 shows the length of the sentence and the number of identifications as isolated utterance. The number of such identifications decreases as the duration of the utterance decreases. There is only one production that is out of its proper position in the sequence: paragraph 132, in which sentence 1 received 21 identifications as having been produced in isolation. In the case of this production, the listeners overwhelmingly identified it as having been produced in initial position (31 votes for 'initial', which happens to be the position in which the speaker produced it). A glance at Table 2 shows that this production had a higher fundamental frequency peak on the initial word than occurred, for example, in paragraph 321 (in 132, the fundamental frequency peak was 185 Hz, whereas in 321 it was 155 Hz). In this case, then, the fundamental frequency cue seems to have overridden the overall length cue.

High fundamental frequency appears associated with judgments that the sentence was produced at the beginning of the paragraph. Table 3 contains also an arrangement of the results in a sequence according to peak fundamental frequency on the first word of the sentence. The fundamental frequency peak shows a similar picture as was observed above for overall duration and identification as isolated production. Higher fundamental frequency peaks are associated with higher numbers of identification as first sentence in the paragraph. Here, too, there is just one instance where a production appears to be out of its proper position; it is utterance 123, where sentence 1 received 22 votes for initial position and 29 votes for production in isolation. A look at Table 2 will show that sentence 1 in paragraph 123 had the second longest duration in the whole set (closest to the duration of the actual production in isolation). In this case, the length cue seems to have overridden the cue provided by the fundamental frequency peak.

Discussion. I would like to come back now to the initial hypothesis--that paragraphs have a definite phonetic structure, and that the linguistic competence of listeners includes the ability to recognize whether sentences have been produced in isolation or as part of a paragraph. Some regularities have indeed emerged from this pilot study. In fact I expected to find these regularities in final position within the paragraph rather than in initial position. I assumed that if there is confusion, it would be between isolated productions and final productions, since I expected the terminal contours to be basically the same and to differ only in degree. To my surprise, the confusion took place between isolated productions and initial productions. On further reflection, however, this seems to make good sense. The terminal intonation contour serves a grammatical purpose: it signals whether the sentence is a completed utterance, and when completed, whether it is a statement or a question. The initial contour thus may be considered relatively free to carry other kinds of information, including information relative to the position of the sentence within a larger structure. In all three sentences, the intonation contour on the beginning of an isolated sentence had a considerably lower fundamental frequency than was found when the sentence occurred in first position within the paragraph. It seems as if knowledge of the length of the stretch of speech to be produced prompts the speaker to start at a higher fundamental frequency level. I speculate that the speaker may anticipate the gradual fall in fundamental frequency during the production of the paragraph. Fundamental frequency cannot fall indefinitely; at a certain frequency, which differs with individuals, the speaker shifts into laryngealization. It seems plausible that the speaker starts at a higher frequency in order to have enough leeway for producing the falling intonation contour without having to shift into laryngealization before the end of the paragraph, or without having to change to a higher frequency in the middle of the paragraph, which might be interpreted as emphasis by the listeners.

Outlook for further research. The pilot study has also pointed out some aspects that should be treated differently in a larger-scale investigation. It seems that the length of individual sentences should be kept down to about the length of the first sentence in this study--approximately 2 seconds, no more. A method should be devised for better control of the semantic aspects of the study. Even though in my judgment all three orderings of the sentences were equally possible and all of them could occur in isolation, I received feedback from some of the subjects that certain orderings were more plausible than others. It is difficult to find sets of sentences that could be permuted in the way I permuted this set of three. I have thought of using jabberwocky--nonsense words complying with English phonotactic patterns--but jabberwocky typically uses connectives and articles in the same way that normal speech does, and normally sentences in a paragraph are connected by the choice of indefinite and definite articles and by deictic adverbs and pronouns. Once an order of sentences is established by the use of 'a', 'the', 'then', and 'this' and 'that', the sentences cannot be permuted any more. Jabberwocky would be useful, however, to guarantee absence of semantic bias.

Summary and conclusion. The study has shown that under certain conditions, listeners are indeed able to agree among themselves whether a sentence was produced in isolation or as part of a larger structure. Some phonetic properties of isolated sentences and paragraphs have been discovered that seem to provide the basis for the decisions made by the listeners. While not in any way exhaustive or conclusive, the results of the pilot study seem to justify a more detailed investigation.

Reference.

Pike, K. L., "A linguistic contribution to composition", Conference on College Composition and Communication, May 1964, p. 87.

Table 1

Judgments made by 30 listeners (2 judgments for each sentence)

Sentence and position	Perceived as occurring				Chi-square	Significance p < (df 3)
	Isolation	Initial	Medial	Final		
1. Language is a composite of form and meaning						
Isolation	29	14	8	9	18.81	.005
Initial 123	29	22	6	3	31.34	.005
Initial 132	21	31	6	2	36.14	.005
Medial 213	19	28	8	5	22.28	.005
Medial 312	18	31	7	4	30.01	.005
Final 231	24	25	9	2	25.74	.005
Final 321	24	12	13	11	7.86	.050
2. If a person tries to study meaning without reference to the formal structure of language, he may end up with no structuring at all.						
Isolation	14	18	18	10	2.95	N.S.
Initial 213	17	15	17	11	1.61	N.S.
Initial 231	6	15	25	14	12.14	.010
Medial 123	11	11	28	10	15.08	.005
Medial 321	10	14	22	14	5.08	N.S.
Final 132	9	14	23	14	6.81	N.S.
Final 312	11	12	19	18	3.34	N.S.
3. Meaning does not occur in isolation, but only in relationship to forms.						
Isolation	22	10	14	14	5.08	N.S.
Initial 312	6	18	23	13	10.54	.025
Initial 321	11	19	13	17	2.68	N.S.
Medial 132	17	18	10	15	1.54	N.S.
Medial 231	4	22	23	11	14.01	.005
Final 123	6	17	29	8	22.01	.005
Final 213	4	10	31	15	26.81	.005

Table 2

Duration and fundamental frequency patterns in the tested sentences

Sentence and position	Duration in msec	Initial word or phrase		Final word or phrase	
		Duration	F_o	Duration	F_o
1. Language - meaning.					
Isolation	2100	439	120-145-140	277	130-100
Initial 123	2050	370	140-175-165	277	105- 85
Initial 132	1950	323	145-185-180	254	105- 90
Medial 213	1935	339	125-180-150	246	100- 90
Medial 312	1885	339	135-205-190	270	110- 80
Final 231	1970	370	125-160-150	254	115- 75
Final 321	1940	347	130-155-120	254	100- 80
2. If a - at all.					
Isolation	6320	277	120-125-115	847	110- 95
Initial 213	6170	123	130-150-140	801	110- 85
Initial 231	6060	123	125-180-165	793	115-140-90
Medial 123	5950	115	110-145-125	816	120-125-lar.
Medial 321	5830	131	125-150-140	793	115-120-80
Final 132	5840	123	120-120-120	816	145-100
Final 312	5875	92	115-150-132	809	105- 95
3. Meaning - to forms.					
Isolation	3930	285	150-170-150	547	130- 95
Initial 312	3755	308	125-200-135	592	125- 80
Initial 321	3840	277	140-210-190	570	125- 90
Medial 132	3465	277	120-200-145	578	115- 80
Medial 231	3840	277	135-200-145	592	125- 75
Final 123	3460	277	130-200-165	539	125-100-lar.
Final 213	3800	300	125-165-120	570	140- 95

Table 3

Sentence 1 - correlation of listener judgments with duration

Production	Overall duration in msec	Listener judgments as production in isolation
Isolation	2100	29
123	2050	29
231	1970	24
132	1950	21
321	1940	24
213	1935	19
312	1885	18

Sentence 1 - correlation of listener judgments with fundamental frequency

Production	Fundamental frequency peak in Hz	Listener judgments as initial in paragraph
312	205	31
132	185	31
213	180	28
123	175	22
231	160	25
321	155	12
Isolation	145	14

DISCUSSION of "The phonetic structure of paragraphs"
LEHISTE

MASSARO:

It is necessary to analyze your results as a function of presentation order in the listening test. Listeners may have judged the sentences relative to the sentences already heard in the test. This normalization process may have been necessary in order to categorize the sentences correctly.

If so, subjects should have performed very poorly in the early part of the listening test and improved during testing. Evidence for a normalization process of suprasegmental shape of a sentence would be analogous to the normalization process in the recognition of vowel quality and duration.

PISONI:

From the work discussed previously, especially DENNIS KLATT's finding, it appears that the segmental information in sentential context is quite different acoustically from the segmental information in isolated words. From the work of Cohen and Lieberman and others it seems reasonable to suppose that the speaker is making every effort in production such that the listener can construct the appropriate semantic representation from the speech signal. Obviously, the differences in segment duration stress and the F_0 pattern in sentential materials are not random occurrences but seem to be part of the speaker's tacit knowledge about his language.

't HART:

I cannot imagine that the observed differences in duration, which amount to only 2 to 4 percent, can constitute cues to the listeners.

LEHISTE:

Most likely the durational differences were not the only cues. But the observed correlation between duration and listener responses is very unlikely to have arisen by chance.

KLATT:

You observe a correlation between overall sentence duration and perception of whether the sentence was spoken in the paragraph or in isolation. 't Hart questioned whether the small (2 to 4%) differences in duration were large enough to be discriminated. One way in which these durational changes could be perceived is if the major durational change is localized to a single portion of the sentence. For example. it is likely that the last syllable or two of the sentence produced in isolation are significantly longer than in sentence context.

COLLIER:
I have two questions regarding your finding that the first sentence in a paragraph shows more listener agreement than the last sentence. First, is it of any use to a listener to know that a sentence is the first one of a paragraph? In other words, should the linguistic competence of a speaker really contain rules that generate cues for paragraph initial position, rather than for paragraph final position? Second as far as the cues themselves are concerned: you interpret higher fundamental frequency at the beginning of a paragraph as the speaker's anticipation of the gradual F_o fall during the production of the whole paragraph. Do we not frequently observe that speakers "reset" their F_o at beginning of each utterance, rather than only once per paragraph?

LEHISTE:
Before I got the results of this study, I would not have claimed that clues signalling the beginning of a paragraph are stronger than those signalling the end of a paragraph. This is simply an empirical finding. I cannot say whether knowledge of these cues is of any use to the listener. As regards my tentative interpretation of the finding that the same sentence is produced with a higher fundamental frequency at the beginning of a paragraph than elsewhere, this was intended to explain the situation found in the present experiment I certainly did not claim that speakers may not "reset" their F_o at the beginning of an utterance; I do believe that speakers would avoid "resetting" in the middle of an utterance since this might be perceived as an unintended contrastive stress.

HUGGINS:
1. In connection with the facilities at Bell Labs for presenting auditory material (e.g. sentences) in any desired order: there is a much cheaper alternative, the Language Master by Bell and Howell. It is a tape recorder whose recording medium is magnetic tape mounted on 3"x8" cards allowing random access.

2. The measurements you have made were mostly at the sentence level - it is also likely that there is paragraphic information at the word level. Sven Öhman and Phil Lieberman have both shown that the less context supports a word in a sentence, the more clearly it is articulated. Similarly, Coker at Bell Labs has shown that the first introduction of a word or topic into a conversation is uttered more clearly (and probably more stressed) than later tokens of the word. Cues of this type are double edged: the first occurrence is marked positively by contrastive stress, later occurrences are negatively marked by reduced (relatively) stress.
3. There is a methodological problem too. A judgment task on a sentence that is repeated is very different to the first-pass processing of the sentence. On repeated listening, it is often possible to invent a setting, or attitude on the part of the speaker, that would account for this having uttered a sentence with <u>any</u> given pattern. On the other hand, the result of interfering with the order of a first pass presentation can be misunderstanding of the meaning or the intent of the speaker.

III. Some Storage Properties of Speech Perception

TEMPORALLY SEGMENTED SPEECH AND "ECHOIC" STORAGE

A. W. F. Huggins
Research Laboratory of Electronics,
Massachusetts Institute of Technology,
Cambridge, Massachusetts 02139

The argument I defend in this paper is that successive events tend to be integrated in perception, if the events are similar and follow each other rapidly, but that they are perceived as separate if the succession is more leisurely. The idea is far from being new - for example, one of the important papers that led to the development of Gestalt Psychology, Wertheimer's (1912) study of the phi-phenomenon (apparent movement) came to exactly this conclusion. Two lines, viewed sequentially in a tachistoscope, were seen as simultaneous if separated by less than 30 msec, but successive if separated by 200 msec or more. At intermediate separations, a single moving line was often seen. Wertheimer's results have been revised and expanded since, but the principle is clear. More recently, Stroud (1955) proposed that psychological time is not continuous, but occurs in discrete "moments." Two events that occur within a single moment are integrated; they can only be perceived as separate if they occur in separate

---

This research was supported by NIH Grant No NS 04332.

moments. Stroud's theory has been hard to sustain, most of the evidence pointing to a continuous, rather than a discrete, perceptual process. But the same principle is present: rapid succession leads to integration, slower succession to separation.

Before we turn to audition, let us consider briefly the implications of this principle in visual perception. We will start with two of the germinal studies in the growth of the information-processing approach to perception and cognition. Sperling (1960) showed that all of the information in a brief visual display was available for a short time after the display was switched off, but that it decayed rapidly. His display was an array of nine or twelve letters, in three rows. The subject was told which row to report by a high, middle, or low pitched tone that followed the display offset. With short delays, any row was correctly reported, but with delays longer than about half a second, only three or four letters in the array could be reported, and there was no further decay. Averbach and Coriell (1961) asked for a single letter to be identified in an array, and used a visually presented bar or circle to point to the required letter, after the array had disappeared. Their results were similar to Sperling's, but with the additional finding that the circular marker sometimes erased the letter it was pointing to. These results, with others (reviewed in Neisser, 1967), suggest that visual input is initially stored in a rich, raw form in "iconic" memory, from which it is read out sequentially for subsequent processing. Information in the store decays quite rapidly, and must be read out before it decays if it is to be perceived. A new visual input interferes retroactively with the processing of earlier inputs.

These findings seem to lend support to ideas like Stroud's, that perception occurs in a blink-by-blink fashion. But the findings may not be correct for all modes of perception. It is possible that discrete processing takes place only when the input is discrete. Since the perceptual apparatus was molded by evolutionary pressures, stimulus situations that do not occur in the real world often provide a way of penetrating the perceptual armor, there being no advantage to the organism of being able to handle these situations. But it may be dangerous to lose sight of the fact that the stimulus is unusual: normal stimuli may be processed differently. Thus the finding that a later stimulus can interfere with processing of an earlier stimulus may not mean that this always happens. If it did, how would we see anything at all? Erasure of the earlier stimulus is obviously appropriate when visual attention is being switched off an old object and onto a new one, but it is not appropriate as long as the focus remains on one object. Although there are some situations that require repeated rapid shifts of fixation - reading is a good example - there are many other activities where changes of fixation are much less frequent, and the interesting things happen during fixation. Any activity that requires visual tracking falls in the second class - for example, watching any object that is itself moving.

Consider the success of the movie industry. Successive frames of a movie do not erase the earlier frames, but are sequentially integrated with them into an ongoing perception of thrill-packed action. The situation is different from that producing erasure, in that successive frames of a movie are very similar. The developing action is continuous. If the interval between exposing successive frames is lengthened, so that the similarity between them disappears, then perception is indeed disrupted. The effect is so striking, it has led to the popular party game of illuminating the party with a slow-running strobe light.

Thus, we can conclude that the perceptual result of following one stimulus with another will depend on the intervening interval, and on the similarity between the stimuli, and on whether the task requires the stimuli to be integrated or separated. Short intervals and similar stimuli will assist their integration into one event, and make it very difficult to treat the two as separate percepts. Longer intervals and dissimilar stimuli will have the reverse effects.

We know much less about auditory processing than about visual processing. Until recently, it has been much more difficult to obtain complete control of the stimulus, but the advent of computers has changed this. Although there are striking differences between audition and vision, the same model of peripheral raw storage followed by sequential read-out has been proposed for audition (e.g. Massaro, 1972). The need for peripheral storage is much more obvious in audition: sound travels past the observer at 1100 feet/sec, so there is no possibility of taking a second look, as there often is in vision, where the objects of fixation tend to be relatively permanent. Secondly, auditory input is already sequential, unlike vision. Sound consists of pressure changes over time, and the short-term spectrum of the sound itself changes over time. One can think of meaningful sound as sequential in both microstructure and in macrostructure. Vision on the other hand is parallel in microstructure, and sequential in macrostructure. I have argued elsewhere (Huggins, 1975) that a storage mechanism that permits stored information to decay over time may not be appropriate for a medium in which time is so inextricably involved in the definition of the stimulus. Decay would produce qualitative as well as quantitative changes in the stimulus.

Many of the arguments made above for vision apply also in audition. Similar events that follow each other rapidly tend to be integrated into a single percept, whereas less similar events are perceived as separate. For example, Miller and Heise (1950) found that two tones alternated at 5 Hz were heard as a single warbling tone when the frequency separation of the tones was small, but when the separation was increased, two separate, interrupted tones were heard. Musicians have known of this effect for centuries: Bach's music is full of examples.

The disruption of auditory perception by lack of continuity is perhaps best exemplified by how difficult it is to report the temporal order of a cyclically repeating set of disparate sounds. If the sounds are as different as a buzz, a hiss, a tone, and a vowel, each sound must last 300-700 msec for listeners to report the sequence correctly (Warren et al. 1969). On the other hand, when four vowel sounds were used (with abrupt boundaries), temporal order was reported correctly when each vowel lasted 125 msec. though not when they lasted 100 msec (Thomas et al, 1970). In the latter case the lack of continuity was much less severe, since all the sounds were from a single speaker, at the same fundamental frequency. The pnenomenon of primary auditory stream segregation (Bregman and Campbell, 1971) provides further evidence that the perception of temporal order is strongly influenced by the similarity of the component tones. One can argue that a large discontinuity in the input forces the perceptual apparatus to treat the sounds on each side as separate, and the rate at which new, unrelated percepts can be handled is limited. Although the sequential sounds in the foregoing studies interfere with each other to the extent that they disrupt perception of the temporal order, subjects report that they hear each of the sounds clearly. Perhaps this is not surprising, since the sounds each last 100 msec or more.

When the sounds are made very short, however, an auditory erasure effect is found that is similar to the visual erasure mentioned above. Thus, Massaro (1972) has reported that 20 msec bursts of tone, followed after a variable silent interval by a masking tone at the same intensity, cannot be reliably identified as being slightly higher or lower in frequency than the masker, when the silent interval is less than 150-200 msec. Similar results have been observed with vowel sounds. These findings, and those outlined above for temporal order, would seem to raise severe problems in explaining how speech is perceived. The different acoustic events produced by the articulation of speech are both brief (often less than 60 msec) and follow each other very rapidly. Why do they not mask each other, and how are they different from other sounds, in that we have no difficulty perceiving their temporal order?

The answer, I argue, is that speech is _not_ to be regarded as made up of a sequence of unrelated events, but rather as a single continuously developing process. Backward recognition masking should only take place when changing the focus of attention between one sound source and another. When auditory attention is fixed on an ongoing sound, even quite large disturbances fail to produce noticeable masking. Listeners make large errors when asked to locate a click within a sentence (Ladefoged and Broadbent, 1960), or even a loud buzz or cough that completely replaces a sound in the sentence (Warren, 1970). On the other hand, when a _silent_ interval is used to replace a sound, no errors are made in locating it. Auditory perception was forced to evolve in such a way as to be able to cope with extraneous noises, which occur frequently in life. But there is no naturally occurring event that produces a short _silence_ in the middle of an ongoing sound. On the contrary, a silence indicates that a sound-generating event has finished. Thus, if a silent interval is artificially introduced into speech, it is immediately perceived as being unnatural, unless it can be integrated into the speech and perceived as a stop (Bastian, 1960).

The introduction of silent intervals does not necessarily interfere with speech intelligibility, however. In an early study of interrupted speech, Miller and Licklider (1950) reported that at rates of interruption above about 10 per second (the speech was switched off for 50 msec every 100 msec) intelligibility of PB words was little affected. They likened the experience to viewing a scene through a picket fence. The result has been repeated with running speech (Cherry and Taylor, 1954; Huggins 1964). Longer silent intervals of course did reduce intelligibility, since substantial parts of the speech wave were suppressed by the interrupting switch. But intelligibility can be degraded by inserting silent intervals even when none of the speech is suppressed (Huggins, 1975). A continuous speech message was "temporally segmented", that is, divided into speech intervals by effectively splicing 200 msec silent intervals into the recorded speech, without discarding any of the speech. Intelligibility declined by about 30% for each halving of speech interval duration below 200 msec. On the other hand, if relatively short speech intervals of 63 msec were separated by silent intervals of variable duration, intelligibility remained low (c. 50%) as long as silent intervals lasted 125 msec or more. But as silent intervals were shortened from 125 msec to 63 msec, intelligibility abruptly recovered to close to 100%, where it remained for shorter silent intervals. We will refer to the latter result, in what follows, as the Gap-Bridging result.

These results fit in exactly with the arguments made above. When silent intervals are short, it is easy to integrate successive similar sounds into an ongoing percept, but hard to treat successive sounds as different. As silent intervals are lengthened, it becomes progressively harder to integrate the successive speech intervals into an ongoing percept, and easier to treat them as separate independent events. If the foregoing interpretation is accepted, the present results agree remarkably well with the study on temporal order in vowels mentioned above (Thomas et al., 1970). Thomas found that temporal order was reported correctly when each of the four cyclically repeated vowels lasted 125 msec, but not when they lasted 100 msec. That is, each vowel could be processed separately when it lasted 125 msec. In the Gap-Bridging result, silent intervals of 125 msec could not be bridged, so successive speech intervals had to be processed separately, and intelligibility was low. On the other hand, when the combined duration of speech and silent interval was 125 msec, the successive speech intervals could be integrated, and intelligibility was high. The results seem highly compatible with an account in terms of a short-term storage mechanism, or a "processing time". As such they are reminiscent of earlier theories put forward to account for the loss of intelligibility that occurs when speech is cyclically alternated between the ears (Cherry and Taylor, 1954). Unfortunately, the processing time argument has been severely undermined by the finding that in both alternated speech, and the very similar temporally segmented speech, the effects are tied to speech rate rather than absolute duration (Huggins, 1964, 1974; Wingfield and Wheale, 1975).

Therefore, before making too much of the agreement between the Gap-Bridging result and temporal order results with vowels, we should explore the parameters of the Gap-Bridging result, and in particular, confirm that the result is not tied to speech rate.

Method

The experimental approach involves separating the duration of the speech intervals from the amount of speech they contain by varying the playback speed of the speech, and measuring the effect of silent interval duration on intelligibility. The speech materials were the same two sets of nine 150-word passages I used in earlier studies. Each of the eighteen passages were temporally segmented in two ways, as diagrammed in Figure 1.

First, a complete 150-word passage was digitized, and stored in the Speech Group's PDP-9 computer. The waveform was then "marked", in unused low-order data bits, to divide it into speech

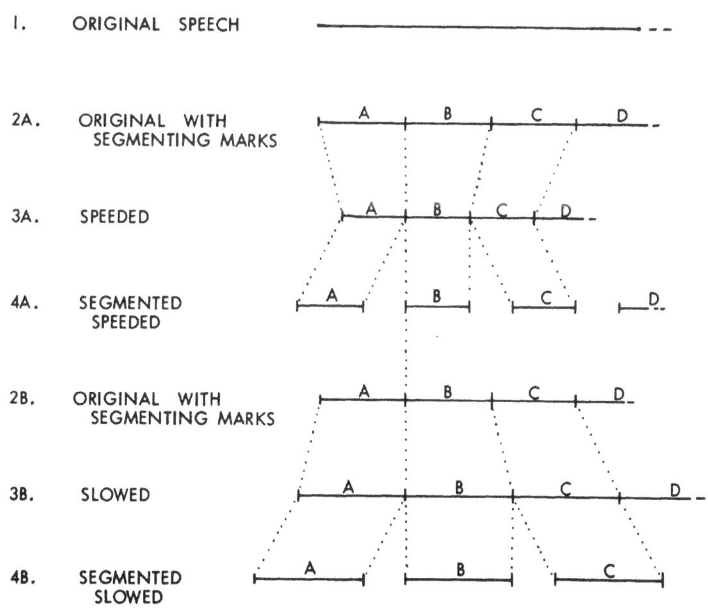

Figure 1: Diagram of the experimental design.

intervals lasting 62.5 msec. Next, the sampling rate was increased by a factor of 1.189 (= $2^{1/4}$), and the silent interval duration was set, say to 125 msec. The stored speech was then played out at the increased sampling rate, and the computer inserted 125 msec of silence whenever it encountered a segmenting mark. Next, the sampling rate was reduced below the rate used for input by the same factor of 1.189. The silent interval duration was reset to 125 msec, and the segmented speech was played out and recorded. The same segmenting marks were used for both versions of a passage, so that corresponding speech intervals consisted of exactly the same waveform. They differed only in their time scale: speech intervals lasted 74 msec in the SLOW-played version, and 52 msec in the FAST-played version. Silent intervals lasted 44 msec in the first two passages of each set, and 52, 62.5, 74, 88, 110, and 177 msec in the remaining seven passages. Twelve subjects each shadowed both sets of passages, one in the SLOW and one in the FAST version, with order of presentation and passage sets appropriately counterbalanced. The subject was told in detail how his shadowing performance would be used to assess intelligibility, and he was asked to try to maximize his score. He was warned against trying to shadow with very short delays, since most subjects find this very difficult, and against long delays, because of the risk of forgetting several stored-up words upon encountering something unintelligible. He was told that what he said did not have to make sense - he would score 50% either if he repeated every word in the first half of the passage, or if he repeated alternate words all through. He was given several minutes practice of shadowing undegraded speech (taken from the same master tape), and the first two passages in each set (44 msec silent intervals) were regarded as practice for segmented speech. The subject's performance was recorded for later scoring: one point was earned for each word correctly repeated from the middle 100 words of each 150-word passage. The remaining 50 words were discarded, from the beginning and the end of the passage, to eliminate start-up and recency effects.

<u>Results and Discussion</u>

The mean shadowing scores are plotted twice in Figure 2, so that their fit to two hypotheses can be compared. If the Gap-Bridging effect is independent of speech rate, then the ABSOLUTE

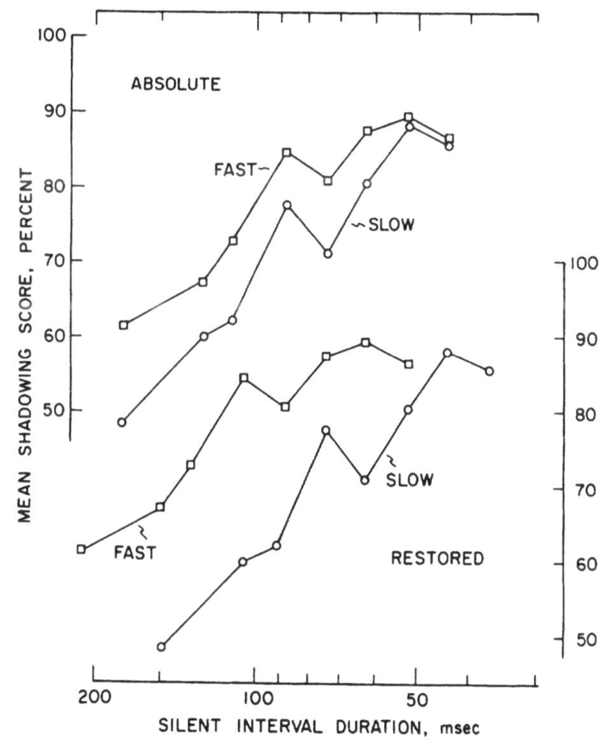

Figure 2: Mean shadowing scores plotted against ABSOLUTE silent interval duration (top), and against RESTORED duration, relative to speech rate (bottom).

duration of the silent intervals is the critical parameter controlling the recovery of intelligibility, and the two sets of data, labelled FAST and SLOW, should be in good agreement at the _top_ of the figure. Here the scores are plotted against the ABSOLUTE duration of the silent intervals, as they were presented to the subjects. If, on the other hand, the critical parameter is the duration of the silent intervals _relative to the duration of the speech events_, then the two sets of data should be in good agreement as they are plotted at the bottom of Figure 2. Here, the scores are plotted against the silent interval durations that would have been heard, had the speech _and_ silent intervals been played back at a speed that RESTORED speech interval duration to 62.5 msec.

Clearly, the two sets of data are in better agreement in the top of Figure 2 than in the bottom. Therefore, if the two options we must choose between are 1) ABSOLUTE: the recovery of intelligibility is independent of speech rate, and takes place at the _same_ silent interval duration when speech rate is increased, or 2) RELATIVE: the critical silent intervals are a fixed proportion of phoneme duration, so that the recovery takes place at shorter silence intervals when speech rate is increased, we must clearly choose the former. However, the two sets of data are not in good agreement with either hypothesis. Inspection of Figure 2 suggests that, rather than the recovery taking place at _shorter_ silence intervals when speech rate is increased, it occurs at _longer_ silence intervals. That is, the FAST data lie to the left of the SLOW data in Figure 2, rather than to its right as would be required by the RELATIVE hypothesis. To test whether the lengthening of the silent intervals from SLOW to FAST data was compensating for the shortening of the speech intervals caused by the change in playback speed, the two sets of data are plotted a third way in Figure 3.

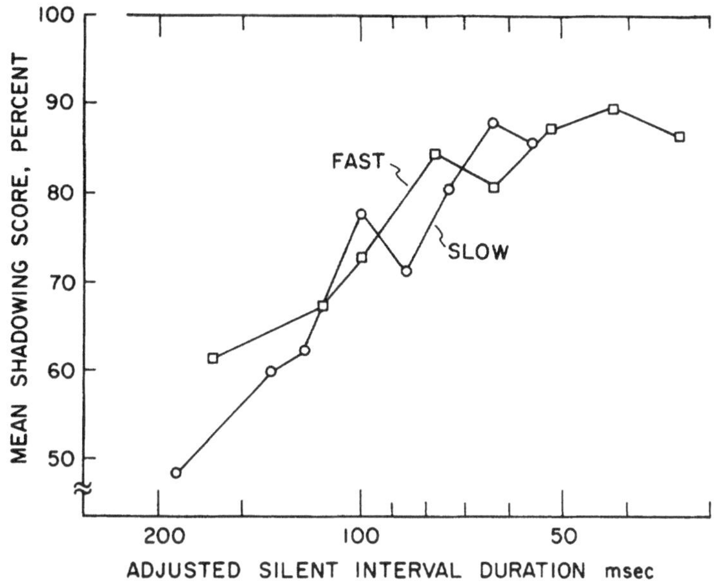

Figure 3: Mean shadowing scores plotted against silent interval duration ADJUSTED so as to equate time from onset to onset of speech intervals.

Here, the two sets of data are plotted against duration in such a way as to equate the duration between successive onsets (or offsets) of speech intervals across the FAST and SLOW versions. In fact, the plotted silent intervals were ADJUSTED, for the FAST data by <u>increasing</u> the true silence duration by the <u>decrement</u> in speech interval duration resulting from playing it FAST, and for the SLOW data by <u>decreasing</u> the true silence duration by the <u>increment</u> in speech interval duration resulting from playing it SLOW. The two sets of data seem to be in better agreement in Figure 3 than in the top of Figure 2, which suggests that the critical parameter controlling the recovery of intelligibility in Gap-Bridging is not the time between the <u>end</u> of a speech interval and the start of the next, but rather the time from the <u>start</u> of a speech interval to the start of the next.

Before exploring briefly the implications of this result, we should note that it was obtained by post-hoc analysis, and therefore needs to be confirmed. Second, the data are not as smooth and regular as might be expected from the results of earlier experiments using the same materials, and it is not obvious whether smoother data would improve or worsen the agreement in Figure 3. It is possible that the inversion present in both sets of data (middle two data points, Fig 2 top) might disappear if the number of subjects was increased. Otherwise, the inversion adds to the need for a confirming experiment, with better counterbalancing of the speech materials.

To return to the result obtained from Figure 3: suppose we are correct in concluding that the critical parameter controlling Gap-Bridging is the time interval between the onsets of successive speech intervals. What are the implications of this finding? The fact that the two sets of data were brought into agreement by plotting them against a single temporal parameter, for two different speech rates, suggests that the recovery of intelligibility may reflect some relatively fixed temporal property of auditory processing, at a stage more primitive than speech, or phonetic processing. There is some evidence that "acoustic" and "phonetic" stages of processing are separate. For example, Wood (1973) showed that reaction time in classifying an acoustic parameter, such as fundamental frequency, is not affected by irrelevant phonetic variation, such as place of articulation on a continuum between /b,d/. On the other hand, reaction time in classifying place of articulation _is_ increased by irrelevant variation in fundamental frequency. Reaction times are fastest of all when phonetic and acoustic variations are correlated (Wood, 1974). Wood argues from these results that acoustic and phonetic processing proceed in parallel, with the results of acoustic processing being available earlier than those of phonetic processing.

If the foregoing arguments are correct, and phonetic processing does not make use of the results of acoustic processing, then any temporal parameter that affects both acoustic and phonetic processing must be more primitive than either.

Secondly, the fact that the temporal parameter underlying the recovery of intelligibility corresponds to the <u>sum</u> of speech and silent interval durations suggests a storage mechanism with a fixed capacity, like Archimedes' bath: the more speech the store contains, the less silence is required to fill it up. I have argued elsewhere that raw "echoic" storage in the auditory system might have some of the properties of a <u>delay line</u> (Huggins, 1975). The foregoing interpretation of the present results appears to be compatible with this concept.

<u>References</u>.

Averbach, E. & Coriell, A.S. (1961) Short term memory in vision. Bell Syst. Tech. J. <u>40</u>, 309-328.

Bastian, J. (1960) Silent Intervals as Closure Cues in the Perception of Stop Phonemes. Haskins Laboratories, Quart. Prog. Rep No. 33, Appendix I.

Bregman, A.S., & Campbell, J. (1971) Primary Auditory Stream Segregation and Perception of Order in Rapid Sequences of Tones. J. Exper. Psychol. <u>89</u>, 244-249.

Cherry, E.C., & Taylor, W.K. (1954) Some further experiments on the recognition of speech, with one and with two ears. J. Acoust. Soc. Amer. <u>26</u>, 554-559.

Huggins, A.W.F. (1964) Distortion of the Temporal Pattern of Speech: Interruption and Alternation. J. Acoust. Soc. Amer. <u>36</u>, 1055-1064.

Huggins, A.W.F. (1974) More Temporally Segmented Speech: Is Duration or Speech Content the Critical Variable in its Loss of Intelligibility. Research Laboratory of Electronics, Quart. Prog. Rep. #114, p 185-193, Massachusetts Institute of Technology, July 15, 1974.

Huggins, A.W.F. (1975) Temporally Segmented Speech. To appear in Perception and Psychophysics.

Ladefoged, P. & Broadbent, D. E., (1960) Perception of sequence in auditory events. Quart. J. Exper. Psychol. 12, 162-170.

Massaro, D. W. (1972) Preperceptual Images, Processing Time, and Perceptual Units in Auditory Perception. Psychol. Rev. 79,124-145

Miller, G. A., & Heise, G. A. (1950) The Trill Threshold. J. Acoust. Soc. Amer. 22, 637-638.

Miller, G. A., & Licklider, J. C. R. (1950) The Intelligibility of Interrupted Speech. J. Acous. Soc. Amer. 27, 167-173.

Neisser, U. (1967) Cognitive Psychology. New York, Appleton-Century Crofts.

Sperling, G. (1960) The information available in brief visual presentations. Psychol. Monogr. 74, No 11.

Stroud, J. (1955) The fine structure of psychological time. In H. Quastler, ed., Information Theory in Psychology, New York, Free Press.

Thomas, I. B., Hill, P. B., Carroll, F. S., & Garcia, B. (1970) Temporal Order in the Perception of Vowels. J. Acoust. Soc. Amer. 48, 1010-1013.

Warren, R. M. (1970) Perceptual Restoration of Missing Speech Sounds. Science, 167, 392-393.

Warren, R. M., Obusek, C. J., Farmer, R. M., & Warren, R. P. (1969) Auditory Sequence: Confusions of Patterns other than Speech or Music. Science, 164, 586-587.

Wertheimer, M. (1912) Experimentelle Studien uber das Sehen von Bewegung. Z. Psychol. 61, 161-265.

Wingfield, A., & Wheale, J. (1975) Word Rate and Intelligibility of Alternated Speech. To appear in Perception and Psychophysics.

Wood, C. C. (1973) Levels of Processing in Speech Perception: Neurophysiological and Information-Processing Analyses. Haskins Laboratories Status Report on Speech SR-35/36.

Wood, C. C. (1974) Parallel Processing of Auditory and Phonetic Information in Speech Perception II: Application of a Probabilistic Model. J. Acous. Soc. Amer. 55, S88.

DISCUSSION of "Temporally segmented speech and "Echoic" storage"
HUGGINS

BERTELSON asked whether localization of the clicks at the two ears was an important part of my dichotically alternated click experiment.

Answer: I think not. I was excited to hear of van Noorden's thesis on coherence of tones. which seems to me have paralleled my result, but in one auditory channel. This is what I have argued in my paper, that integration (coherence) is facilitated by short silence intervals, and by similarity of successive stimuli.

SUMMERFIELD asked me to justify the penultimate line of my paper, that suggested the concept of a delay line. In a delay line, information enters the line at the point, travels down it without degradation, and on reading the end, is suddenly lost.

My answer: for a detailed discussion, I refer you to my Perception and Psychophysics paper (1975, in press). There, I describe an earlier experiment with alternated clicks. Subjects heard a sequence of clicks presented alternately to the left and right ears. They adjusted a binaural train to equate the perceived rates. At slow rates they matched the total rate of pulses into the head; at fast rates they matched the rate in one ear of the alternated train. The explanation I proposed was that with long inter-pulse intervals, successive pulses are perceived independently of each other, but as the interval is shortened, successive pulses in one ear begin to interact. The change-over takes place as one-ear interpulse interval is shortened from 200 msec to 100 msec. If echoic memory was a delay line of 200 msec, there would never be more than one pulse in the line when pulses were separated by more than 200 msec, in one ear, and never less than two pulses in the line when interpulse interval was shorter than 100 msec, with proportional changes in between. I also attempted to apply the explanation to my results with temporally segmented speech.

PREPERCEPTUAL AUDITORY STORAGE IN SPEECH RECOGNITION

Dominic W. Massaro and Michael M. Cohen
University of Wisconsin, Dept. of Psychology, Madison, Wisconsin, U.S.A.

Given that the acoustic stimulus for speech perception is extended in time and that perception cannot be immediate, it seems necessary to postulate a preperceptual auditory storage that holds the first part of the sound pattern until it is complete and perception has occurred. The duration of this preperceptual auditory storage places an upper time limit on the sound patterns functional in speech recognition. A recognition masking task has been developed to study the properties of preperceptual auditory storage and the temporal course of the speech perception process. In this task, a short speech stimulus is preceded or followed after some variable silent interval by a second sound. Both sounds are presented at a normal listening intensity. A number of studies have shown that the second sound interferes with the perception of the first if the second sound is presented before recognition of the first is complete. Backward masking results have shown that speech perception is not immediate but requires time for the synthesis of the sound pattern held in preperceptual auditory storage. The present studies evaluate some of the properties of preperceptual auditory storage and the primary recognition process. The fact that a second sound can interfere with perception of a first sound even if the sounds are presented to opposite ears locates preperceptual auditory storage at a central rather than a peripheral level. A first sound can interfere with a second sound if the sounds occur within roughly 80 msec. whereas the second interferes with the first out to an intersound interval of roughly 250 msec. This asymmetry supports the idea that the second sound overwrites and replaces the first in preperceptual storage when they occur outside the short 80 msec. period of temporal integration. Similarity effects do not appear to be critical in the masking paradigm whereas the effectiveness of the interference of one sound on another may be dependent on their acoustic structure.

INTRODUCTION

An information processing model appears to be particularly well-suited to the dynamic aspects of speech perception, the topic of this symposium. An information processing model is structured along the temporal (dynamic) dimension. A sequence of processing stages is assumed to occur between some observable stimulus and some observable response. Each processing stage has some information available to it and makes some transformation of this information available to the next processing stage. The model presented in Figure 1 has been utilized in our research in auditory information processing and speech perception. The model distinguishes four processes or functional components (circles): feature detection, primary recognition, secondary recognition, and recoding. Corresponding to each process there is a structural component (boxes) which represents the information available to that

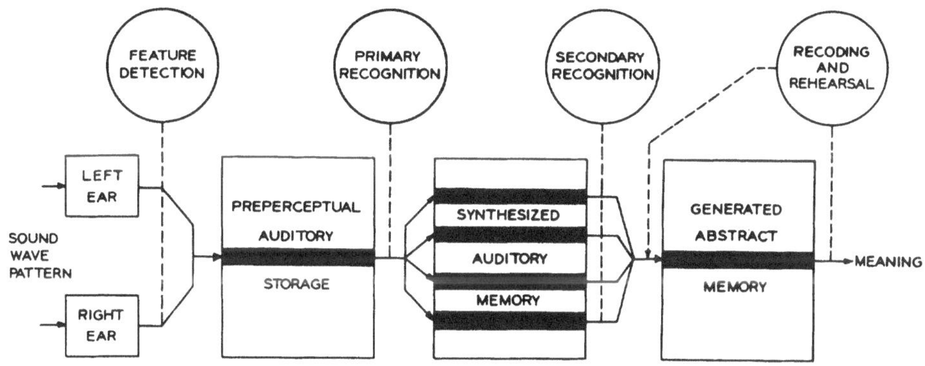

Fig. 1. Structural and functional components of auditory information processing model.

stage of processing. The structural component defines the nature of the information at a particular processing stage whereas the functional component corresponds to the psychological operations or processes at that stage of processing. Although the onsets of the stages are necessarily successive, the processing at one stage may overlap in time with the processing at another stage. In this paper, we focus on the properties of preperceptual auditory storage and the primary recognition process.

The feature detection process transforms the sound wave pattern into acoustic features that are stored centrally in preperceptual auditory storage. The features are described as acoustic since we assume that there is a positive relationship between the nature of the auditory signal and the information in preperceptual storage. Given that a speech sound is temporally extended, its acoustic features can be detected at varying times during or after the speech sound. Furthermore, different features might require different amounts of time for feature detection. Accordingly, the features cannot enter preperceptual storage simultaneously, but must be integrated across some short period. We also assume that there is a certain amount of background noise in the feature detection process. Since this background noise fluctuates over time, a nonexistent feature may be erroneously detected or a feature actually represented in the stimulus may be missed.

Preperceptual auditory storage holds the features passed on by the detection process for a short time after a sound is presented. The primary recognition process resolves the featural information producing a percept in synthesized auditory memory. The phenomenological outcome of primary recognition is perceiving a particular sound of a particular loudness and quality at some location in space. In speech perception, preperceptual auditory storage accumulates the acoustic features of a speech stimulus until the sound pattern is complete. Primary recognition occurs during and possibly after this time in order to produce a synthesized auditory percept. A second pattern does not usually occur until the first pattern has been perceived. However, if the second pattern is presented soon enough, it should interfere with recognition of the first pattern. By varying the delay of the second pattern, we can determine the duration of preperceptual auditory storage and the temporal course of the recognition process. The experimental task is referred to as a backward recognition masking paradigm (Massaro, 1972). Pure tones differing in frequency, intensity, waveform, and spatial location have been employed as test items in the backward masking task.

We have also used speech sounds in the task (Massaro, 1974). In one study, the consonant-vowel (CV) syllables /ba/, /da/, and /ga/ were used as test and masking stimuli. Only enough of the syllable was presented to make it sound speech-like. The 42-msec. syllables had 30 msec. of CV transition plus 12 msec. of steady state vowel. On each trial, 1 of the 3 syllables was presented followed by a variable silent interval before presentation of a second syllable chosen from the same set of 3 syllables. The subject's task was to identify the first syllable as one of the 3 alternatives. He was told to ignore the second syllable, if possible. The speech sounds were presented at a normal listening intensity.

Figure 2 plots the observed results in terms of discriminability (d') values for each of the three test alternatives. The d' measure of signal detection theory provides an index of how well the subject discriminates a given test alternative from the other test alternatives in the task. This measure allows the experimenter to correct for any decision bias the subject may have under a particular experimental condition. The d' values are computed from the hit and false alarm probabilities. The probability of identifying a syllable correctly is designated as a hit, and responding with that syllable to any other syllable alternative is designated a false alarm. For example, the probability of responding ba given the test alternative /ba/, P(ba|ba), would be a hit, whereas the probability of responding /ba/ to /da/ or /ga/, P(ba|da or ga), would be a false alarm. The d' value given by these two independent probabilities indexes the discriminability of the syllable /ba/.

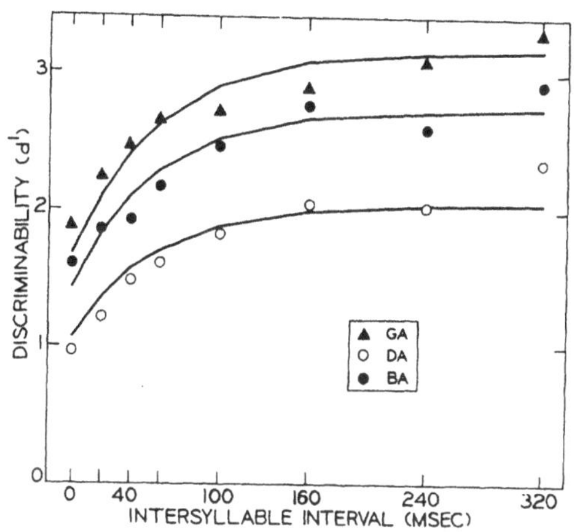

Fig. 2. Discriminability (d') values for the three test syllables as a function of the silent intersyllable interval. (After Massaro, 1974, Experiment IV.)

Figure 2 shows that correct identification of the first speech sound increased with increases in the silent interval between the 2 sounds. These results show that recognition of the consonant phoneme was not complete at the end of the CV transition or even at the end of the short vowel segment of the sound. Syllable recognition required perceptual processing after the speech sound was terminated. The second speech sound interfered with perception if it was presented before recognition was complete. These results support the idea that the speech sound is held in preperceptual auditory storage while processing takes place. A second sound interferes with any further resolution of the first sound.

The purpose of the present report is to provide further information about the properties of preperceptual auditory storage, and the temporal course of the primary recognition process. One question is whether preperceptual auditory storage occurs centrally receiving inputs from both ears or is ear specific. Figure 1 shows that preperceptual storage occurs centrally in our model after the input from the two ears is combined in the auditory system. Huggins (1974), on the other hand, has argued "that immediate preperceptual auditory storage occurs separately for the two ears" (p. 9). An answer to this question might be found in the effectiveness of a masking sound when it is presented to the opposite ear of the test sound presentation; the first experiment compares monaural and dichotic backward masking of consonant-vowel sounds.

A second question is whether the second sound replaces the first in preperceptual storage or is simply integrated with the first. If two sounds are integrated in preperceptual storage, the second sound should not interfere with perception of the first anymore than the first interferes with perception of the second. The second sound should dominate, however, if it overwrites and replaces the first in preperceptual storage. Massaro (1975) presents evidence that two pure tones are integrated only if they occur within 60 to 80 msec of one another. The integration time of speech sounds may differ from that found with pure tones, however. This second question must be answered by a comparison of backward and forward masking in which the masking syllable either follows or precedes the test syllable presentation. If a second syllable that falls outside the short period of integration overwrites and replaces the first syllable in preperceptual auditory storage, then backward masking should be much more effective than forward masking. In contrast, if the two syllables are simply integrated, forward masking should be as effective as backward masking. The integration idea also predicts that the similarity relationship between the test and masking sounds should be critical in the recognition task. According to the integration notion, if the second sound shares features with the first sound, less masking might occur since the integration of identical features will not disrupt identification performance. In contrast, if the second sound overwrites and replaces the first in preperceptual auditory storage, the primary recognition of the first sound may terminate with the presentation of the second sound so that the similarity relationship between the two sounds will not be critical. The second and third experiments study the backward and forward masking of consonant-vowel syllables as a function of the similarity of the test and masking syllables.

Experiment I

This experiment asks whether preperceptual auditory storage is central or ear specific. If preperceptual storage exists after the inputs from the two ears are combined in the auditory system, then dichotic masking should be as effective as monaural masking. In contrast, no backward masking should occur in the dichotic situation if there is separate and independent preperceptual storage for each ear. Consonant-vowel syllables were utilized as test and masking syllables in a backward masking paradigm. The test and masking syllables were either presented to the same ear (monaural) or different ears (dichotic).

Procedure

Ten undergraduates fulfilling a course requirement were tested for five days. Three CVs were employed as both test and masking stimuli. The CVs were /ba/, /da/, and /ga/ (/a/ as in box). The two formant CVs were synthesized at Haskins laboratory and recorded digitally for playback during the experiment. Enough of the CV syllable was presented so that it sounded speech-like. The duration of the CV syllables was 42 msec. and they were presented at 76 dB SPL. On each trial, one of the three CV syllables was presented followed by one of the three syllables. The first test syllable could be presented to the right or left ear and the second masking syllable could also be presented to the right or left ear independently of the test syllable presentation. The silent intersyllable interval lasted 0, 20, 40, 60, 100, 160, 240, or 320 msec. Each observer recorded his decision by pressing one of three pushbuttons labeled "GA-1", "DA-2", and "BA-3", respectively. Following the 1.5 sec response interval, feedback was provided by presenting the digits 1, 2, or 3 visually for 500 msec. The intertrial interval was 1.5 sec. Presentation of the three test alternatives, the three masking alternatives, the location of the test syllable, the location of the masking syllable, and the inter-syllable intervals was completely random. The alternative /ga/ could occur with probability 2/8 and /ba/ and /da/ occurred with probability 3/8 each. All other procedural details were the same as in Experiment IV of Massaro (1974).

On the first day, observers identified the test syllable without the masking syllable present for 300 trials. In the second session, the masking stimulus was presented. On Days 2, 3, 4, and 5, the experiment proper was carried out. There were two sessions of 300 trials each per day. Only the results of the last three days were included in the data analysis.

Results

Figure 3 shows that backward masking was observed under both the monaural and dichotic conditions, $F(7,63) = 28$, $p < .001$. Performance improved 28% with increases in the silent intersyllable interval for the best 5 subjects. The worst 5 subjects show a similar but much flatter function. There is some indication that the dichotic presentation actually produced more backward masking than the monaural presentation, $F(1,9) = 18$, $p < .005$. The dichotic and monaural functions seem to converge to a common asymptote for the best 5 subjects whereas the poorest subjects perform better under the monaural presentation at all masking intervals. The reason for the poorer performance is not readily apparent. It is possible that subjects occasionally

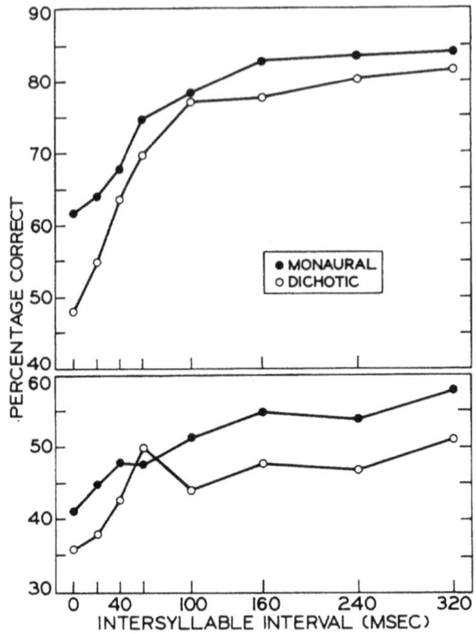

Fig. 3. Percentage of correct recognitions as a function of the intersyllable interval for the monaural and dichotic conditions. The top and bottom panels give performance for the best five and poorest five subjects, respectively.

reported the masking syllable in the dichotic case because of a problem in determining the temporal order of the two sounds. One cannot argue, however, that the backward masking observed under the dichotic presentation was due only to a temporal order problem. Massaro (unpublished) observed the same kind of backward masking with dichotic presentation when the subjects knew in advance the ear of the test sound presentation. The present results then, argue for a central preperceptual storage located after the inputs from the two ears are combined.

The next two experiments provide a comparison of backward and forward masking while simultaneously manipulating the similarity between the test and masking syllables. The first study and previous studies (e.g., Massaro, 1974) did not allow a straightforward test of similarity effects since the masking syllables were drawn from the same set as the test syllables. In this case, the d' measure of performance did not provide a measure of masking as a function of similarity, although it gave information about the effectiveness of the different masking stimuli (cf. Massaro, 1974). In order to evaluate similarity effects directly, the next two experiments employed the CVs /ba/ and /ga/ as targets with the CVs /ba/, /ga/, /bi/, /gi/, white noise, and patterned noise as masking sounds.

Experiment II

Procedure

Ten female and eight male students participated in the experiment for five consecutive days.

Each trial started with the presentation of the test stimulus, a 40 msec. CV that could be either /ba/ or /ga/. On seven out of eight trials a 40 msec mask was presented. The mask could be one of the CVs /ba/, /ga/, /bi/, or /gi/ with a probability of .125 or white noise (WN) or patterned noise (PN) with a .25 probability. The observers had 1.5 sec. to make their responses. The observer recorded his response by pressing one of two buttons labeled "BA" and "GA." Following the response interval, feedback indicating the correct response was presented by displaying for 500 msec. either "BA" or "GA" on two light emitting diode displays (Monsanto #MDA-III). The intertrial interval was 1.2 sec.

Four observers could be tested simultaneously in sound insulated rooms. All experimental events were controlled by a PDP-8L computer. The stimuli were generated under computer control by a digital to analog converter (Digital model VC8-L). These signals were filtered by a bandpass filter (KHRON-HITE model 3500R), which attenuated the acoustic signal below 20 Hz and above 5000 Hz. Then the signals were amplified (McIntosh model MC-50) and presented binaurally over matched headphones (Grason-Stadler type TDH-49). The stimuli were presented at a normal listening intensity (about 75 dB SPL).

Two sessions of 300 trials each were given each day. On the first day the test CVs were presented without masking stimuli. On the second through fifth days the experiment proper was carried out. All experimental conditions were completely random in a given session and, except for the restrictions on mask occurrence noted above, were programmed to occur equally often.

The CV stimuli used in this experiment were originally produced at Haskins Laboratory. Portions 98 msec. long of the two formant CVs /bi/, /ba/, /gi/, and /ga/ were digitized at a rate of 5000 samples per sec. by a computer controlled analog to digital converter (Digital model AD-08B) and placed in four adjacent 490-word sections of computer memory. During the experiment the first 400 words of a given section were played back at a rate of 10,000 words per sec. giving a stimulus of 40 msec. The first 20 msec. of each CV stimulus was a transition; the last 20 msec. was a steady state vowel. Spectrograms of the CVs are shown in Figure 4. WN was produced by assembling in one section of memory 400 words chosen randomly from the four CVs. PN was produced by randomly selecting two 200 word segments from the four CVs. A new sample of WN or PN was taken for each presentation of these stimuli in the experiment. Spectrograms of three samples of PN and one sample of WN are also shown in Figure 4.

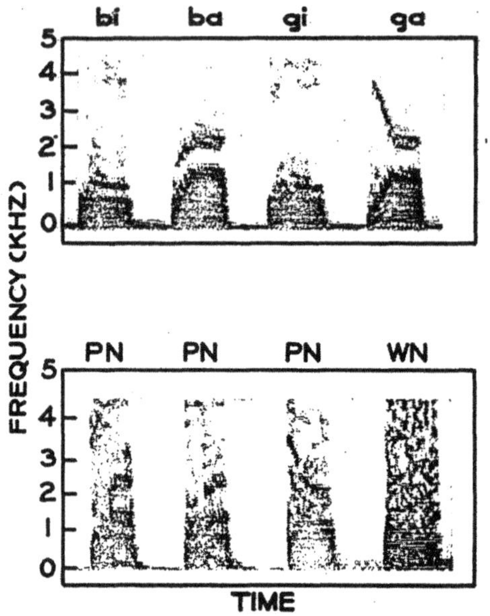

Fig. 4. **Spectrograms of the sounds used in** Experiments II and III. See text for description.

Results

In order that performance be measured in a manner independent of possible decision bias the results were analyzed in terms of the signal detection metric d'. With this metric the interal presentations of the conditions, called "BA" and "GA" are considered to be expressable as statistical values with normal distributions on an underlying subjective dimension. The value of d' is then defined as the separation of the means of the "BA" and "GA" distributions in standard deviation units, thus providing an index of the discriminability of the two targets (Massaro, 1975).

Rather than compute a d' for each subject, the response frequencies were first pooled over subjects at each experimental condition. The d' values were then derived from the response probabilities computed from these pooled observations. Figure 5 plots the observed d' values as a function of the masking stimulus and the silent interval between the test and masking sounds. The top panels give the results for the present backward masking experiment. The statistical analyses were carried out on the percentage of correct responses for each subject at each of the experimental conditions.

As in earlier studies, overall performance improved with increases in the ISI, $\underline{F}(6,102) = 31$, $\underline{p} < .001$. Discriminability for the no-mask condition was 3.63. The

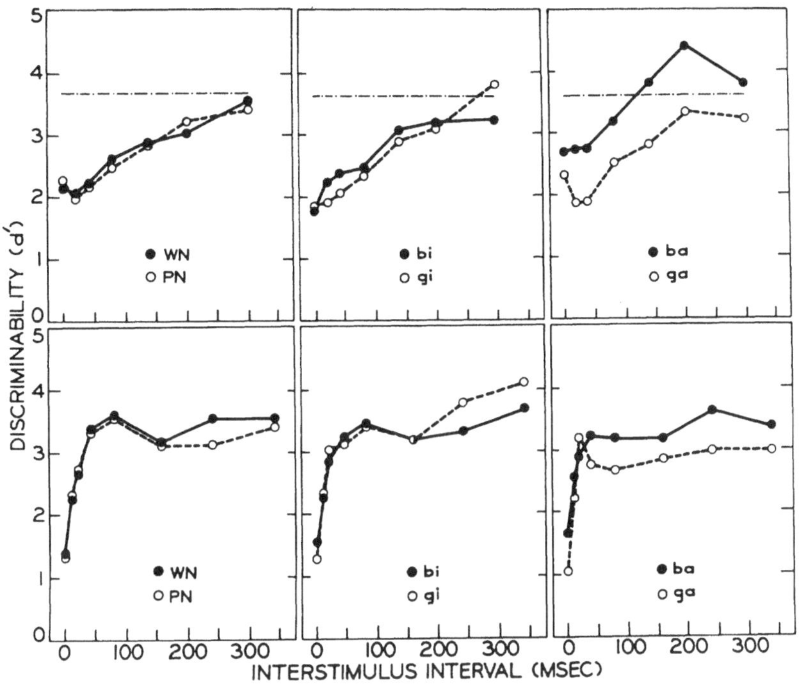

Fig. 5. Discriminability (d') values as a function of the interstimulus interval and the masking stimulus. The top and bottom panels give performance under backward and forward masking, respectively. The broken line in the top panels give performance under the no masking condition.

figures show that performance asymptoted at about 300 msec. A comparison between overall performance when the ISI was 300 msec. and the condition when no mask was presented was nonsignificant, $F(1,17) \approx 1$.

The only difference in performance as a function of mask, $F(5,85) = 12$, $p < .001$, was due to the fact that the /ba/ mask produced less overall masking. A specific comparison of performance with the /ba/ mask against performance with all other masks showed that /ba/ was a less effective mask, $F(1,85) = 54$, $p < .001$. Specific comparisons for the noise masks were not significant, $F(1,85) \approx 1$. These comparisons show that masking by WN or PN does not differ, overall, from masking by CV syllables.

The results show that the quantitative similarity between the test and masking sounds was not critical in the masking paradigm. The white noise and pattern noise were as effective as the syllables /bi/ and /gi/ in terminating processing of the test syllable. This evidence reinforces the idea that the aftercoming mask effectively overwrites and replaces the test syllable in preperceptual auditory storage, thereby terminating processing of the target. In speech, however, noise bursts occur quite frequently (e.g., /s/, /f/, /t/), and thus the stimuli used as masks in this study are still somewhat related in character to the targets. The logical possibility exists that if a mask of less similarity (e.g., a tone) were employed, masking would not have occurred. This question bears further experimentation.

The masking function for the /ba/ mask was elevated above the rest. Massaro (1974, Exp. IV) obtained similar results and suggested that the effect might be due to the fact that /ba/ is lacking in energy at high frequencies (cf. Figure 4). The stimulus /ba/ did serve as a masking stimulus, however, simply not to the extent of the other masks. The results show that similarity effects may not be critical in the masking task, although one sound may be a more effective mask than another because of the acoustic properties of the sounds. The masking sounds /ba/ and /ga/ must be equally similar to the test sounds /ba/ and /ga/ so that similarity cannot account for the differences found with the /ba/ and /ga/ masks. Accordingly, future studies will have to distinguish between similarity effects and overall masking effects in the masking paradigm. A pure tone may produce less masking of a CV syllable, not because of their similarity relationship, but simply because a tone has its energy concentrated at a narrow part of the frequency spectrum.

Wolf (1974) also looked at similarity effects in a backward recognition masking paradigm. The test stimuli were the syllables /da/, /ga/, and /ka/ and the masking syllables were /ba/, /pa/, and /ta/. In her analysis of the results, Wolf computed the probability of responding correctly to a given stimulus alternative as a function of its similarity relationship to the masking stimulus. The problem with this analysis is that it does not allow the experimenter to distinguish between similarity effects on the perceptual system and/or on the decision system (Massaro, 1975). Subjects may simply be more biased to respond with a test alternative that is similar to the masking stimulus even though the similarity of the masking stimulus was not critical in the perceptual resolution of the test syllables. One can still, however, evaluate similarity effects in Wolf's study utilizing a signal detection analysis.

For example, one can compute d'_{dg} as an index of how well /da/ is discriminated from /ga/. There are 2 independent ways to compute d'_{dg} and the two computations should give similar estimates of d'_{dg} if the underlying assumptions of the analysis are correct (Massaro & Schmuller, 1975). In one case, P(da|da) would be a hit and P(da|ga) would be a false alarm. In the second case, P(ga|ga) would be a hit and P(ga|da) would be a false alarm. One can then ask if the 2 estimates of d'_{dg} are equivalent except for random error and if the discriminability of /da/ and /ga/ in the task is influenced by the similarity between the test and masking sounds. We carried out this analysis on the pooled results of the five backward masking experiments of Wolf (1974). These results are presented in Table 1.

This analysis fails to reveal any large effects of similarity between the test and masking stimulus. For example, the discrimination of /da/ from /ga/ is given

Table 1

Hit and false alarm probabilities and d' values for the 3 possible test stimulus pairs as a function of the masking stimulus. The probabilities were computed from the pooled results of 5 backward masking experiments carried out by Wolf (1974).

Test (A-B)	Mask	P(A|A)	P(A|B)	d'_{AB}	P(B|B)	P(B|A)	d'_{BA}
da-ga	ba	.77	.30	1.26	.64	.20	1.20
da-ga	pa	.76	.27	1.32	.54	.14	1.18
da-ga	ta	.68	.35	.85	.39	.13	.85
ga-ka	ba	.64	.39	.64	.38	.06	1.25
ga-ka	pa	.54	.20	.94	.61	.19	1.16
ga-ka	ta	.39	.16	.72	.61	.26	.92
da-ka	ba	.77	.24	1.45	.38	.02	1.74
da-ka	pa	.76	.19	1.59	.61	.10	1.56
da-ka	ta	.68	.23	1.21	.61	.19	1.16

when /ba/ and /pa/ are masking stimuli. The masking syllable /ba/ shares the voiced feature with /da/ and /ga/ whereas /pa/ does not. If similarity was critical in backward masking, performance might be expected to differ in these two cases. In contrast, the d' values were 1.26 and 1.20 in the case of a /ba/ masking stimulus and 1.32 and 1.18 in the case of the masking stimulus /pa/. In general, no similarity effects are apparent although /ta/ produced more overall masking than either /ba/ or /pa/.

As mentioned earlier, the results of the first experiment can be interpreted in two ways. In terms of our model, a short speech sound is stored in preperceptual storage, so that primary recognition can occur. A second sound overwrites and replaces the information about the first in storage terminating the recognition processing of the first sound. Rather than replacing the first sound in storage, however, it is possible that the second sound is simply integrated with the first decreasing the signal-to-noise level of the first syllable. That is to say, it is possible that two sounds are mixed or integrated when they arrive within 1/4 of a sec. of each other. One test between these two interpretations has been to compare the results of backward masking to those found in forward masking, when the masking sound precedes the test sound presentation (Massaro, 1972, 1974, 1975). Massaro (1975) has argued that two sounds can be integrated if they occur within 60 or 80 msec; after this time the second sound replaces the first in preperceptual auditory storage. Therefore, we predict that forward masking should not occur beyond silent intervals of 60 to 80 msec. In contrast, if integration was entirely responsible for the backward masking results, forward masking should occur out to the same intervals as backward masking.

Experiment III

Procedure

Ten male and eight female students at the University of Wisconsin participated in the experiment. Each trial started with the presentation of a 40 msec. mask, which could be one of the CVs, /ba/, /ga/, /bi/, or /gi/, with a probability of .125 or WN or PN with a probability of .25. After a silent ISI of 0, 10, 20, 40, 80, 160, 240, or 340 msec. the test stimulus, a 40 msec. CV that could be either /ba/ or /ga/ was presented. The observers made their response during a 1.4 sec. interval following presentation of the target. Following the response interval the correct answer was displayed for 500 msec. All other procedural details and data analysis were identical to those described in the backward masking experiment.

Results

The bottom panels of Figure 5 display discriminability, d' values as a function of ISI and masking stimulus. Overall performance improved with increases in ISI, $F(7,119) = 54$, $p < .001$, asymptoting at about 40 msec. as opposed to the asymptote at about 300 msec. found for backward masking. Again, the similarity relationship between the test and masking sounds was not critical although overall performance was better with a /ba/ than /ga/ masking stimulus. The differences between the /ba/ and /ga/ masking stimulus persist even after performance is asymptotic where the masking sound supposedly no longer influences the perceptual processing of the test sound. This result suggests that some of the difference between the /ba/ and /ga/ masking stimuli might be accounted for by some decision bias or criterion variability not accounted for in the signal detection analysis.

Figure 6 provides a direct comparison between backward and forward masking by plotting the results as a function of stimulus onset asynchrony, the time between the onsets of the test and masking sounds. The figure shows that there is actually more, or as much, forward masking than backward masking when the onsets of the two sounds are presented within the short 60 to 80 msec. period of integration. If the two sounds fall outside of this period, no forward masking occurs although significant backward masking occurs out to roughly 1/4 of a sec. It should be noted that there were individual differences in quantitative performance. Figure 7 displays results for 6 typical observers from the backward and forward masking experiments. The substantial asymmetry of the forward and backward masking functions argues against a pure integration interpretation of backward masking of speech sounds. To what extent these results are relevant to the processing of continuous speech is the goal of future research.

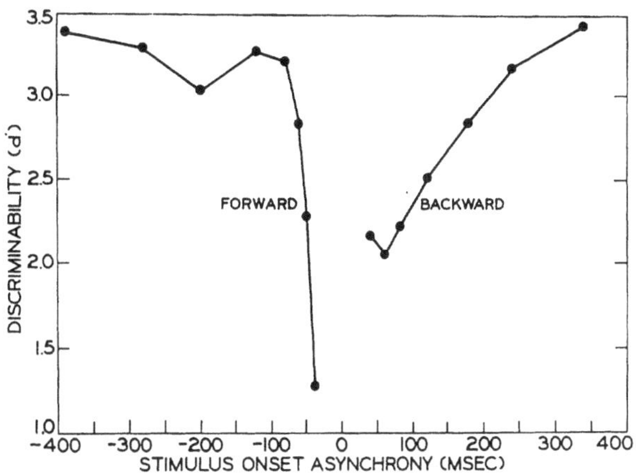

Fig. 6. Discriminability (d') values as a function of stimulus onset asynchrony, the time between the onsets of the test and masking sounds.

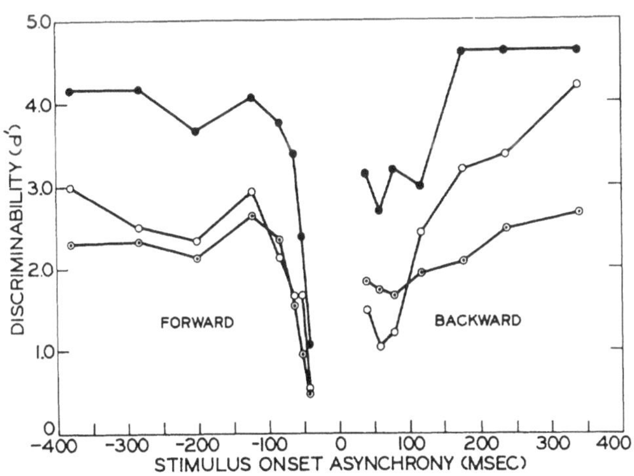

Fig. 7. Discriminability (d') values for 6 typical observers in the backward and forward masking experiments.

DISCUSSION

Given that the interpretation of the present experiments has been presented in the introduction and results sections, it seems worthwhile to direct the discussion section at some general issues in recognition masking experiments. These issues are best presented in discussion of some recent experiments of backward masking. Dorman, Kewley-Post, Brady, and Turvey (1973) have carried out a series of vowel masking studies taking issue with the hypothesis that vowels require time for perceptual processing. They want to argue that vowels readily identified in isolation are not susceptible to backward masking. In the first experiment, the natural vowels /i/, /ɛ/, and /ʌ/ as in heat, bet, and up were followed by a synthesized sound pattern with two formant frequencies at 489 Hz and 1690 Hz. One problem is that the natural vowels would have had energy at high frequencies whereas the masking stimulus did not. Our present studies show that a masking stimulus may not be completely effective if its frequency range is limited relative to the test stimuli. Therefore, masking may not have been complete in this situation weakening the conclusions that can be drawn. A direct comparison between naturally-spoken vowel masks and the synthesized-sound masks is needed. There is some evidence in Dorman et al.'s Experiment II that, in fact, the natural vowel /o/ as produced by the same speaker of the test vowels produced more masking than the synthesized 2-formant sound even when the synthesized sound was made 20 dB more intense than the test vowels. First, 5 of the 10 subjects showed significant masking with the mask /o/ whereas only 2 of another 10 subjects showed masking with the synthetic-sound mask. (Both groups of 10 subjects had previously achieved perfect accuracy on a practice sequence without a masking stimulus present.) Second, of the subjects who showed masking, significantly more masking was achieved with the /o/ mask than the synthesized sound mask (cf. Figure 2 in Dorman et al.).

When the experimenter brings a natural phenomenon into the laboratory, he must intervene and manipulate the situation before the experiment captures what he is after. Dorman et al. were interested in how vowels were recognized in real speech, i.e., whether perceptual processing time was required for their identification. In the actual experiment, however, subjects were only required to identify 3 vowels and the acoustic structure of each of the vowels was exactly constant throughout the experiment. The vowels were also presented under optimal conditions in a noise free situation. At first glance, one might expect that the task of the subjects in the situation would be much easier than in normal speech. Observers can do amazing things when they put on a set of headphones in a sound-attenuated chamber. More importantly, to identify the vowels correctly, the observers did not have to resolve the vowels to the degree that may be required in normal speech when many more alternatives are possible from a wide variety of speakers.

Table 2 presents typical formant frequencies of the first three formants for male speakers. To identify the vowel /i/, it would be sufficient for the observer

Table 2

Typical formant frequencies for a male speaker with a fundamental frequency of 120 Hz and the formant frequencies of the synthesized sound pattern used as a mask by Dorman et al. (1973).

	Vowel			
Formants	/i/	/ɛ/	/ʌ/	mask
F_0	120	120	120	120
F_1	270	530	640	489
F_2	2290	1840	1190	1690
F_3	3010	2480	2390	

to pick up either a fairly low F_1 or a high F_3 without really identifying the vowel as such. Analogously, energy in the range of 1200 Hz is sufficient for the subject to identify /ʌ/ correctly. If we consider the vowel stimulus as a composite of acoustic features, then subjects require fewer of these features in the experimental task than in the processing of real speech. This is exactly analogous to asking an observer to distinguish between the lower case letters o, t, and m, as opposed to identifying the letters as one of the 26 possible. In the first case with only three alternatives, the gross shape of the letter would be sufficient for recognition. Much more of the letter would have to be resolved when any of the 26 letters could occur. In both the vowel recognition and letter recognition tasks, then, one could not make the jump from experiment to speech perception or reading.

Regardless of the applicability of experiment to the real world, researchers have realized that an experimental question cannot be asked if performance in the experimental situation is insensitive to the changes in the independent variable. In tachistoscopic studies of the temporal course of letter recognition experimenters have always manipulated the figure-ground relationship of the display in order to put performance in the range in which it could be influenced by the independent variables of interest. Pilot work is usually carried out to set the duration and figure-ground contrast of the display so that the subject transmits information in the middle of the range between chance and perfect performance. Traditionally, conditions have been set so that the subject averages 75% correct in a two-alternative forced-choice task.

Dorman et al. selected their observers in all of the experiments on the basis of perfect performance in a training task. Subjects who made identification errors in the training task were eliminated from the study. Therefore, given only 3 discriminable alternatives and this selection criterion, it is not surprising that all of Dorman et al.'s studies are plagued by ceiling effects. Subjects are at perfect performance at a 200 msec. intervowel interval as they are at the 500 msec. interval. Does this result mean that 300 msec. additional processing time did not influence the perceptual experience of the observers? No! Subjects *may* have resolved more of the test vowel with a 500 msec. than a 200 msec. silent interval, but there would be no way to observe this fact since performance was at the ceiling of 100 percent. Although the observer knew more about the vowel in one case than the other, he knew enough about it in both cases to respond accurately.

Given the inevitable individual differences in an experimental task, Dorman et al. observed more of a ceiling effect for some observers than others. However, Dorman et al. concluded that the differences between subjects reflected "the outcome of two qualitatively different modes of vowel processing" . There is no justification for this conclusion. It was pointed out earlier that experimenters must manipulate the situation to get performance in the appropriate range in each experiment. This task is also required with different subjects. Subjects differ in the simplest task imaginable and a ceiling for one subject may be a floor for another. These individual differences in no way show different modes of processing. The experimenter must first equate the subjects on the basis of performance level and then ask if they differ in the way that they are influenced by his independent variable. Pisoni's (1973) goal was to compare the temporal course of consonant recognition to that of vowels. When the sets of CV syllables and vowels were constructed, there was no way to insure that identification performance would be at the same level in the two cases. Accordingly, if the vowels were more discriminable from one another than the consonants, the masking function for vowels might asymptote at perfect performance faster than the masking function for consonants. However, this in no way leads to the conclusion that vowels are perceived faster than consonants in real speech. Pisoni should have constructed vowel and CV syllable sets so that both sets could be recognized at 75% correct with maximal processing time. Given that identification of the two sets of sounds asymptoted at the same 75% level, the masking functions of these stimuli could be compared to see if the masking function would reach asymptote faster in the one case than in the other.

Some students of speech perception have utilized Dorman et al.'s and Pisoni's experiments either to question the validity of backward masking in speech perception or to support certain ideas of speech processing. For example, Studdert-Kennedy interprets these results to mean that "both the degree of masking and the critical ISI (interstimulus interval) for release of masking vary with target (consonant or vowel), size and range (acoustic or phonetic) of target set, target and mask energy, relations between target and mask structure (acoustic or phonetic), and individual listeners, many of whom show no masking whatever." (P. 23) Our analysis of these same results indicates that conclusions similar to those of Studdert-Kennedy are not justified. We hope that the present work and discussion will encourage and stimulate investigators to utilize the recognition masking paradigm to explore the unmapped territories in the dynamics of speech perception.

Acknowledgement

This research was supported, in part, by U.S. Public Health Service Grant MH-19399 and the Wisconsin Alumni Research Foundation.

References

Dorman, M., Kewley-Post, D., Brady-Wood, S., & Turvey, M. Forward and backward masking of brief vowels. Haskins Laboratories Status Report on Speech Research, 1973, 33, 93-100.

Huggins, A. F. On perceptual integration of dichotically alternated pulse trains. Journal of the Acoustical Society of America, 1974, 56, 3, 939-943.

Massaro, D. W. Preperceptual images, processing time, and perceptual units in auditory perception. Psychological Review, 1972, 79, 124-145.

Massaro, D. W. Perceptual units in speech recognition. Journal of Experimental Psychology, 1974, 102, 199-208.

Massaro, D. W. Experimental Psychology and Information Processing. Chicago: Rand-McNally, 1975.

Massaro, D. W., & Schmuller, J. Visual features, preperceptual storage and processing time in reading. In D. W. Massaro (Ed.), Understanding language: an information processing analysis of speech perception, reading, and psycholinguistics. New York: Academic Press, in press.

Pisoni, D. B. Perceptual processing time for consonants and vowels. Journal of the Acoustical Society of America, 1973, 53, 1, 369. Also appears in Haskins Laboratories Status Report on Speech Research, SR 31/32, 1972, 83-92.

Studdert-Kennedy, M. Speech perception. Haskins Laboratories: Status Report on Speech Research, SR-39/40, 1974.

Wolf, C. D. G. An analysis of speech processing: some implications from studies of recognition masking. Unpublished dissertation, Brown University, 1974.

DISCUSSION of "Preperceptual auditory storage in speech recognition"
MASSARO

GILBERT:
The experiments which you report seem to have little to do with "speech" but do demonstrate certain general resolving powers of the auditory system.

MASSARO:
I believe that understanding auditory processing can only illuminate speech processing. Although categorical perception and the differences between consonants and vowels were believed to be unique to speech, recent work has shown that these phenomena are more easily understood in terms of auditory information processing without reference to speech qua speech (cf. Massaro, D.W. (Ed.), Understanding Language: On information processing analysis of speech perception reading and psycholinguistics, New York: Academic Press, 1975).

LEVELT:
I like your stressing the importance of making process models of speech perception. For that reason I tried to make a process model explaining your forward and backward masking results. which you were able to refute on the basis of data that had not yet been available to me. However, my motivation remains, since I still do not understand what, in terms of processing, is the exact meaning of your concepts "integration" and "overwriting". Could you clarify these points?

MASSARO:
The concepts of integration and interruption are necessary to describe the results of recognition masking. Elsewhere I have reviewed the operations of these processes in both visual and auditory information processing (cf. Massaro, D.W., Experimental Psychology and Information Processing, Chapters 17, 18, 21 and 33).

KLATT:
Your data suggest that the perception of a consonant in a VC environment might be easier than in a CV environment, all else being equal. In language, CV patterns (e.g. at the beginning of an utterance) bear at least as much if not more information than VC patterns (e.g. at the end of an utterance). Why should this be the case?

An interesting answer can be suggested in the case of plosive consonants. The formant patterns for a voiced plosive in CV and VC environments are nearly mirror images, but the CV environment includes an additional cue, the plosive burst. It is also possible that special feature detectors exist for processing the rapid spectral changes that accompany a sudden onset of energy (Stevens, M.I.T. Q.P.R.. about 1973). In part your data motivate the advantages of postulating the existence of this type of detector.

MASSARO:
In fact, given CV and VC sounds with mirror image formant patterns and no bursts. the VC is perceived better than the CV. As you suggest, the burst must provide an additional cue in CV environments.

RECOGNITION MEMORY FOR VOICES

Edward C. Carterette and Alan Barnebey
Department of Psychology and Brain Research Institute
University of California
Los Angeles 90024

Introduction.

Nowadays the study of coding processes in human memory is a large central topic of psychology. How is information perceived, selected, transformed, abstracted or elaborated and finally put in memory? A recent review of the work on encoding processes has been given by Melton and Martin (1972). In this book Liberman, Mattingly and Turvey contrast speech codes and memory codes. For them speech is a special process having a code which is complex, lawful, natural, efficient and resistant to distortion. Memory codes are not special but are relatively arbitrary and generally inefficient. Liberman et al suggest that improving memory codes is likely to require a recoding into a form like a language code which is the class of codes to which speech belongs. An interesting question which arises is, how hard or easy is it to tell one speech-encoding system, or voice, from another? If a voice which will later be heard again is stripped of its semantical, grammatical and contextual constraints so as to lose its specialness of speech except as a carrier, are its abstracted properties laid down in a speech code or a memory code? The answer is important in the biology of survival, and also in our own human society which is held together by voice communication. The day is near when men and machines will talk fluently to each other. And even if it were not, the answer is interesting because whatever the evidence for or against, it is widely held that a voice can be recognized as familiar from a brief fragment of speech.

A primitive biological instance is the ability of the pallid bat (*Antrozous pallidas*) to recognize the voice of its infant (Brown, 1973). If separated from its mother a baby bat emits a continous series of isolation calls of about 60 msec duration. An infant behind a screen some distance away elicited responses from its own mother on 95 per cent of trials and on 65 per cent of the trials the mother flew to a position opposite its infant against the screen. Other females responded 28 per cent of the time but never flew to the infant. At any given age a bat's vocal signature has a characteristic pattern of cadence and temporal patterning of frequencies. Only the temporal patterning of frequencies remains fairly constant as the bat matures. Since a mother was able to recognize its infant's voice as the infant matured, it appears that vocal tract properties rather than intonation pattern carries the distinctive features used by the mother. Brown reports the work of another investigator who found that infant bats (5 to 23 days old) chose their own mothers 15 out of 20 times in a Y-maze. This hit rate probably underestimates performance since an infant bat may not hear during its first week of life.

Speech perception and automatic recognition.

Not so much work as one might expect has been done with perceptual aspects of voice recognition, though of course speech perception and automatic recognition are topics of very great activity, as well attested by the four volumes of the Stockholm Speech Communication Seminar (1974). The work on speech perception concentrates on the microstructure of phonemic elements and has been a rich source of hypotheses about the nature of the mechanisms which encode and decode speech (see, e.g., Liberman, Cooper, Shankweiler, and Studdert-Kennedy, 1967).

Automatic speech recognizers are eclectic in choosing their basis of data and hypotheses for they aim at getting the job done efficiently and cheaply. Most use short-term spectra of the acoustic signal or features of the time-varying waveform. Measured patterns are correlated against stored patterns of words or phonemes. Flanagan's (1972) succinct review of this work makes very clear that we can learn very little from it of how humans process speech or recognize its linguistic elements. Much the same can be said for automatic recognition or verification of speakers although this is a more tractable problem. The general problem of speech recognition is very difficult because the semantical, grammatical and contextual constraints which are so easy for humans to use have not yet been realized as machine algorithms in any deep way. These constraints do not hamper machines which identify a speaker, or verify a speaker as one of a class, the class of familiar speakers, for example.

Two illustrations depict some of the methods and the high rate of success in recent work in automatic speaker recognition. Wolf (1972) sought a set of efficient acoustic parameters closely related to voice features which distinguish speakers. Those parameters found useful included pitch period, features of the spectra of vowel and nasal consonants, slope of glottal source spectrum, word duration and voice onset time. Speakers were wrongly verified only 2 per cent of the time "When only 17 such parameters were used..." (p. 2044). Identification from among a set of 21 speakers was perfect. Atal (1974) used a number of different parametric abstractions of speech which he got from a model wherein a speech sample is predicted as a linear combination of past speech samples. From 12 predictor coefficients obtained each 50 msec from speech sampled at 10 kHz, such speech parameter functions as impulse response, autocorrelation and cepstrum yielded 98 per cent speaker identification (10 speakers) for .5 sec of speech. Verification accuracy was 98 per cent for 1 sec of speech, and text-independent identification was 93 per cent for 2 sec of speech. Verifying and identifying were based on the distance between test and reference parameter vectors weighted by the pooled intraspeaker covariance matrix.

We conclude that voice recognition algorithms are very accurate and may be based on many different aspects of the speech waveform.

Recognition memory for voices.

Recognition of pictures is remarkably accurate even when hundreds are compared (Nickerson, 1965; Shepard, 1967) and Shepard found that even after a delay of one week, memory for pictures was nearly the same as that for words and sentences tested immediately after presentation. In contrast, some studies find that auditory information decays within a second or two (see, e.g., Morton, 1970). But others find that some acoustic properties of speech are retained up to 8 seconds (Cole, 1972). Murdock and Walker (1969) conclude that prelinguistic auditory information was remembered for at least 5 to 10 seconds. The acoustic properties of speech can be retained for relatively long periods of time. Pollack's (1959) subjects heard a test word from a fixed set of words in noise and then were asked to choose that word from a reduced set. Performance on the reduced set was better than on the larger fixed set, even when 15 seconds intervened between the test word and presentation of the list of alternatives. Carterette (1960) verified Pollack's result using consonant-vowel sounds. And Crossman's (1958) similar findings held for delays of 40 seconds.

These results and the fact that a familiar voice is recognized with quick ease demonstrate that we can store apparently *nonverbal* stimulus attributes over time periods which are substantially longer than those predicted on the current view of memory as a three-stage model consisting of (1) a sensory register, (2) a short-term store where verbal tagging and rehearsal occurs and before being placed in (3) a long-term store. One is led to conclude that sensory attributes of a stimulus survive and are stored in parallel with verbal attributes as Posner (1972) argues. Diana Deutsch (1974) reviews numerous findings, from which she concludes that memory for tonal pitch is highly specific, complexly organized and enduring. She believes it "likely that similar characteristics will be demonstrated in other nonverbal memory systems."

Little is known about the processing of nonlinguistic and non-semantic sounds at the three assumed stages of the standard information processing model. Is there an echoic store? How are sounds coded for storage? Does the echoic store participate in retrieval? If so what is the form of the information and how does it get into the store? Is recognition memory for sounds such as voices or speech passages or environmental sounds as good as it is for pictures? What features of a voice do we use in deciding whether or not we have heard it before?

We suspected from some earlier recognition memory experiments (Carterette and Barnebey, 1974) that the difficulty of recognizing voices might be underestimated. This is because contexts and expectancies play large roles and the opportunities for false alarms are not always assessed. Using immediate testing Stevens, Williams, Carbonell and Wood (1968) obtained probabilities of about 90 per cent when either 4 or 16 of 32 test spondees were spoken by unfamiliar speakers. Hit rates were high and false alarm rates low and we estimated the value of $\underline{d}'$ at about 2.75. Listeners could freely compare test reference items, however, so that no burden whatever was placed on memory. McGehee (1937) found that a speaker who was heard once reading 57 words of text was recognized correctly among four unfamiliar others with an accuracy of 83, 81, and 13 per cent after 1 day, 1 week, and 5 months, respectively. This performance is nearly as good as for pictures in recognition memory tasks (Shepard, 1967). Of course McGehee's subjects could have used many cues beyond those belonging to the vocal tract. In particular the first speaker was more practiced in subsequent sessions.

Perceptual vs physical dimensions of voices.

What aspects of a voice are used in recognizing a speaker? One way to find out is to discover the structure of the listener's perceptual space. The next step would be to map the perceptual space onto the physical space. The perceptual space of voices has been described by means of semantic differential scales and factor analysis. Voiers (1964) extracted the four factors, clarity, roughness, magnitude and animation whereas Holmgren (1963) extracted the four factors intensity, quality, pitch and rate. These same workers correlated perceptual and acoustical factors. Using measures on physical properties such as intensity, pitch, period, speaking rate, they found only a few scales to be correlated. The implication is that the union of the perceptual and physical spaces is quite imperfect but this could arise from poor choices of perceptual or physical factors.

After having identified speakers by use of a multiple-choice test the subjects of Clarke and Becker (1969) defined a set of scales on subjective dimensions of pitch, pitch variability, rate, clickiness, sibilance and breathiness. The voices were rated on these scales and a decision function (Euclidean distance) on the ratings was used in identifying the voices. The decision function was also applied to measures of pitch period, spectral energy and duration. Pitch and rate were correlated with pitch period and duration. Rate of correct identification on the test (65 per cent) was higher than by decision function (61 per cent).

We conclude from the above discussion that several different sets of base vectors can be found which will span, or almost span, the perceptual or physical spaces but only a partial mapping of a perceptual into a physical space has so far been found.

It was our hope to induce a useful perceptual space from the nonmetric multidimensional scaling (MDS) methods which in contrast to factor analysis require very few assumptions. Indeed we need only assume that the form of the function relating the data to Euclidean distances is monotone. The idea is to start with a confusion matrix of the listener's responses with no preconception about the physical dimensions of the stimuli. By means of MDS algorithms (Shepard, 1972) we identify a purely psychological structure or space. If we are successful in this then we turn to psychophysical experiments and to patterns of measurements made on the physical stimuli in an attempt to relate the perceptual and physical structures. For examples of fruitful use of such an approach, see Shepard's (1972) extraction of the underlying structure in the cases of some classical auditory data on vowels (Barney and Peterson) and consonants (Miller and Nicely).

A perceptual space for voices.

1. Recording and playback. Nine male voices were recorded in an anechoic room directly onto a disk store of a computer via a 10-bit analog-to-digital converter at a rate of 9000 samples/sec. Each voice uttered the 1-sec phrase /who are we/. Voices were played for listeners directly from the disk via a 12-bit digital-to-analog converter after low-pass filtering at 4000 Hz cutoff. Listeners heard the voices over a single, amplified channel which fed a loudspeaker of research quality. Listening was in an anechoic room with the source at 1.5 m distance from the ears at which the Sound Pressure Level (re .0002 microbar) was about 67 dB.

2. Stimuli and responses. Triples of voices were heard, one voice at a time, in each of the 84 sets of C^9_3. For a given triple voices were assigned randomly to keys and each voice could be heard as many times as wished and in any order by pressing a V key. Subject's first task was to press the key lying between the pair of V keys associated with the pair of voices which appeared most similar (V_iSV_j). Immediately following, his second task was to press the key lying between the most dissimilar voices ($V_i\bar{S}V_j$). The keys were a triangle of six on a silent typewriter keyboard and were masked off from other keys. A value of 2 was recorded for the most similar pair, a value of 0 for the most dissimilar pair, and the value of 1 was assigned to the remaining pair.[1] Fifteen subjects were run, 8 male and 7 female.

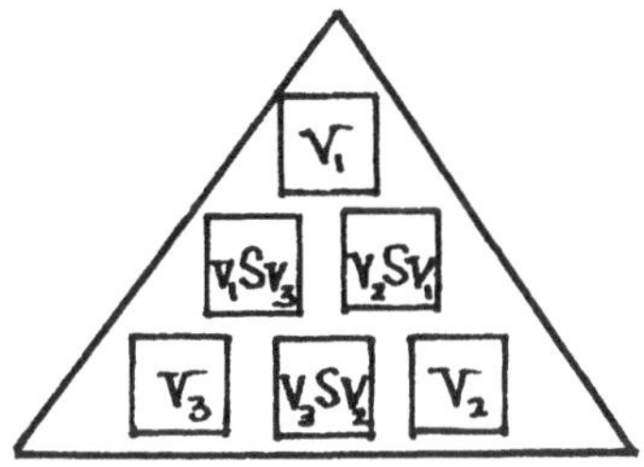

3. Response analysis.[2] The resultant similarity matrices were subjected to an MDS analysis by means of the Shepard/Kruskal/Carroll MDSCAL and INDSCAL computer programs (see Shepard, 1972 or Wish and Carroll, 1974). Analyses were made of the composite data of the 15 subjects and also of the data of each resulting from the triadic voice comparisons. The final stress values were .02 and less than .01 for 3 and 4 dimensions in the composite case. Individual subjects for the most part required 3 dimensions with values of stress less than .010. The subject with most stress (.20) in 3 dimensions had a stress value of .05 in 4 dimensions. A plot made from the output of the MDS programs is shown in Figure 1 for the 3-dimensional solution on composite data of 15 subjects.

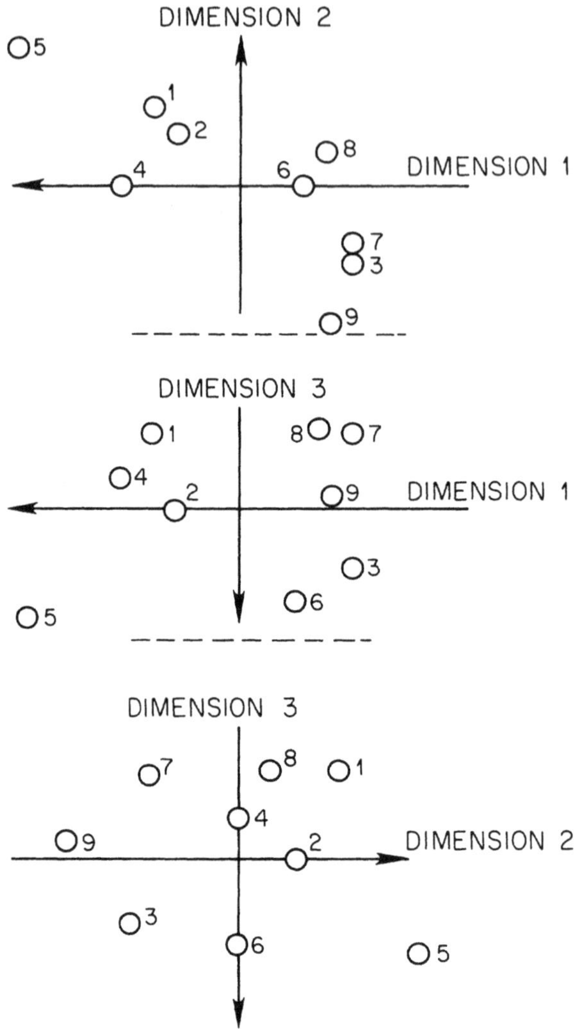

Figure 1. A composite perceptual space locating 9 male voices in three dimensions. Tentatively, Dimension 1 is identified as fundamental voice frequency F_0, Dimension 2 as intensity, and Dimension 3 as variability of F_0 or intonation pattern. Arrowheads point in the direction of increase. The similarity matrix of the combined data of 15 subjects from a triadic voice comparison task was used to obtain solutions based on a multidimensional Euclidean model of distances. Final stress of the 3-dimensional MDS solution was .0097.

4. Stimulus analyses. Let the sequence X_i (i = 1, 2,...., N) be the points sampled from the time waveform, that is to say, the sequence of numbers specifying a brief segment of a voice. After forming the differences $D_i = X_i - X_{i-1}$, the N-1 D_i were weighted by a Hamming time window, normalized, and then subjected to a DFT (Discrete Fourier Transform). The DFT when weighted by 20 log(f) is known as the power spectrum, P(f), and is measured in decibels (dB).

The power spectra are shown in Figures 2 and 3 as the first and third columns of the accompanying plots. After normalization, these power spectra were subjected to an inverse DFT, which is in the time domain. Such a spectrum is called, after Tukey, a cepstrum (say "kepstrum"). The cepstra are shown as the 2d and 4th columns. In any column the signal length analyzed was 28.44 msec (256 samples, at 9000 samples/sec). Since the signal is a real function its power spectrum and cepstrum are symmetric about the middle points. Therefore the spectrum is limited at 4500 Hz and the pitch period of the rightmost cepstral value gives an F_0 of 70 Hz which is (1/14.22) msec.

It is convenient to think of the cepstrum as a time spectrum. Thus events at the left (low time) have a short period, whereas those toward the right (at high time) have a longer period. This is a neat way of finding and tracking the pitch period of a voice. As an example, in Figure 2 Voice MM02 ranges between 137 and 107 Hz for /u/ and /i/ whereas in Figure 3 Voice MM05 has a range of 175 to 94 Hz.

For tracking formant location we used linear prediction coefficients, both stationary (Markel, 1972) and nonstationary (Atal and Hanauer, 1974). Based on methods of Akaike (1970) we have also been able to estimate variances of estimated formant locations as well as estimating the order of the autoregressive process, that is, the optimal number of LPC coefficients. These results will be reported elsewhere (Ionnides, Carterette, and Barnebey, 1975). In addition, nonlinear distances between voices were computed from the linear prediction coefficients (Atal, 1974) and cepstra from different voices were cross-correlated.

These analyses were used as a basis for labeling the dimensions of the perceptual space. For the composite data the 3 dimensions may be identified tentatively with pitch period, power, and variation of pitch period or intonation pattern. Although the orientation of points in individual spaces differed from that of the composite space, the same subgroups of voices emerged within a space.

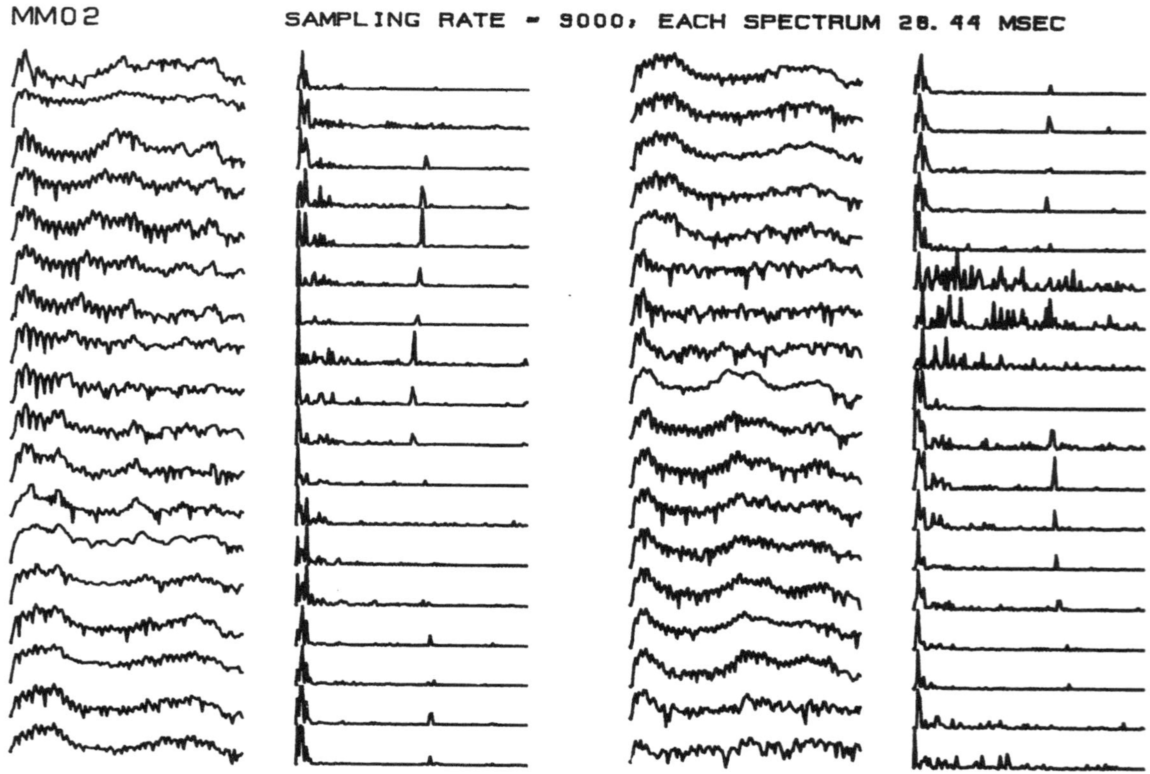

Figure 2. Power spectra (1st and 3d columns) and corresponding cepstra (2d and 4th columns). From top down each plot is based on consecutive 28.44 msec segments of speech. Voice MM02 is speaker AB.

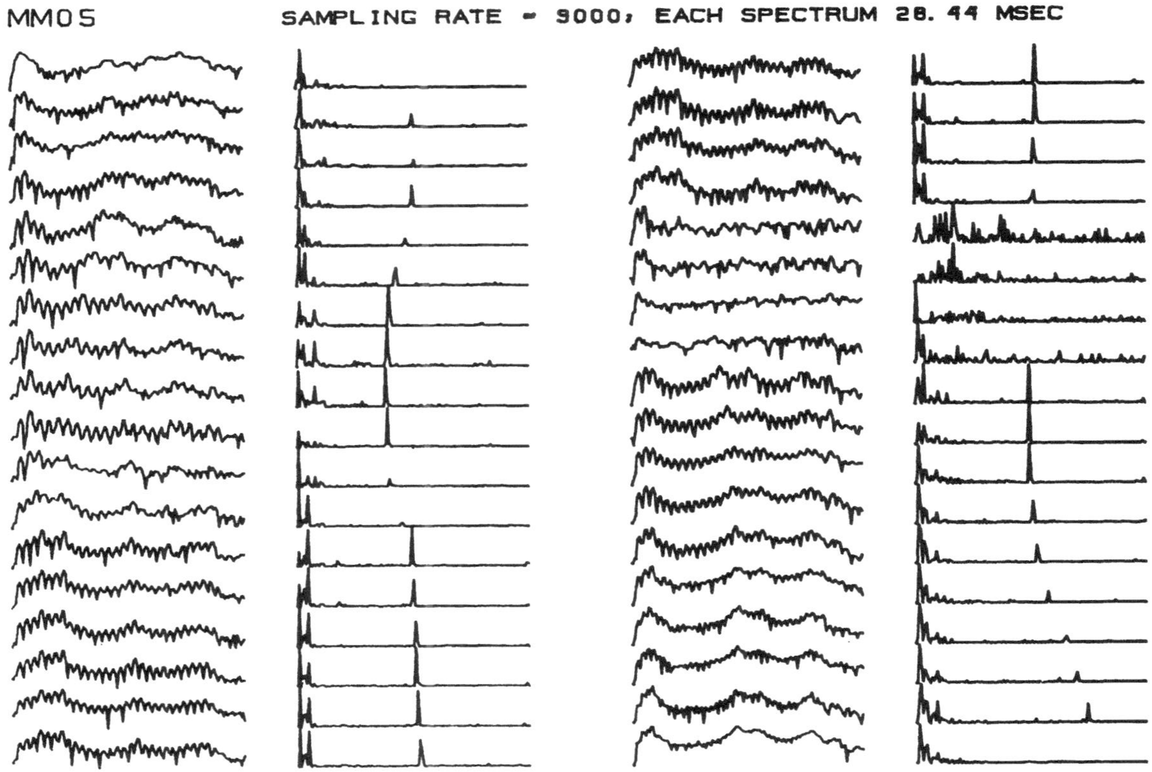

Figure 3. Power spectra (1st and 3d columns) and corresponding cepstra (2d and 4th columns). From top down each plot is based on consecutive 28.44 msec segments of speech. Voice MM05 is speaker DN.

An experiment on recognition memory for voices.

This experiment may be thought of as a more primitive auditory analog to experiments on recognition memory for pictures.

1. Procedures.

Twenty different males whose ages ranged from 22 to 55 each spoke two one-second passages into a B&K condensor microphone. The passages were digitized via a 10-bit analog-to-digital converter at 10,000 samples/sec and filed on the disk store of a computer. The passages were /a year ago/ and /you all be/. The second passage was chosen because the three vowels map out the extrema of the vowel triangle in F1 vs F2 formant space and effectively give the length of the vocal tract. Playback was also at 10,000 samples/sec, through a 12-bit D/A converter and a 4.5 kHz smoothing filter into binaural Koss ESP-9 electrostatic earphones.

In a practice session every listener heard each of 6 different voices randomly selected from the 20 /a year ago/ passages. Following a light signal after a 5-sec delay these 6 "old" voices were heard randomly interspersed with 6 "new" voices. The subject's task was to decide whether or not each of the voices belonged to the first 6 voices. He pressed a key labeled old or new. Instructions were to respond as quickly as possible but to be accurate. The time between voices was approximately 1300 msec and each voice lasted about 960 msec.

In the main experiment the passages /you all be/ were heard by a given subject in only one of the 12 conditions of (delay) x (number of voices). Nominal delays were 0, 15, or 45 sec. The actual delay was 1300 msec longer. The number of voices in the recognition set was 4, 6, 8, or 16, of which exactly half (2, 3, 4, or 8) were new voices.

Ten different listeners served in each of the 12 conditions, for a total of 120 listeners of whom about half were men and half women. After 48 listeners had been run (4 in each condition) the mean latencies for those in the 0-delay condition appeared to be greater than those of listeners in the 15-sec and 45-sec delay conditions. Therefore the last 6 of the 10 listeners in the 0-delay conditions were run at a 5-sec delay. A variance analysis showed no difference between latencies of listeners in the 0-sec and 5-sec delay groups, so their data were combined and the label 0-seconds kept.

2. Results.

(a). P(C).

Table I shows the mean proportion of correct responses P(C). The trends are easier to see in Figure 4. The best performance is for

Number of Voices in the Recognition Set	0	15	45	overall
4	.800	.800	.775	.792
6	.750	.733	.683	.722
8	.750	.591	.577	.639
16	.601	.608	.657	.622
mean	.725	.683	.673	.694

Table I. Shows mean proportion of correct responses, P(C), based on the weighted means of correct recognitions and correct rejections. Each cell mean is based on the means of 10 listeners and each marginal mean is based on 30 (rows) or 40 (columns) listeners.

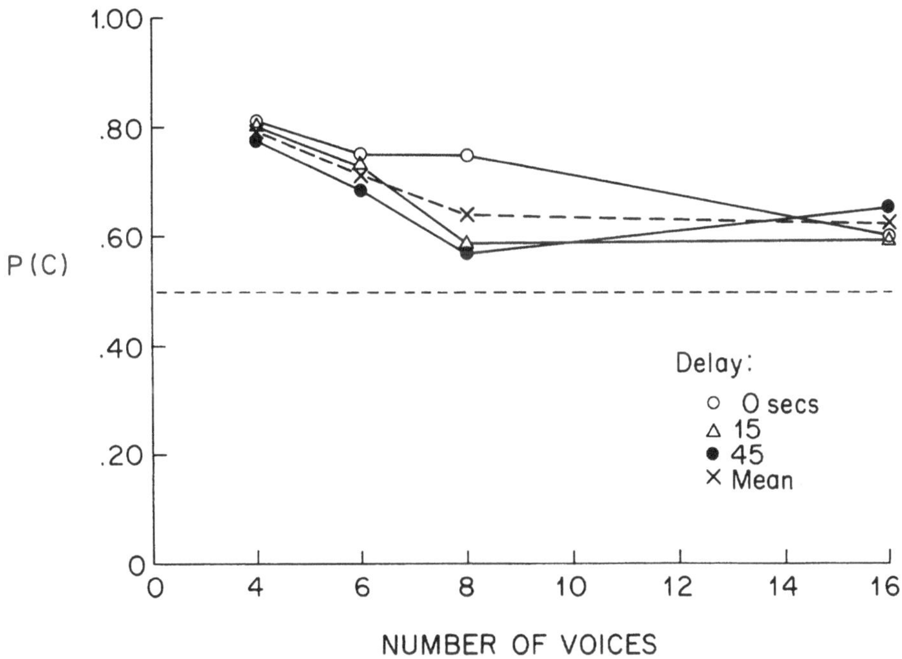

Figure 4. The probability of a correct response, P(C), is plotted on the ordinate as a function of the Number of Voices in the recognition set, plotted on the abscissa. The dashed horizontal line at P(C) = .5 represents chance performance. Three different delays, 0, 15, and 45 seconds, as well as the mean of these three delays, are shown.

0-delay which yields .80 for 4 voices and falls to .60 for 16 voices, a linear decline of less than 2 per cent per voice. For 6 voices all delays give about .80 but for 8 voices 15- and 45-second delay conditions P(C) is some .17 below that at 0-sec delay. For 16 voices P(C) is the same at all three delays.

This analysis is based on variance analyses[3] of the first runs of all 120 listeners. Only the main effect of number of voices was significant, $Prob[(F_{3,180} = 5.05)] < .003$. For 8 voices the mean P(C) at 0-sec delay was significantly (at $P < .05$) higher than mean P(C)'s at 15-sec and 45-sec delay. No other contrasts were tested. A similar analysis was made of the mean P(C) data of the 60 listeners who were tested on a total of 5 replications. The number of voices had a significant effect, $Prob[(F_{3,59} = 8.31)] < .0004$, but none of the effects of delays, replications, or their interactions were significant at the $P < .09$ level, including effects which could be fit by polynomials up to order 4.

(b). <u>Latency</u>.

The latencies of responses (Table II) differ significantly across the four categories [$F_{3,180} = 6.96$, Prob $(F) < .0002$] and differ between but not within the categories of hits (L_{11} and L_{22}) or false positives (L_{12} and L_{21}).

The effect of Delay on latencies was significant [$F_{2,60} = 6.43$, Prob $(F) < .003$]. Latencies at 0-sec and 15-sec delays were not significantly different from each other but both differed from latencies at 45-sec delay ($p < .05$). A cursory glance at the mean latencies averaged over response categories at each delay suggest an interaction between number of voices and delay or between response categories and delay. However, the analysis showed no interactions because $F_{6,60} < 1$ and $F_{6,180} < 1$. In fact none of the possible interactions were significant at the level $p < .08$. Neither did the number of voices in the response set have a significant effect on mean latencies [$F_{3,60} = 1.69$, Prob(F) = .180].

Briefly, in the case of latencies there is an influence of Delay but not of the Number of Voices, just the reverse of the results for P(C). Generally, latencies increase with Delay and without regard for the number of voices. The mean latencies for delays of 5 and 15 sec do not differ from each other but differ significantly ($p < .05$) from the mean latency at 45 sec.

Latencies of hits (L_{11} and L_{22}) are shorter than those of false positives (L_{12} and L_{21}), which suggests that the response is governed by its conditional membership in a category--not merely whether it is

		DELAY (seconds)		
		5	15	45
4 voices	L_{11}	1715	2097	2215
	L_{12}	2002	2291	2279
	L_{21}	2030	2060	2207
	L_{22}	1663	1737	1951
	mean, $\bar{L}..$	1853	2046	2163
6 voices	L_{11}	1876	1863	2184
	L_{12}	2087	1740	2568
	L_{21}	2116	2089	2841
	L_{22}	2144	1760	2306
	mean, $\bar{L}..$	2056	1863	2475
8 voices	L_{11}	1775	1578	1942
	L_{12}	1820	1790	2048
	L_{21}	2115	1567	2165
	L_{22}	1834	1618	2030
	mean, $\bar{L}..$	1886	1638	2046
16 voices	L_{11}	1678	1718	2049
	L_{12}	1697	1792	2241
	L_{21}	1752	1750	2357
	L_{22}	1830	1825	2132
	mean, $\bar{L}..$	1741	1771	2195
Grand mean (Delay)		1884	1830	2220

overall mean	L_{11}	L_{12}	L_{21}	L_{22}
	1885	2031	2085	1904

Table II. Each entry in a voice-by-delay subtable is the mean latency in msec and is based on the means over 5 trials of 6 listeners. A subscripted value L_{ij} is the latency of an i-response to a j-stimulus where, as before, 1 = old, and 2 = new. Thus in the first column at a 5-sec delay the 1st and 4th rows are latencies of hits (1715 and 1663 msec) and the 2nd and 3rd rows are the corresponding latencies of false positives (2002 and 2030 msec).

old or new.

(c). Sensitivity and criteria.

Table III shows hits and false alarms averaged over delays. The most interesting feature is that the hit rate and corresponding latency appears to be almost independent of the number of voices in the recognition set whereas the false alarm rate increases steadily from 4 to 16 voices while the corresponding latency decreases. In signal detection

Number of Voices in the Recognition Set	Pr(o/o)	Pr(o/n)	ℓ(o/o)	ℓ(o/n)
4	.75	.17	1975	2733
6	.76	.31	2037	2183
8	.71	.37	1861	1941
16	.69	.45	1864	1917
mean	.73	.33	1934	2194

Table III. Column two shows the proportion of correct recognitions, Pr(old/old) and false recognitions Pr(old/new) averaged across delays of 0, 15, and 45 seconds. Columns four and five show the corresponding mean response latencies, ℓ(.). Each entry in the table is based on data from 30 subjects.

terms the listener is maintaining a constant hit rate by relaxing his criterion as the signal-to-noise ratio worsens. The mean $\underline{d}'$ values associated with 4, 6, 8, and 16 voices are about 1.8, 1.2, 0.9 and 0.6, and are shown as solid symbols in Fig. 5.

Figure 5 also shows the 12 pairs of hit and false-alarm rates in ROC space for each of the voice-delay conditions. The tendency to keep Pr("old"/old) constant and allow Pr("old"/new)to vary more widely is easy to see. A crude table of $\underline{d}'$ values is shown here:

		DELAY IN SECONDS		
		0	15	45
NUMBER OF VOICES IN RECOGNITION SET	4	2.1	1.8	1.5
	6	1.5	1.3	1.0
	8	1.7	.7	.4
	16	.8	.6	.8

Discussion.

The pattern suggested, which is statistically significant only for the effect of voices and only at Prob $[F_{2,6} = 3.81] < .09$, is one in which sensitivity is greatest for the smallest number of voices and shortest delay, and is least for the largest number of voices and greatest delay. The data also suggest an interaction between Delay and Number of Voices. Delay has a large effect on $\underline{d}'$ in the case of 4 voices but almost none in the case of 16 voices. A possible interpretation is that with a small recognition set a pure auditory memory may be sufficient to keep sensitivity high for very short delay (0 sec) with some use of recoding or rehearsal. Some subjects have spontaneously told us what is confirmed in our own experience as listeners,

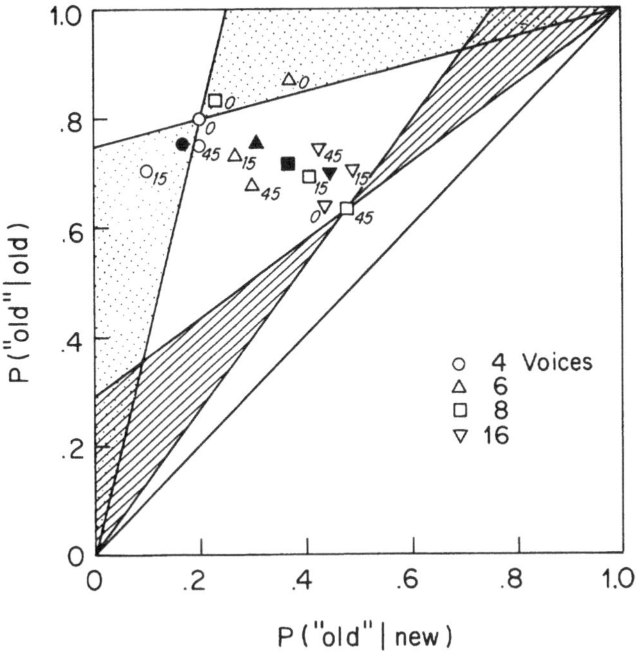

Figure 5. Shows the hit rate, P("old"/old) on the ordinate plotted as a function of the false-alarm rate, P("old"/new) on the abscissa. Each point is based on the mean of 4 subjects. The number next to the open points refers to delay in seconds. The solid points are mean values averaged over the three delays, for each of the four voice conditions. The regions bounded by lines emanating from (0,0) and (1,1) and passing through the 4 voice/0-sec delay and 8 voice/15 sec delay points define degrees of performance which are discussed in the text.

that they "rehearse" the voices. This is done not by active vocal imitation but by imaging, e.g., the articulatory positions, the tension in the diaphragm, the tension of the throat, and the required repositioning of chin and lips in order to achieve a low or high pitch (lengthening or shortening of the vocal tract). With a large recognition set this strategem is not useful and one resorts to remembering some more abstract features of the voice such as its pitch and pitch range or intonation pattern.

The point in Fig. 5 for 4 voices/0-sec delay represents the best performance and the point for 8 voices/45-sec delay represents the worst in terms of $\underline{d}'$ values. A simple graphical procedure may be derived for comparing points in ROC space with few assumptions about the underlying process (Norman, 1964). The ROC domain (the unit square) may be partitioned into four regions such that any arbitrary performance level represented by a single point may be related to that of any other point in the region. With respect to the point for 4 voices/0-sec delay a point lying above, below, or within the stippled region represents a superior, inferior or ambiguous performance, respectively. Similar performance regions are defined by the striped region fanning out from 8 voices/45-sec delay. Generally speaking the graphical analysis suggests that the larger the number of voices in the recognition set and the greater the delay the worse the performance as measured by *d*'. The *d*' values based on means (solid symbols in Figure 2) lie between the extremes (as required by the mean value theorem!) and show clearly the tendency for hit rate to remain constant at the expense of an increasing false-alarm rate.

These data are sparse but they do seem to validate the task as a useful one for revealing properties of recognition memory for an important class of sounds. The task tends to equalize subjects with respect to coding strategies. This is at least because in such brief and highly similar utterances there is no role for semantical and grammatical features and little role for gross intonation patterns. Identification is not required and the speakers are not known to the subjects, facts which probably reduce to a minimum elaborate or idiosyncratic memory schemes.

Our findings lead us to conclude that attributes of unfamiliar voices must survive for a considerable length of time. We cannot say that storage of purely sensory attributes is sufficient. We did show, in the triads experiment, that listeners may have relatively simple perceptual spaces of three dimensions in which voices could be disposed. Listeners may be able to make use of such spaces. It is an open question whether the attributes can be stored in such a space in immediate sensory form or in a mediated verbal code. An attractive solution is that the sensory output is a small set of sensory attributes which are stored in parallel with verbal attributes.

Acknowledgments.

We thank Dr. George Ionnides of the Institute of Geophysics and Planetary Physics for a helpful reading of the manuscript and for discussions of the subtleties of spectral analysis. We thank Cheryl Grossman and Glen Sampson, our laboratory assistants, for running subjects and help with programming, and we thank Esther Zack for preparing the manuscript. Financial assistance has been received from the National Institute of Mental Health, The Ford Motor Company and The Regents of The University of California.

References.

Akaike, H. Statistical predictor identification. Ann. Inst. Stat. Math., 1970, 22, 203-217.

Atal, B. S. Effectiveness of linear prediction characteristics of the speech wave for automatic speaker identification and verification. J. Acoust. Soc. Amer., 1974, 55, 1304-1312.

Atal, B. S. and Hanauer, S. L. Speech analysis and synthesis by linear prediction of the speech wave. J. Acoust. Soc. Amer., 1974, 55, 637-655.

Bower, G. H. and Holyoak, K. Encoding and recognition memory for naturalistic sounds. J. Exptl. Psychol., 1973, 101, 360-366.

Brown, Patrica E. Vocal communication and the development of hearing in the Pallid Bat (Antrozous pallidus). Ph.D. Dissertation, Department of Biology, University of California, Los Angeles, 1973.

Carterette, E. C. The constant-ratio rule applied to confusions among consonants. Paper presented to the Society for the Promulgation of d' at the Meeting of the Acoustical Society of America, Cleveland, October, 1959. Unpublished, 1960.

Carterette, E. C. and Barnebey, A. Recognition memory for voices. Paper read to the Psychonomic Society, November, 1974, Boston, Mass.

Clarke, F. R. and Becker, R. W. Comparison of techniques for discriminating among talkers. J. Speech and Hearing Res., 1969, 12, 747-761.

Cole, R. A. Different memory functions for consonants and vowels. Cognitive Psychology, 1973, 4, 39-54.

Crossman, E. R. F. W. Discussion on the paper by J. Brown in The Mechanization of Thought Processes. National Physical Laboratory Symposium. London: HMSO, 1958.

Deutsch, D. Experiments on short term memory for tonal pitch, and their implications for theories of nonverbal memory. In Deutsch, D., and Deutsch, J. A. (Eds.). Short term memory. New York: Academic Press, 1974.

Flanagan, J. L. Speech analysis, synthesis and perception. New York: Academic Press, 1965.

Ionnides, G., Carterette, E. C., and Barnebey, A. Estimates of the order of coefficients and variance of formant locations in linear predictive coding analyses of speech. In preparation.

Liberman, A. M., Cooper, F. S., Shankweiler, D. P., and Studdert-Kennedy, M. Perception of the speech code. Psychological Review, 1967, 74, 431-461.

Liberman, A. M., Mattingly, I. G., and Turvey, M. T. Language codes and memory codes. In Melton, A. M. and Martin, E. (Eds.), Coding processes in human memory. Washington, D. C.: V. H. Winston & Sons, 1972, 307-334.

Markel, J. D. Digital inverse filtering--a new tool for formant trajectory estimation. IEEE Trans. Audio Electroacoust., 1972, AU-20, 129-137

McGehee, F. The reliability of the identification of the human voice. Journal of General Psychology, 1937, 17, 249-271.

Melton, A. W., and Martin, E. (Eds.). *Coding Processes in Human Memory*. Washington, D. C.: V. H. Winston & Sons, 1972.

Morton, J. A functional model for memory. In D. A. Norman (Ed.) *Models for human memory*. New York: Academic Press, 1970.

Murdock, B. B., Jr., & Walker, D. K. Modality effects in free recall. *J. Verb. Learn. Verb. Behav.*, 1969, 8, 665-676.

Nickerson, R. S. Short-term memory for complex meaningful visual configurations: A demonstration of capacity. *Canadian Journal of Psychology*, 1965, 19, 155-160.

Norman, D. A. A comparison of data obtained with different false-alarm rates. *Psychological Review*, 1964, 71, 243-246.

Pollack, I. Message uncertainty and message reception. *J. Acoust. Soc. Amer.*, 1959, 32, 1500-1508.

Posner, M. I. Coordination of internal codes. Paper presented to the 8th Carnegie Symposium of Cognitive Psychology. Pittsburgh, Pa., May, 1972.

Shepard, R. N. Recognition memory for words, sentences and pictures. *J. Verb. Learn. Verb. Behav.*, 1967, 6, 156-163.

Shepard, R. N. Psychological representation of speech sounds. In David, E. E., Jr., and Denes, P. B. (Eds.) *Human Communication : A Unified View*. New York: McGraw-Hill, 1972. pp. 67-113.

Speech Communication Seminar Proceedings, Stockholm, August, 1974. 4 Volumes. New York: Wiley, 1975.

Stevens, K. N., Williams, C. E., Carbonell, J. R., and Woods, B. Speaker authentication and identification: A comparison of spectrographic and auditory presentations of speech material. *J. Acoust. Soc. Amer.*, 1968, 44, 1596-1607.

Torgerson, W. S. *Theory and Methods of Scaling*. New York : Wiley, 1958.

Voiers, W. D. Perceptual Bases of speaker identity. *J. Acoust. Soc. Amer.*, 1964, *36*, 1065-1073.

Wish, M. and Carroll, J. Douglas. Applications of individual differences scaling to studies of human perception and judgment. In Carterette, E. C. and Friedman, M. P. (Eds.), *Handbook of Perception*, Vol. II. New York: Academic Press, 1974, 449-491.

Wolf, J. J. Efficient parameters for speaker recognition. *J. Acoust. Soc. Amer.*, 1972, 51, 2044-2056.

Notes.

1. Strictly speaking this is the _method of triadic combinations_. From the _pair_ of judgments we deduce _three_ relations about V_1, V_2, and V_3. Thus from V_1SV_2 and V_2SV_3 we get (1) V_1 is more like V_2 than V_3, (2) V_2 is more like V_1 than V_3, and (3) V_3 is more like V_1 than V_2. The _complete method of triads_ allows judgments of the relations to be obtained directly but requires each possible triad to be presented thrice. Thus one asks for each V_i whether it is most like V_j or V_k $(i \neq j \neq k)$. Within either method the aim is the same, to obtain a scale of comparative distances from the observed differences of distances. See Torgerson (1958) for a complete discussion.

2. Another group of 15 listeners heard the same voices in all possible combinations of pairs and judged the similarity of each pair on a 9-point scale. These data were also subjected to INDSCAL analysis. The resulting spaces of 3 or 4 dimensions do not appear to be related in any obvious way to the spaces obtained by the method of triads.

3. Computations were made at the University of California, Los Angeles, Campus Computing Network. The BMDP2V programs analyze variance and covariance, including repeated measures, and do orthogonal decompositions as well. This last feature estimates the best polynomial fit to trends of main effects or their interactions. See _BMDP Biomedical Computer Programs_, Berkeley, Los Angeles and London: University of California Press, 1975.

IV. Short-Term Context Effects, Dichotic Listening, Speech Perception in the Very Young

PERCEPTION OF TIME-VARYING RESONANCE FREQUENCIES IN SPEECH AND NON-SPEECH STIMULI

Hiroya Fujisaki and Sotaro Sekimoto

Department of Electrical Engineering and
Research Institute of Logopedics and Phoniatrics
University of Tokyo
Bunkyo-ku, Tokyo, 113 Japan

Abstract

Time-varying formant frequencies constitute essential cues for the perception of phoneme sequences in connected speech. This paper investigates the conditions under which such formant transitions elicit perception of two contiguous phonemic segments in a syllable-like unit. Based on the analysis of formant transitions in natural speech, synthetic speech stimuli were generated with various values of magnitude, rate, and duration of formant transitions. Discrimination tests of dynamic and static stimuli indicated the existence of perceptual extrapolation of targets that underlie formant transitions. Results of discrimination tests on non-speech stimuli with similar formant transitions suggested that the extrapolation was to a large extent auditory, and thus was not specific to perception of speech stimuli. On the other hand, identification tests of dynamic and static speech stimuli clearly indicated the short-term context effect in perception of connected segments, which was quantified as the amount of temporary shift in the threshold for phonemic judgment due to perception of the immediately preceding segment. Vowels, semivowels and stop consonants were compared with respect to the magnitude of the context effect which they exert on the perception of the following segment.

1. Introduction

It is a well-known fact that the acoustic characteristics of a vowel in connected speech are often quite different from those of the same vowel uttered in isolation. Namely, formant frequencies of a vowel in a dynamic context display trajectories that almost invariably undershoot their target values. This phenomenon is called vowel reduction since in most cases the influence of the context reduces the distinctive characteristics of individual vowels, and often tends to neutralize them (Lindblom, 1963).

For the most part, the phenomenon is ascribable to the inertial movements of all the articulatory organs, constraining realization of the intended articulatory gestures only in an asymptotic manner. Thus, in connected speech where the commands for a sequence of phonemes follow each other, the intended articulatory gestures and the resulting acoustic characteristics of a phoneme are scarcely complete before they are followed by changes for the next phoneme, or they are truncated by the end of an utterance. The range of such inter-phoneme interference is not necessarily limited to the immediately adjacent phonemes, but can extend over several phonemes (Benguerel and Cowan, 1974). In particular, the influence of a vowel is commonly observed to extend beyond the adjacent consonant to the next vowel.

It is certainly not appropriate to ascribe the entire process of coarticulation to the inertial movements of articulatory organs, and there exist instances where an interpretation in terms of reorganization of articulatory commands seems to be more appropriate. It has been shown, however, that the process of coarticulation between successive phonemes in connected vowels can be approximated in the formant frequency domain in terms of a critically-damped second-order linear system, which represents the combined effects of neural, muscular and motional characteristics that intervene between articulatory commands and the observed acoustic characteristics (Fujisaki et al., 1973).

On the other hand, such variabilities of acoustic characteristics of vowels in dynamic contexts seem to be removed by the perceptual process. Namely, the perceptual invariance of a vowel is maintained by a mechanism which removes the coarticulatory effects, or, more properly, utilizes them to infer the underlying targets which are not attained. In other words, the articulatory undershoot seems to be compensated for by the perceptual process. Although the exact characteristics of such a mechanism have not been precisely formulated, there exist experimental results to indicate that the compensation is almost complete if a vowel is presented along with its immediately neighboring vowels (Kuwahara and Sakai, 1972).

Experimental studies of such perceptual processing of time-varying formant frequencies have been conducted by several investigators. Brady et al. (1961) used test stimuli generated by periodically exciting a tuned circuit whose resonance frequency was varied linearly upwards or downwards over a range of 500 Hz with the duration of 20 msec or 50 msec. They asked subjects to match the test stimuli with comparison stimuli having stationary but adjustable resonant frequencies, and found a consistent tendency for subjects to place the resonant frequency of the comparison stimulus near the terminal value of the time-varying resonance of the test stimulus. Lindblom and Studdert-Kennedy (1967) used synthetic speech stimuli with five formants of which the first three were controlled parabolically to simulate vowels in /w-w/ and /j-j/ contexts, and investigated the influence of such contexts on the phoneme boundary on the /u/-/i/ continuum defined by the formant frequency values at the point of closest approach to the vowel target. In most of the subjects, they found a strong context effect that shifted the phoneme boundary toward the locus of the adjacent semivowel, supporting the view that the undershoot in articulation is compensated for by perception. Hiki et al. (1968) analyzed the context effect in the identification of a vowel pair, and tried to formulate a dynamic model of vowel perception in terms of the conditional probability of identification. Arai et al. (1971) used the method of triad comparison to find a stationary vowel that was perceptually equivalent to the dynamic vowel contained in /bVb/, /bV/, or /Vb/ syllables, and formulated the process in a psychological space constructed by multi-dimensional scaling. More recently, Kanamori et al. (1974) used the method of adjustment to find similar correspondences between stationary and dynamic vowels.

Although these previous studies served to clarify the problem and supplied some useful evidence for further investigation, the process of dynamic perception of vowels is still far from being elucidated. Namely, the following questions are still open:

(1) Is the perceptual processing for time-varying resonances the same for both speech and non-speech stimuli?
(2) How does such processing relate to the so-called context effect that is known to exist between a pair of stationary sounds?

2. Problem

As mentioned above, most of the previous studies have investigated perception of the vowel in symmetrical CVC syllables, which involved two-way transitions. Since, however, the minimal context for a vowel is an open syllable where a one-way transition elicits perception of two contiguous phonemes, the investigation of vowel perception in such an environment seems to be essential. Furthermore, the shapes of formant transitions adopted in previous studies were either linear or parabolic and were not good approximations to those observed in real speech. Consequently, the present study concentrates on the perception of synthetic stimuli characterized by one-way formant transitions which are close approximations to those found in /V_1 V_2/ or /CV/ combinations.

The first problem is to investigate the perceptual processing of such time-varying characteristics. As indicated by some of the previous studies, subjects are reported to be capable of finding the subjective equality between a dynamic stimulus with time-varying resonances and a static stimulus with stationary resonances both in speech sounds and in non-speech sounds. If the same transition of resonance frequency, shared by speech and non-speech stimuli, is perceived as equivalent to the same stationary resonance frequency in a static speech stimulus and a static non-speech stimulus, the result may be regarded as evidence for the existence of a mechanism for compensating the articulatory undershoot at the level of auditory processing rather than phonetic processing. If, on the other hand, the equivalence between dynamic and static stimuli is found to be different for speech and for non-speech, then at least a part of the mechanism should be operating at the phonetic level. A series of discrimination tests have been conducted to elucidate this problem.

The second problem to be investigated is the so-called context effect in dynamic stimuli. The context effect is generally defined as the change in phonemic identification of a speech sound due to its context (Fry et al., 1962). It is, however, not clear whether the apparent shift in the phonemic identification is caused by a change in the auditory or phonetic representation while the criterion for phonemic judgment remains unchanged by the context, or it is caused by a change in the judgment criterion itself. Identification tests on various dynamic and static stimuli have been conducted to find a solution to this problem.

3. Method

It has been shown that close approximation to formant trajectories of natural utterances of $/V_1V_2/$ or $/CV_2/$ can be realized by the step response function of a critically-damped second-order linear system (Fujisaki et al., 1973). If the target values of the nth formant frequency of the first and the second vowels in a vowel pair are denoted respectively by F_{n1} and F_{n2}, and the transition to the second vowel is assumed to be initiated at an instant t_2, the approximated trajectory of the nth formant $F_n(t)$ is given by

$$F_n(t) = F_{n1} + (F_{n2} - F_{n1}) \left[1-\left\{1+\alpha_{n1}(t-t_2)\right\} \exp\left\{-\alpha(t-t_2)\right\}\right] u_{-1}(t-t_2),$$

where we assume that the articulatory gestures for the first vowel are initiated well in advance so that its target formant frequencies are realized at the onset of the utterance. This model has been demonstrated to be capable of approximating formant transitions not only in $/V_1V_2/$ combinations, but also in /CV/ combinations where the consonant phoneme is a semivowel or a voiced stop consonant (Fujisaki et al., 1975).

In order to investigate the perception of the first formant transition in $/V_1V_2/$, the vowel /u/ was selected as V_1 and the target value for V_2 was selected on the /u/-/a/ continuum, where only the first formant frequency F_1 was varied, keeping all the higher formant frequencies constant. Corresponding non-speech stimuli were generated by using only the first formant transitions, removing all other formant structures. Likewise, in investigating perception of the second formant transition, the target value for V_2 was selected on the /u/-/i/ continuum, where only the second formant frequency was varied.

On the basis of analysis of natural utterances, the duration of the stationary portion was always kept at 80 msec, while the transition was truncated at points where 95, 80, or 70% of the total excursion was attained, producing three different dynamic stimuli corresponding to durations of the transient portion of 118, 76, or 62 msec, respectively. An example of the first formant transition is shown in Fig. 1. Target formant frequencies of /u/, /a/, and /i/ adopted in the present study are given in Table 1, while the fundamental frequency is held constant at 130 Hz. The speech stimuli were generated by computer simulation of a terminal-analog synthesizer, while non-speech stimuli were generated by using only one formant characteristic of the synthesizer. The synthesized signal was normalized by the overall maximum amplitude, and was read out at 10 kHz with an accuracy of 10 bits, converted into analog waveform and recorded for off-line experiments. In the listening tests, the stimuli were presented to subjects in an anechoic room through a loudspeaker. The subjects were three male adults with normal hearing.

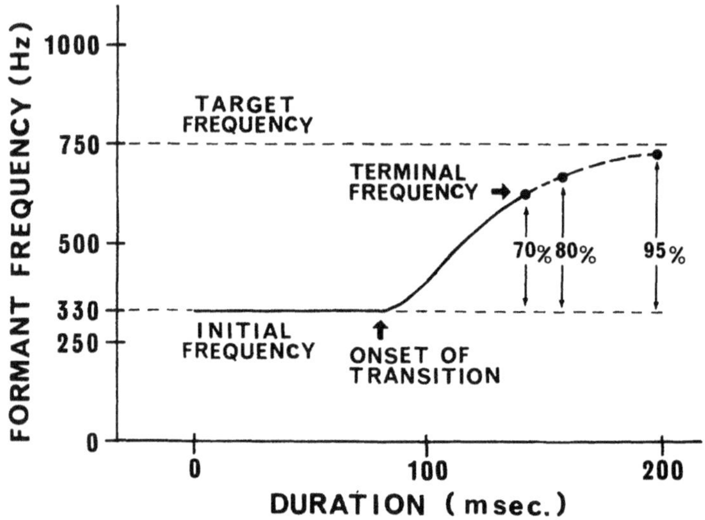

Fig. 1. Formant transition of dynamic speech and non-speech stimuli, approximated by step response of a critically-damped second-order linear system.

Table 1. Target formant frequencies of vowels adopted in synthetic speech stimuli.

	F_1	F_2	F_3	F_4	F_5	
/u/	330	1250	2750	3500	4500	Hz
/a/	750	1250	2750	3500	4500	Hz
/i/	330	2250	2750	3500	4500	Hz

4. Perceptual Equivalence of Dynamic and Static Stimuli

Preliminary listening tests of the dynamic stimuli indicated that it might actually be possible to find the point of subjective equality on the continuum of static stimuli generated by replacing the time-varying resonance by a stationary one. In the case of speech stimuli, paired comparison between the second vowel in a dynamic stimulus and a static stimulus can be made to determine which of the two sounds more like /a/. In the case of non-speech stimuli, on the other hand, comparative judgment of dynamic and static stimuli is also possible regarding which of the two sounds higher.

The experiments to find the perceptual equivalence between dynamic and static stimuli were based on the A-X procedure, in which a dynamic stimulus (A) is followed by one of seven static stimuli (X) whose first formant frequencies are selected in equal steps within the range from 80 Hz below to 40 Hz above the target of the formant transition of the dynamic stimulus. The duration of the static stimuli was 160 msec, and the interval between A and X was 1 sec, while a response interval of 4 sec was inserted between diads. A brief tone marked every 10 diads. The discrimination test was performed for each of the three dynamic speech stimuli and three dynamic non-speech stimuli. Each test consisted of a randomized sequence of 70 diads preceded and followed by 5 dummies, and each subject took at least four tests on each of the dynamic stimuli.

Because of individual differences in subjects' performance, the test results were analyzed individually. In the case of speech stimuli, the percentage of times that a static stimulus is judged to be more like /a/ than the dynamic stimulus, was obtained for each of the static stimuli, and was approximated by a cumulative Normal distribution, whose mean and standard deviation were calculated by the method of least-mean-squared error with Müller-Urban weighting. The mean corresponds to the point of subjective equality of the dynamic stimulus on the continuum of static stimuli, while the standard deviation serves as an index for the accuracy of discrimination. Results for non-speech stimuli were also analyzed by the same method.

Results of these discrimination tests are summarized in Fig. 2, where the abscissa shows the terminal frequency of the truncated formant transition of dynamic stimuli and the ordinate shows the formant frequency of static stimuli that are judged to be subjectively equal to dynamic stimuli. The solid and broken lines respectively indicate the averaged results of three subjects for speech and non-speech stimuli, and the vertical line indicates the standard deviation due to individual differences. These results indicate that formant transitions, when truncated at 70% or 80% of the total excursion, are perceptually extrapolated and are equated by a stationary formant frequency that is close, if not exactly equal, to the target of the transition. Furthermore, no significant differences are found between the results for speech and non-speech stimuli, suggesting that the processing for such extrapolation occurs at the auditory level. Results obtained for the first formant transition from /u/ to the phoneme boundary between /u/ and /a/, and for the second formant transition from /u/ to /i/, also indicate the same tendency as found in the /u/-/a/ transition.

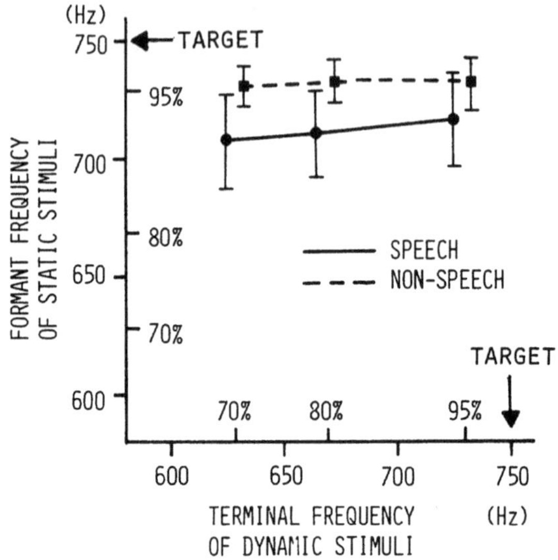

Fig. 2. Subjective equality between dynamic and static stimuli.

5. Identification of Dynamic Speech Stimuli.

While the context for a vowel in connected speech inevitably involves formant transitions, the effects of such a dynamic context on vowel identification have scarcely been studied systematically. The following experiments were conducted to obtain quantitative estimates of the effects of magnitude and rate of transition as well as of duration of the preceding segment upon the phonemic identification of a vowel.

The stimuli used in the identification test possessed formant transitions similar to the one shown in Fig. 1, but their target values were selected at points in 15 equal steps on the entire /u/-/a/ continuum.

The first experiment examined the effect of the magnitude and duration of the transition on the identification of a dynamic stimulus. Three sets of stimuli, corresponding to formant transitions terminated at 95, 80, and 70% of the total excursion, were generated and used for identification tests. These stimuli could be identified either as /uu/ or as /ua/, and subjects were asked to write down only the second vowel. The stimuli were randomized and presented every 4 sec, and a brief tone was inserted at every 10 stimuli. The test was repeated until each stimulus was identified at least 20 times by one subject, and the results were analyzed individually.

The percentage of times that the second vowel was identified as /a/, was calculated for each stimulus, and was approximated by a cumulative Normal distribution, whose mean and standard deviation were calculated by the same method used in processing the results of the discrimination tests. The mean and the standard deviation are considered as the phoneme boundary and an index for the accuracy of identification, respectively.

Phoneme boundaries thus obtained for dynamic stimuli are compared with those for static stimuli in Fig. 3. The abscissa indicates stimulus conditions, and the ordinate shows the phoneme boundary represented by the target frequency of formant transition. For the three conditions of formant transition, the phoneme boundaries are found to remain almost constant, suggesting that perceptual extrapolation of incomplete formant transition in each case produces almost the same auditory representation for the second vowel. Furthermore, comparison with phoneme boundaries for static vowel stimuli indicates that the effect of context is contrast in two subjects (HF and KI), while it

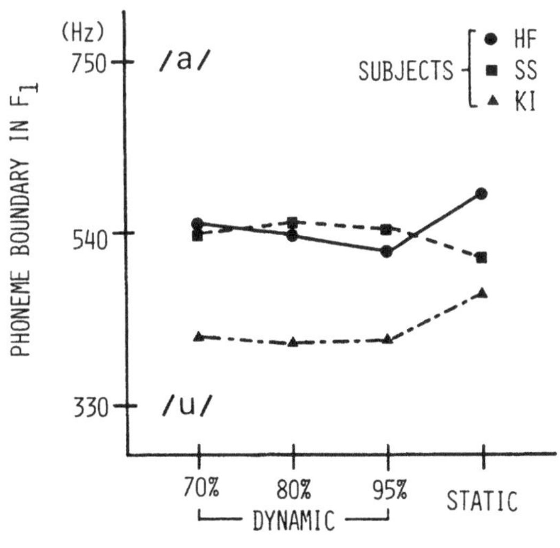

Fig. 3. Phoneme boundary between /u/ and /a/ in target value of F_1 for dynamic and static vowels.

is assimilation in one subject (SS). The results of all subjects, however, clearly indicate the short-term shift in categorical judgment for the second vowel which cannot be accounted for by the perceptual compensation found in discrimination tests.

The second experiment examined effects of various changes in the dynamic context on the identification of a vowel. The formant transition was terminated at the point of 95% of the total excursion, while the duration of the stationary initial segment was selected at 120, 80, 40, and 0 msec, resulting in four sets of stimuli for identification. In the first three conditions, the context is identified as /u-/, while it changes into /w-/ in the fourth condition. Furthermore, a fifth set of stimuli was generated by increasing the rate of formant transition by a factor of 5 while removing the initial segment, to produce a /b-/ context.

The results are shown in Fig. 4, again with those for the isolated static vowel. Since little variation was found among phoneme boundaries for the first four sets of stimuli, only results for the two extreme conditions, corresponding respectively to /u-/ and /w-/ contexts, are shown. The fact that such a large change in context duration does not appreciably affect the phoneme boundary suggests that the origin of the context effect observed in these cases may not be auditory, but may rather be phonemic. It is also interesting to note that the semivowel /w/ and the vowel /u/ are almost equivalent in their roles as context in the identification of a dynamic vowel, while a marked difference is found in the contextual role of the voiced stop consonant /b/ in at least two of the subjects.

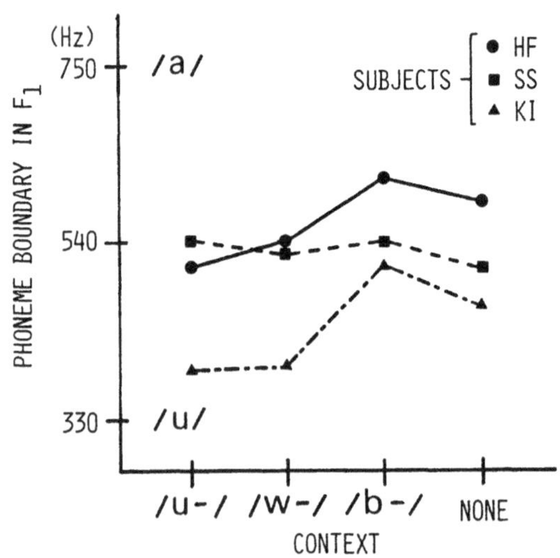

Fig. 4. Influence of various contexts on phoneme boundary between /u/ and /a/.

6. Summary and Discussion

The results of the present study have demonstrated that at least two distinct processes are involved in the perception of a vowel in a dynamic context. Namely, the extrapolation of incomplete formant transitions at the auditory level, and the short-term adaptive change of category boundaries at the level of phonemic decision.

Our findings on auditory perception of moving formants are generally in line with those of Brady et al. (1961) on non-speech stimuli, but our results made clear that subjects do not necessarily perceive the terminal frequency of a truncated transition, but are capable of approximately extrapolating the target frequency, not only in non-speech stimuli, but also in speech stimuli. The discrepancy may be partly due to the fact that in Brady's experiments, the transition of resonance frequency was not exactly truncated, but was controlled by a trapezoidal function, so that damped oscillation of the terminal frequency might have been continued at the end of the stimuli.

Our results on the effects of dynamic context in vowel identification are also consistent with those of a previous study by Lindblom and Studdert-Kennedy (1967) in that the effect of /w-/ context is contrast in most subjects. The two studies, however, differ with respect to the effect of an increased rate of formant transition; a twofold increase in transition rate did not change the phonemic context, but resulted in an increase of contrast effect in their study, whereas in the present study a fivefold increase in transition rate was found to change the phonemic context from /w-/ to /b-/, and resulted in a decrease of the contrast effect. Whether the discrepancy is due to the change in the phonemic context, or due to the lack of a monotonic relationship between the rate of transition and the magnitude of the context effect, awaits further investigation.

Though both of these processes seem to be responsible for the perceptual compensation of articulatory undershoot in connected speech, their relative roles are found to be subject to individual differences. While the characteristics of auditory extrapolation were quite similar for all three subjects, the effects of dynamic context on categorical judgment were rather dissimilar, being contrast in two subjects and assimilation in another. This fact is also in agreement with the results of Lindblom and Studdert-Kennedy (1967), which indicate that the effect of /w-w/ context was contrast for most of their subjects but assimilation for the rest of them. Whether the observed individual variability in context effect is also observed in perception of speech in a larger context, or it is an artefact produced by the experimental paradigm adopted, also calls for further study.

Finally, it is quite important to know to what extent these perceptual processes are mediated by articulation, although its role was only implicit in the above interpretation of the obtained results. For instance, if reference to articulation or its underlying motor commands is playing a major role in the perceptual compensation of time-varying acoustic characteristics, better performance will be expected along dimensions where articulatory undershoot is more likely to occur. Further work is in progress to elucidate this point.

References

Arai, S., Y. Kanamori, H. Kasuya and K. Kido, Model for the perception of vowels in connected speech, Report of Technical Committee on Electroacoustics, IECE Japan, No. EA 70-27 (1971).

Benguerel, A-P. and H.A. Cowan, Coarticulation of upper lip protrusion in French, Phonetica, 30, 41-55 (1974).

Brady, P.T., A.S. House and K.N. Stevens, Perception of sounds characterized by a rapidly changing resonant frequency, J. Acoust. Soc. Am., 33, 1357-1362 (1961).

Fry, D.B., A.S. Abramson, P.D. Eimas and A.M. Liberman, The identification and discrimination of synthetic vowels, Language and Speech, 5, 171-188 (1962).

Fujisaki, H., M. Yoshida, Y. Sato and Y. Tanabe, Automatic recognition of connected vowels using a functional model of the coarticulatory process, J. Acoust. Soc. Japan, 29, 636-638 (1973).

Fujisaki, H., y. Sato, Y. Noguchi and T. Yamakura, Automatic recognition of semivowels in word context, J. Acoust. Soc. Japan, 31 (1975).

Hiki, S., H. Sato, T. Igarashi and J. Oizumi, Dynamic model of vowel perception, Reports of 6th ICA, Tokyo (1968).

Kanamori, Y., Y. Shigeno and K. Kido, Dynamic perception of vowels, Reports of Autumn Meeting, Acoust. Soc. Japan (Oct. 1974).

Kuwahara, H. and H. Sakai, Perception of vowels and C-V syllables segmented from connected speech, J. Acoust. Soc. Japan, 28, 225-234, (1972).

Kuwahara, H. and H. Sakai, An experiment on the phoneme boundary locations in the dynamic perception of synthetic vowels, J. Acoust. Soc. Japan, 31, 18-23 (1975).

Lindblom, B., Spectrographic study of vowel reduction, J. Acoust. Soc. Am., 35, 1773-1781 (1963).

Lindblom, B. and M. Studdert-Kennedy, On the role of formant transition in vowel recognition, J. Acoust. Soc. Am, 42, 830-843 (1967).

Thompson, C.L. and H. Hollien, Some contextual effects on the perception of synthetic vowels, Language and Speech 13, 1-13 (1970).

DISCUSSION of "**Perception of time varying characteristics of speech and non-speech sounds**"
FUJISAKI

KLATT:
The auditory extrapolation that you observe was not seen by others who have studied the perception of brief stimuli that contain a rapid spectrum change (e.g. Nabelek, Nabelek and Hirsh). They find that a subject will match either the average pitch or the final pitch of the brief glissando depending on whether the spectrum of the stimulus has one or more peaks, but there are no conditions of stimulus duration or rate of frequency change under which auditory extrapolation is consistently observed. The difference between their experimental paradigm and yours seems to be that their subjects are told to treat the stimulus as a unitary percept while your subjects assume that the stimulus consists of two sequential events, a steady state harmonic complex followed by a transition to a second (truncated) steady state. Under these conditions, extrapolation is probably not a strictly auditory (i.e. peripheral) process. Instead it is likely governed by central intellectual analyses.

HAGGARD:
Heinz et al. (IEEE, Au-16, 1968, 107-111) failed to find perceptual extrapolation for very simple glide stimuli. I wonder if your result is to be explained by some variability in whether the non-speech controls are in fact sometimes processed as speech. In particular a single formant approximation to /u/ and /a/ is quite good. This would perhaps overestimate the purely auditory component.

FUJISAKI:
The work of Heinz, Lindblom and Lindqvist you mentioned was not on perception of time-varying formant frequencies, but on perception of pitch of a tone signal whose frequency was changing linearly with time, so that it is not possible to compare their results with those of the present study, although their work is not irrelevant to ours.
The tasks of subjects are also different. In their study, a subject was asked to adjust the frequency of a steady pure-tone to match the

overall pitch of the transient signal, whereas in our study subjects compare the quality of the second sound of a time-varying signal with that of a stationary stimulus that follows the transient signal.

It is true that the non-speech stimuli adopted in my study retains some speechlike quality, since they are parts to the speech stimuli (consisting of single formant transition). This cannot be avoided if we want to generate stimuli that share some essential characteristics with the speech stimuli. For the present investigation, therefore we consider a stimulus as non-speech as long as subjects cannot consistently identify it as sequence of phonemes.

COARTICULATION OF LIP ROUNDING AND ITS PERCEPTION

André-Pierre Benguerel and Sharon Adelman
Division of Audiology and Speech Sciences
University of British Columbia
Vancouver, B.C., Canada V6T 1W5

Abstract

This study investigates the perceivability of coarticulated lip rounding in French. Nine utterances containing the clusters /kstr/, /rstr/, and /rskr/ followed by one of the vowels /i/, /y/, or /u/ in all possible combinations, were truncated at 4 different points before the vowel. Subjects were asked to predict the identity of the missing vowel.

Results show that when segments up to and including at least half of the final consonant of the cluster are present, subjects correctly identify the missing vowel well above chance level. Several individuals were able to identify the vowel even when presented with shorter versions of the utterances. No significant difference in performance was found between French and English subjects, nor between subjects with and without phonetic training. Perceivability of individual features of the missing vowel is discussed.

It is concluded that coarticulatory effects due to lip rounding (as well as to horizontal place of articulation) provide perceivable information at a level significantly above chance, and that this information may be used by the perceptual mechanism as an aid in speech sound identification.

Introduction

In a previous study of coarticulation of upper lip protrusion in French (Benguerel and Cowan, 1974), it has been found that the protrusion movement due to the rounding of the second vowel of a group - VCC... CV - starts as early as the first consonant of a cluster of 4-6 consonants. In some cases, it even starts during the vowel preceding the cluster. The acoustic signal will reflect these coarticulatory effects. Can the listener make use of coarticulatory information contained in the speech signal?

Moll and Daniloff (1971) have shown that velopharyngeal opening in CVN and CVVN sequences (where N is a nasal consonant) almost always begins during the CV transition. To test the perceivability of this coarticulated nasality, Ali <u>et</u> <u>al</u>. (1971) spliced out the final consonant and its VC transition from English CVC and CVVC utterances, in which the final consonant was sometimes a nasal and sometimes not. Subjects were asked to identify the missing consonant as nasal or non-nasal. Results showed that nasal stimuli were correctly identified significantly above chance level.

Lehiste and Shockey (1972) tested the perceivability of vowels removed from VCV utterance (where C is a stop consonant). Öhman (1966) had previously shown that VC and CV transitions in such an utterance are influenced by the transconsonantal vowel. For the perceptual test, VCV utterances were cut in two during the consonant closure. Subjects were asked to identify the missing initial or final vowel. Though Lehiste and Shockey noted the same coarticulation effects spectrographically for their utterances as did Öhman, they found that these contextual effects were not sufficient for identification of the deleted segment. Nor was enough information present in the spliced utterances to identify a feature of the deleted phone, such as high/low or front/back; incorrect responses did not tend to share a feature with the correct response.

The present study looks at utterances in which cues for a certain vowel are known to exist several phones preceding that vowel. Subjects were asked to predict the upcoming vowel after hearing only part of the utterance.

Procedures

Speakers

Three male native speakers of French recorded the utterances used for this experiment. Speaker 1 was born in Lausanne, Switzerland.

Speaker 2 was born in Grenoble, France. Speaker 3 was born in Albi, France. All three had been in North America for several years.

Corpus

Each speaker recorded fifteen utterances, each utterance at least twice. Each utterance contained one of the consonant sequences /kstr/, /rstr/, /rskr/, followed by one of the three vowels /i/, /y/, /u/ in all possible combinations. Benguerel and Cowan (1974) have found that for utterances of this type, upper lip protrusion most often begins with the approach to the first consonant in the cluster, if the cluster is followed by the rounded vowel /y/ or /u/.

One set of 9 utterances including one instance of each of the three clusters followed by each of the three vowels, was chosen from each speaker, on the subjective criteria of voice clarity, absence of noise, similarity of intonation patterns of utterances containing the cluster, and presence of all phonemes in the cluster. Spectrograms and mingograms were made of all utterances, for reference in the editing process.

Editing of Utterances

The utterances were edited on a PDP-12 computer with a set of programs written by Lloyd Rice at U.C.L.A. The speech signal was filtered at 6 kHz and digitized with a 10-bit analog-to-digital converter at a 12 kHz sampling frequency.

Each utterance was truncated at four different points, producing the four groups of stimuli shown in Table I. The shortest stimuli (Group IV) were truncated after the release of the third consonant (either a /t/ or a /k/) of the cluster.

With 3 speakers, 9 utterances per speaker and 4 truncation points per utterance, a total of 108 test items was produced.

The order of taping items was randomized with respect to speaker, cluster, and vowel. The tape contained three practice items, followed by the 108 test items and 2 buffer items.

TABLE I

Groups of edited stimuli with consonant clusters /kstr/, /rstr/, and /rskr/. Each sample as described above has 3 versions, one of which originally had the following vowel /i/, one /y/, and one /u/,

GROUP I	Truncation immediately after the final consonant of the cluster	/ladɛkstr/ /lavɛrstr/ /lamɔrskr/
GROUP II	Truncation in the middle of the final consonant	/ladɛkstɍ/ /lavɛrstɍ/ /lamɔrskɍ/
GROUP III	Truncation immediately after aspiration of the third consonant	/ladɛkst^h/ /lavɛrst^h/ /lamɔrsk^h/
GROUP IV	Truncation immediately after release of the third consonant, before aspiration	/ladɛkst/ /lavɛrst/ /lamɔrsk/

Utterances for cluster /kstr/

la dextre imitable

la dextre universelle

la dextre outragée

Utterances for cluster /rstr/

l'averse tribale

l'averse truquée

l'averse troublée

Utterances for cluster /rskr/

l'amorce criptique

l'amorce cruciforme

l'amorce croupissante

<u>Subjects</u>

A group of 10 native French speaking adults and a group of 10 native English speaking adults participated in the experiment.

Four females and six males made up the French speaking group. They had been in North America for 3 to 14 years. Seven subjects were from France, two from Switzerland and one from Haiti.

Six females and four males made up the English speaking group. All subjects had had approximately 4 years of high school French in Canada.

Test Procedure

The subjects were seated, one at a time, in a soundproof room with the experimenter. The test tape was presented over headphones at a level of 60-70 dB SPL.

Subjects were instructed in writing to listen to each utterance and to mark the missing vowel on an answer sheet. English subjects were first asked whether they were familiar with the vowels as represented in French orthography. The experimenter then pronounced each vowel in isolation for the English subjects.

Included in the instructions were the nine whole utterances from which the edited versions had been taken, (cf. Table I). Subjects were told that the vowels were represented in approximately equal proportion on the test (that is, that each vowel appeared approximately 1/3 of the time). Guessing was strongly encouraged.

Each English subject was asked whether he preferred to hear the numbers in French or English, and each chose French.

Measures of Perceivability

Two measures were used for the analysis of the data: relative transmission (T_{rel}) and correct score (S).

The relative transmission is a measure of covariance between input (the stimulus), and output (the subject's response) (Miller and Nicely, 1955). This measure was used to describe the amount of transmissible information available in the truncated utterances, and is given by

$$T_{rel}(x;y) = \sum_{i,j} p_{ij} \log_2 \frac{p_i p_j}{p_{ij}} \Bigg/ -\sum p_i \log_2 p_i$$

where the input variable is x, with any one input x_i having the probability p_i, and the output variable is y, with any one output y_j having the probability p_j. The symbol p_{ij} represents the probability that a particular input x_i will elicit the particular response y_j. The more consistently a response can be predicted from the stimulus, that is, the better the transmission of information, then the closer T_{rel} is to a value of 1. If the transmission of information is poor, then stimulus and response are unrelated, and T_{rel} has a value near 0.

The distribution of relative transmission values and the distribution of correct scores for a series of computer-generated random responses were calculated.

These two distributions were based on tests of 27 items for comparison with each group of edited utterances, as there were 27 items of Group I utterances, 27 items of Group II utterances, and so on.

Results

Correlation Between Relative Transmission (T_{rel}) and Score (S)

Because the relative transmission is a measure of information transmission, and not necessarily correct information transmission, a high T_{rel} value does not always reflect a high score. For example, if a subject consistently responds with the vowel /i/ when the stimulus is /y/, he is receiving predictable information from the stimulus. Although he misinterprets this information consistently, he obtains a high value of T_{rel} (assuming other responses are also highly predictable from their stimuli). If a subject responds in a random manner, using no information from the stimulus, S will be at chance level, for example in the test described here, distributed around 9 (out of a maximum of 27), whereas T_{rel} will be relatively low. The better a subject performs on the test, the better one expects T_{rel} and S to correlate.

Identification of the Missing Vowel

Responses were tabulated in 3 x 3 confusion matrices, one matrix per group of items and per subject. There were therefore 4 matrices per subject, each with 27 items.

T_{rel} and S were calculated for each matrix. Figure 1 shows mean

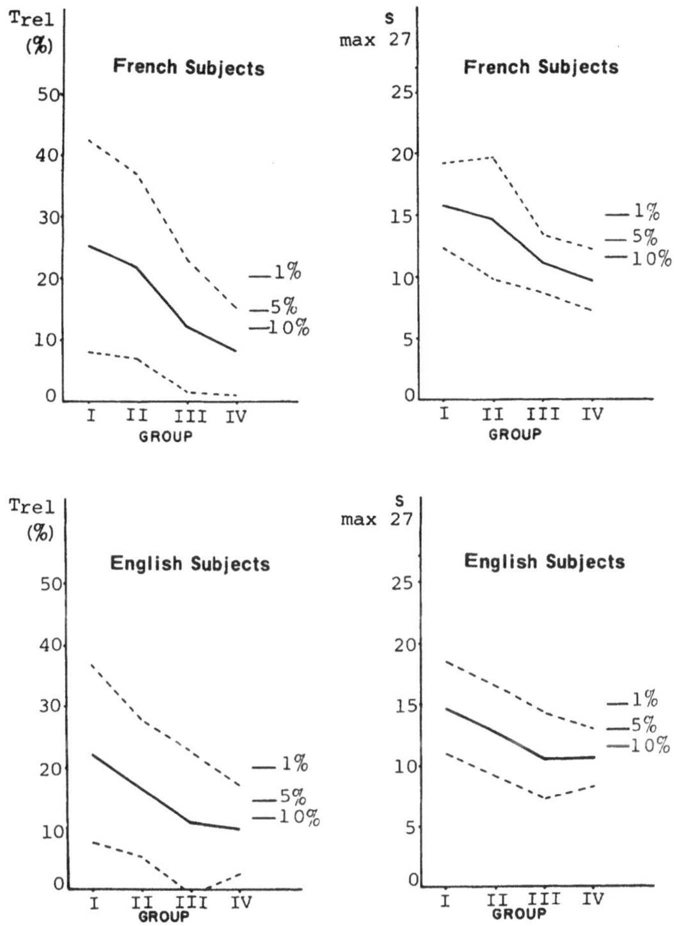

Figure 1. Mean values of T_{rel} and S, plus or minus one standard deviation, for each group of items.

T_{rel} and S values for each of the four groups of items, displayed separately for French and English speakers. T_{rel} and S values one standard deviation about the mean are also shown. Levels that one subject would obtain by chance 1%, 5%, and 10% of the time (obtained from the computed distributions) are also shown in Figure 1. As can be seen in that figure, all subjects showed a downward trend in both T_{rel} and S, from Group I to Group IV. That is the farther from the vowel the utterance was truncated, the less able subjects were to correctly identify the vowel. An analysis of variance showed a significant treatment effect among the groups of items for both T_{rel} and S for both French and English subjects. This effect was significant at 0.001 level for French subjects and at the 0.05 level for the English subjects.

Results show a great deal of individual variation. Figure 2 shows

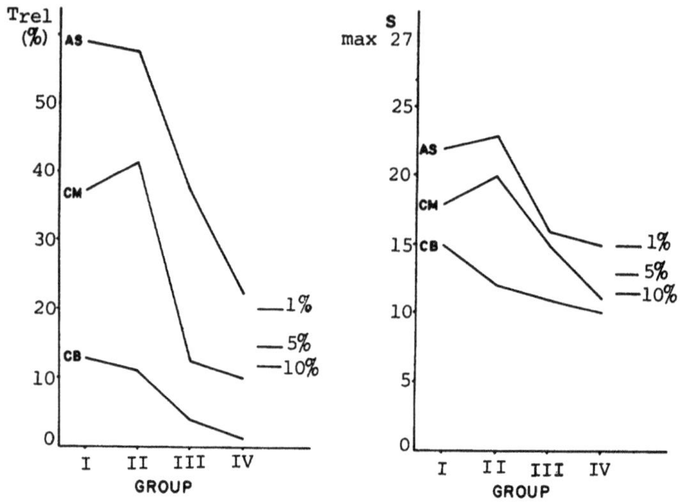

Figure 2. T_{rel} and S values for each group of items for three selected subjects.

T_{rel} and S values for each group of utterances, for 3 different subjects. Levels that one subject would obtain by chance 1%, 5%, and 10% of the time are indicated. Subject AS scored consistently higher than any other subject. She was a native French speaker who was a teacher of French, but had had no formal phonetic training. Subject CM was a female native English speaker who had had some phonetic training. Subject CB was a male native speaker of French and one of the speakers on the test; he had also had some phonetic training. Because of her high performance relative to other subjects, subject AS was retested. On the second run of the test she maintained her high levels, scoring slightly higher than she had on the first run.

Results as shown in Figures 1 and 2 indicate that, for several of the item groups, subjects were able to identify the missing vowel above chance levels. For example, on the average, English subjects obtained T_{rel} and/or S values higher than those one subject would obtain by chance 5% of the time, for items of Group I, and French subjects, on the average, obtained similarly high levels for items of Group I and II. Several individuals of both languages performed well above the 5% chance levels for Groups I and II, and 5 individuals did so for Group III. In general though, Group III and IV performances were at the level which one subject would obtain by chance 80% of the time. It is interesting to note that individual variations were so great that some subjects were able to identify the vowel for Group III and IV items better than others were able to identify vowels for Groups I and II.

Discussion

French utterances containing the sequence $-C_1C_2C_3C_4V-$ were truncated at four points before the vowel, as shown in Table I. From the above results, in particular Figures 1 and 2, one sees that the segments up to and including C_4 contain information about the following vowel that is utilizable in the perception process. This information is not restricted to the C_4V transition, since the vowel can be correctly predicted when segments up to only the middle of C_4 are heard. Though on the articulatory level the influence of the vowel is apparent as far back as the first consonant of the cluster or earlier, the information present in C_3, C_2 and C_1 is not by itself sufficient to identify the upcoming vowel. It is not obvious whether the perceivable information is restricted to C_4 or whether it is the cumulative information present in the whole cluster up to and including at least half of C_4 which is used in perception. However, because coarticulation due to the vowel may begin by the first consonant of the cluster, it seems likely that several segments, and not just C_4, contain information which, when it is all available to the listener, can be used in the perception process, but when only early segments are available, is not perceptually sufficient. However, there are great individual differences, and one subject at least was able to consistently predict the upcoming vowel above the level she would obtain by chance 5% of the time even for items of Group IV.

The above results show that the perceivability of coarticulated information does not seem to be related to the listener's native language, even though one of the vowels employed in the study (/y/) does not have any close equivalent in English. Such findings are in agreement with the findings of Stevens _et al_. (1969), that the listener's linguistic background, be it English or Swedish, did not affect his ability to make subphonemic distinctions, even among vowels that were not present in his language. Though the pattern of coarticulation may be language-dependent, its perception does not seem to be.

Phonetic training did not affect the test results. The subjects with phonetic background in general were not better able to identify the missing vowel than the phonetically naive subjects. This suggests that the ability to make use of coarticulatory information does not depend on specific training. However, large individual variations in test performance indicate that not all listeners make the subphonemic distinctions necessary to predict the missing vowel.

Without exception, all subjects reported that they found the test difficult. Most felt certain they had performed badly (though they may or may not have), and that they had guessed a large proportion of the time. The fact that subjects thought the test was a difficult one and that they had "only guessed" does not necessarily mean that the perceptual mechanism, to a large extent working subconsciously, could not handle the task. However, only one of the twenty subjects consistently gave an indication of his confidence in each of his responses, as was suggested in the instructions. This seems to indicate that the task of identifying the vowel was sufficiently difficult to impede subjects from making the further decision of how confident they were in each response. This may mean, that though the vowel was perceivable to an extent, use of coarticulatory information is not a process used in everyday speech perception.

Most subjects could not describe the strategy they had used in responding. However, several subjects were seen to repeat the test item subvocally two or three times before choosing their response.

The fact that subjects can use subphonemic coarticulatory information to identify an upcoming vowel does not mean that the perception process necessarily incorporates this ability. But it is possible that in unfavorable conditions, such as a noisy environment or a large amount of information having to be processed quickly, coarticulatory effects are used as cues by the perceptual mechanism. Use of such redundant cues would facilitate correct identification of any one speech sound. It is clear that some coarticulatory effects provide significantly perceivable information to the listener.

References

Ali, L., Gallagher, T., Goldstein, J., and Daniloff, R. (1971). "Perception of Coarticulated Nasality," J. Acoust. Soc. Amer. 49: 538-540.

Benguerel, A-P., and Cowan, H. A. (1974). "Coarticulation of Upper Lip Protrusion in French," Phonetica 30: 41-55.

Lehiste, I., and Shockey, L. (1972). "On the Perception of Coarticulation Effects in English VCV Syllables," Working Papers in Linguistics No. 12, The Ohio State University, 78-86.

Miller, G. A., and Nicely, P. E. (1955). "An Analysis of Perceptual Confusions Among Some English Consonants," J. Acoust. Soc. Amer. 27: 338-352.

Moll, K. L., and Daniloff, R. (1971). "Investigation of the Timing of Velar Movements During Speech," J. Acoust. Soc. Amer. 50: 678-684.

Öhman, S. E. G. (1966). Coarticulation in VCV Utterances: Spectrographic Measurements," J. Acoust. Soc. Amer. 39: 151-168.

Stevens, K. N., Liberman, A. M., Studdert-Kennedy, M., and Öhman, S. E. G. (1969). "Cross Language Study of Vowel Perception," Language and Speech 12: 1-23.

DISCUSSION of "Coarticulation of lip rounding and its perception"
BENGUEREL and ADELMAN

KLATT:
One might speculate on the origin of the negative result (no ability to guess next vowel in truncated vowel-plosive-vowel sequences) of LEHISTE when testing perceptual importance of ÖHMAN's observation of vowel-vowel coarticulation across a consonant and your more-positive result. The acoustic cues available in the VCV case occur primarily during the rapid formant motions associated with VC closure. In your case the acoustic cues that allow the listener to predict the vowel quality following the /r/ include more-static spectral cues including shifts in the spectral quality of the /s/ and in the formant pattern for the /r/. The target formants for the /r/ may shift as a function of both anticipatory lip rounding and some anticipatory tongue-body coarticulation. Perhaps these cues are available in precategorical acoustic storage for decoding purposes whereas the rapid spectrum change cues needed by the LEHISTE listeners may not be available in PAS.

HAGGARD:
One possible index of the perceptual importance of these effects would be in a reaction-time experiment. You could ask subjects to monitor the vowel phoneme /y/ in Dr. BENGUEREL's examples and see if speakers (or stimuli synthesised on an articulatory basis) with a large degree of anticipatory rounding might give a shorter reaction time.

WINGFIELD:
You have previously shown reliable articulation differences in as early as C_1, dependent on the character of the final vowel. Your present result suggests an influence of C_4 on predicting the final vowel, but not C_1, C_2 or C_3. I am curious to know how much variability in C_1 is shown by different speakers, and whether speaker variability might determine the degree to which C_1 differences influence prediction of the terminal vowel.

BENGUEREL:
The onset of upper lip protrusion, in $V_1C_1C_2C_3C_4$ V-rounded sequences, varied from the end of V_1 to the middle of C_2, and our data included

six speakers. For a given speaker on the other hand, repeatability was very high, and the range of variation in this case, was considerably smaller.

DICHOTIC SIGNS OF SPEECH MODE LISTENING

by

Charles I. Berlin and John K. Cullen, Jr.

Louisiana State University Medical Center

Department of Otorhinolaryngology

Kresge Hearing Research Laboratory of the South

1100 Florida Ave. - Bldg. 164

New Orleans, Louisiana 70119 (USA)

INTRODUCTION

In this presentation we should like to make two major points:

1. The use of dichotic listening may help to isolate the critical acoustic elements used in identification of some syllables.
2. Dichotic studies in temporal lobectomees, hemispherectomees, and a patient with a single-functioning medial geniculate body may help to identify some of the physiologic mechanisms by which that identification takes place.

In order to support these points, some historical background is necessary. In previous work (Cullen, Berlin, Hughes, Thompson, and Samson, 1975) we showed that as one listens dichotically to CV stops, the performance of both ears is well below 100%. If we degraded only one of the channels in intensity, bandwidth, or signal-to-noise ratio, the accuracy in the unaffected channel approached 100% (see Figure 1). This suggested to us that the *release* from dichotic interference in one channel was related to the reduction of critical information in the other channel; thus, if there were *no* critical elements in one channel, the second channel would be identified nearly perfectly. Conversely, if there *were* important elements in the other channel, the second channel would be identified at levels *below* 100% as a function of the information weighting of these elements.

It is this "interference" aspect of dichotic listening which we wish to emphasize, rather than the often-reported ear advantages associated with dichotic presentation of speech and non-speech signals. Let us clarify our terminology for the dichotic interference experiments we will report:

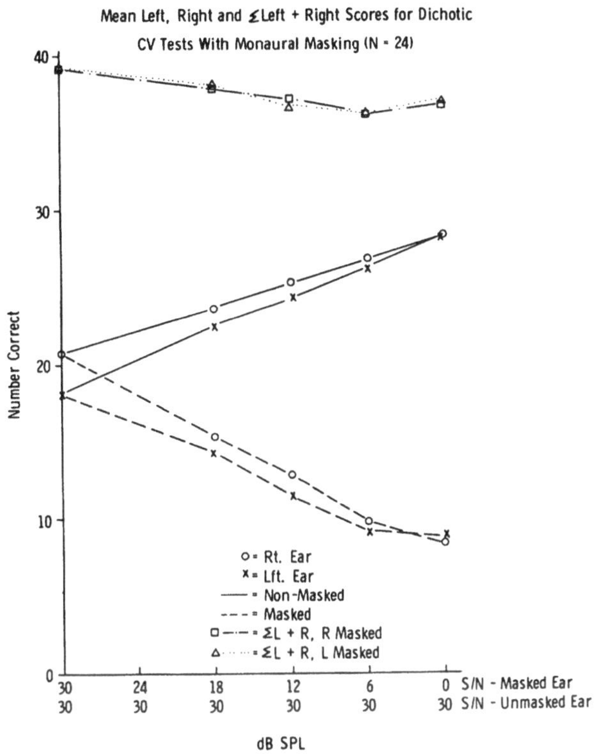

Figure 1. A sample of the performance of right and left ear when one channel is degraded. In this case, the degradation took place by adding noise in one channel. Similar results have been obtained by filtering one channel, or reducing the intensity of one channel. Top trace reflects the sum of the performance of both ears.

TARGETS

For the experiments reported here, natural or synthetic CV syllables (the six English stops, pa, ba, ta, da, ka, and ga) served as targets. We specified the onset and offset points of the signals and matched amplitudes on the two channels, using digitized samples and computer control. Subjects were asked to listen for the targets only, and report back which of the CV's they perceived.

THE CHALLENGES

Figure 2 shows spectral analyses of the challenges used in the experiments. These challenges are labeled bleats, CV's, transitionless CV's, sweeps (upward and downward, or horizontal), and CV masks. The bleats are second formant transitions taken from CV's synthesized by

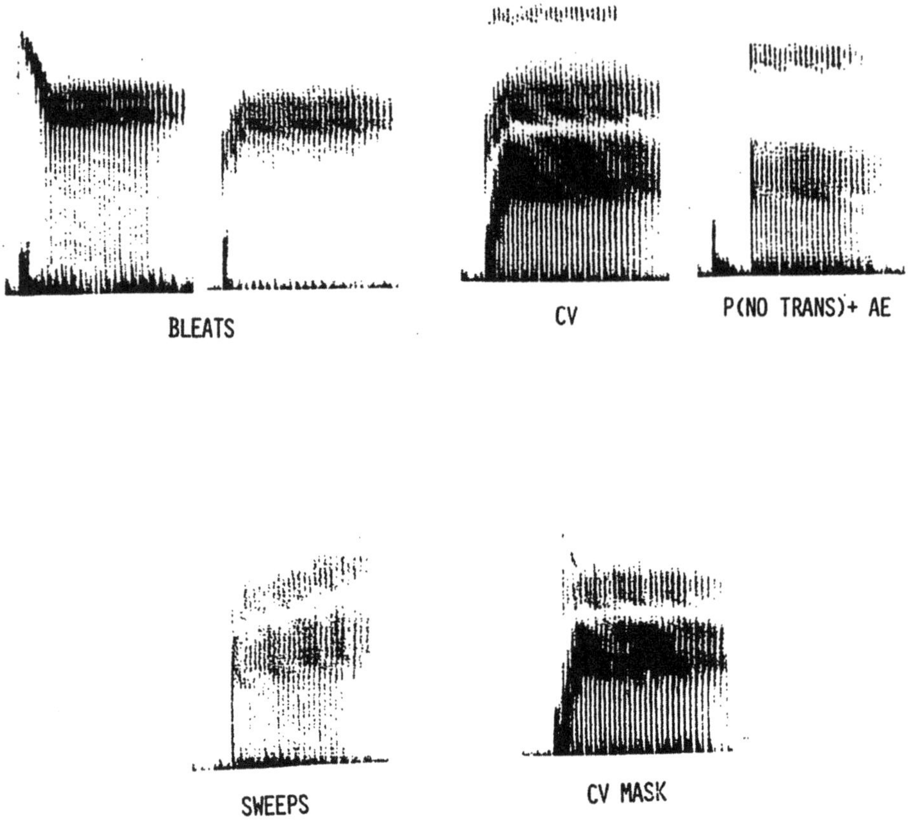

Figure 2. Spectrographs of the "challenges."

Haskins Laboratories. The CV's are natural speech and/or synthetic tokens stored in digital form for computer generation of test tapes. Transitionless CV's are modified, voiceless synthetics used as part of a larger set of studies (Cullen, 1975). The sweeps are 90 msec, first, second, and third formant-like glides swept upward, downward, or held constant. The CV-mask is the normalized sum of three synthetic, voiced stops + /a/ tokens (Mirabile and Porter, 1975).

SUBJECTS

Subjects were normal-hearing, right-handed adults, who had no history of brain trauma or hearing loss and were native speakers of English.

RESULTS

Figure 3 shows the performance of the right and left ears, re-

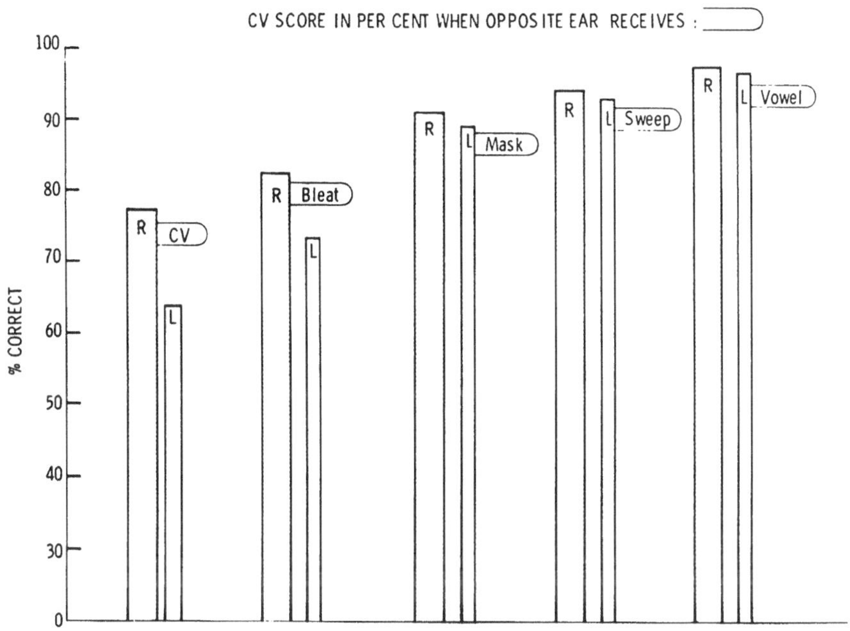

Figure 3. Right- and left-ear performance of normals when the opposite ear received one of the "challenges." The nature of the "challenges" is listed in the ⊃.

spectively, when subjects listened to targets paired with challenges in the opposite ear. The bar group is labeled to indicate the type of challenge (Berlin, Porter, Lowe-Bell, Berlin, Thompson, and Hughes, 1973; Cullen, 1975; Mirabile and Porter, 1975). The CV's and bleats generate suppression of targets for both the right and left ears. In addition, the right ear is more resistant to suppression than the left. The transitionless CV challenges (Figure 4) also generate a right-ear advantage but show somewhat less interference than a normal CV challenge. Sweep, vowel, and the CV-mask challenges produce considerably less interference with the speech targets.

The target-challenge interference seen in this series of dichotic experiments does not represent simple acoustic masking. This fact can be deduced from the observation that monotic presentation of the targets and challenges produces completely different interference patterns. Whereas the mask, sweep, and vowel challenges were the *least*

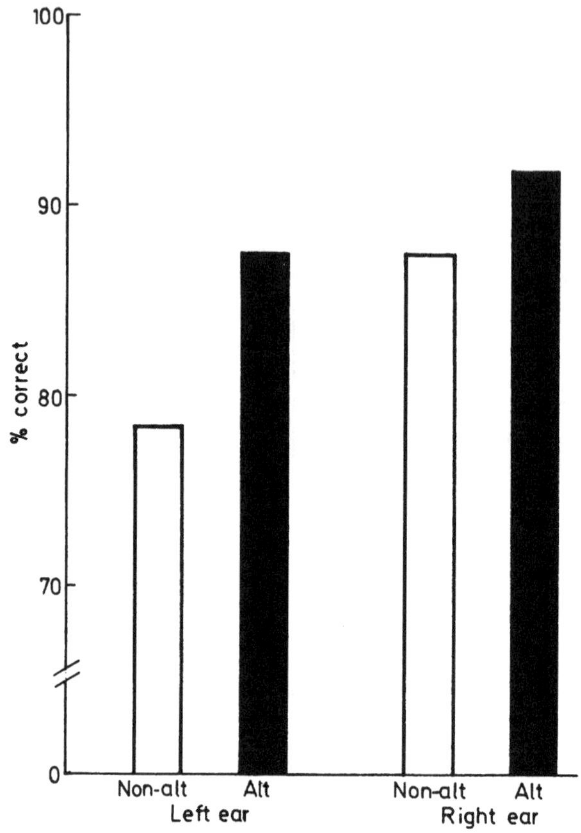

Figure 4. Left- and right-ear performance when the opposite ear received the voiceless CV (non-alt) versus a CV with the transition removed (alt).

effective in producing dichotic interference, they are the *most* effective maskers monotically.

In summary then, some challenges give us considerable monotic (but not dichotic) interference in the perception of the paired speech target. We have suggested that the degree of *dichotic* interference may be indicative of the informational weight placed on the attributes of the challenge in speech processing.

THE USE OF PATIENTS WITH BRAIN LESIONS

While the technique we are proposing has potential for yielding additional data on what is important in the acoustic speech stream, can the same basic paradigm be used to provide insight into the processes and structures involved in speech decoding? We believe it can, for there exists a fairly large population of patients with circumscribed lesions involving telencephalic structures which affect performance on dichotic, but not monaural, tasks. Figure 5 displays our

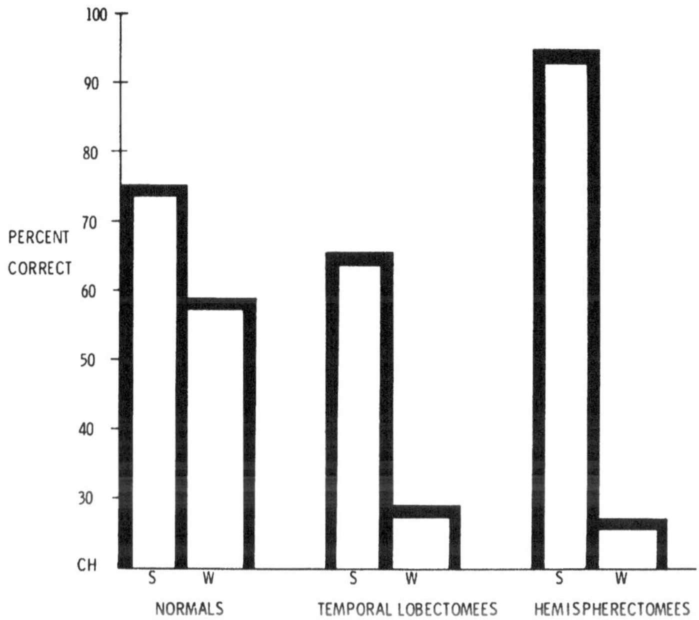

Figure 5. Strong or ipsilateral ear performance in normals, temporal lobectomees, and hemispherectomees. S = the right ear for normals and ipsilateral ear for patients.

observations for patients with temporal lobectomies and hemispherectomies. The essentials of these observations may be summarized as follows:

1. The ear contralateral to the side of resection usually performs very poorly relative to the ear ipsilateral to the resection. The level of the weak-ear performance is also well below that seen for either ear in normal subjects.
2. The performance of the ear ipsilateral to the side of resection generally appears to have been "released" from interference by the information presented through the resection-contralateral ear. This release seems similar to that shown in normals when one channel is degraded.

3. Release for hemispherectomees is nearly complete, while the temporal lobectomee demonstrates only a partial release.
4. Patients with section of the corpus callosum and anterior commissure behave similar to hemispherectomees. However, it is generally the *left* ear which performs poorly.

A dichotic flow model presented elsewhere (Cullen et al., 1975), implies some form of "gating mechanism" in the auditory nervous system which reduces the transmission of ipsilateral input. Figure 6 shows

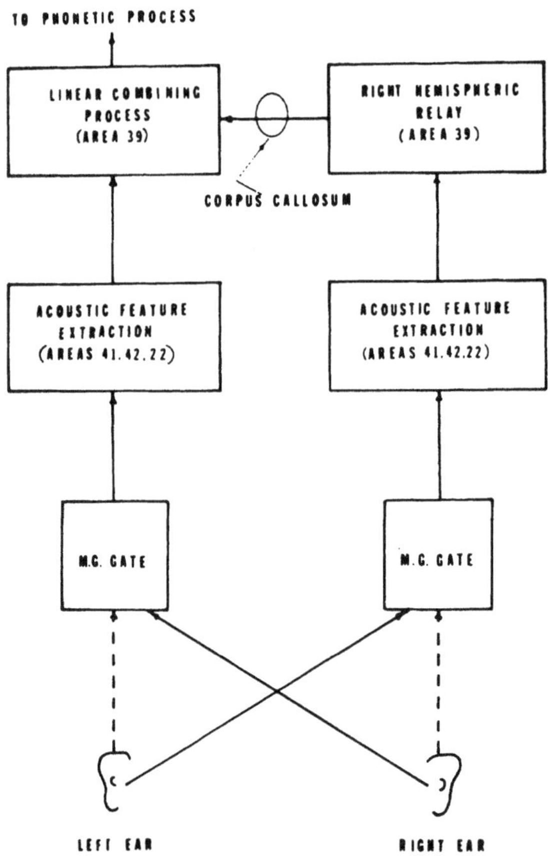

Figure 6. A flow model of dichotic listening (Cullen, 1975).

that model. We would like to show data which suggest that the gate might in part be related to the medial geniculate body. The model has deliberately omitted interactions at lower auditory system levels (cochlear nuclei, olivary complex, lateral lemnisci, inferior colliculi) to concentrate on possible higher level interactions in the admittedly unreal listening case of dichotic, competing syllables. Other workers have shown that in cats, the medial geniculate contains units which pass both ipsilateral and contralateral information, until impinged upon simultaneously; then only contralateral information is passed (Aitken and Webster, 1972).

We have recently seen a patient with a thalamic level lesion presumably confined to the region of the medial geniculate as indicated by no apparent damage to the lateral geniculate, thalamic somatosensory nuclei, corpus callosum, or fibers of the internal capsule.

Mrs. C. R. has a presumed lesion in the parathalamic area, approximately at the brachium of the inferior colliculus. She has been seen twice for auditory studies on 1/23/75 and again on 5/28/75. Figure 7 shows her monaural intelligibility scores of the left and right ears on the two testing dates. The dichotic portion of the graph shows that she had almost complete suppression of her left ear on 1/23/75 with improved performance in both ears on 5/28/75 after chemo- and radiation-therapy. The next figure (8) shows the monotic performance of each of her ears with bleats, noise, vowel, and sweeps competing with the CV in the *same* ear. Keep in mind this patient probably had only one functional medial geniculate body. Monaural and monotic performance for C.R. is comparable to normal subjects on similar tasks. However, the dichotic interference patterns are quite different from those observed for normals (compare Figures 1, 3, and 9). These data support the hypothesis of interference between challenges and targets, and point to the possible role of the medial geniculate body in effecting ipsilateral-contralateral ear information separation for cortical processing (Cullen et al., 1975). Notice that vowels and sweeps had *great* effect on her performance as monotic challenges, but much less effect as dichotic challenges. Thus, acoustic masking and the related VIII N and basilar membrane patterns at the peripheral level can not entirely account for the dichotic effects.

We are not prepared to fully explain the nature of these interference results, but it is clear that the spectral content of the individual challenges seems to have differential effect on monotic vs. dichotic listening.

Dichotic studies involving some of the challenges have been extended to persons having other focal brain lesions (Figure 10). These data also show characteristic patterns for particular lesion groups, from the standpoint of performance changes. We feel these differences,

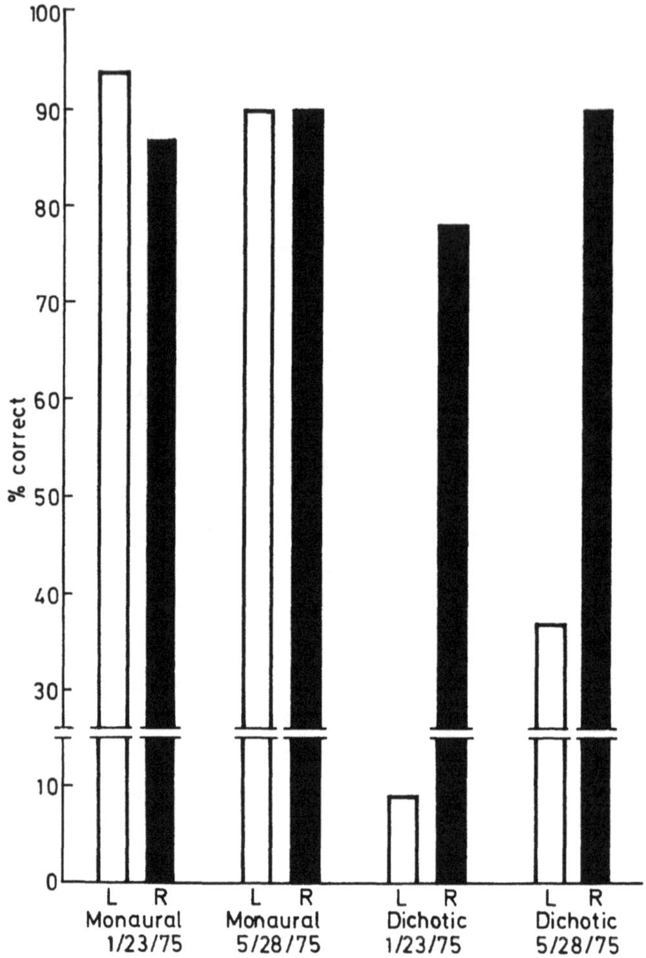

Figure 7. Test-retest data for monaural and dichotic CV's -- patient with medial geniculate lesion.

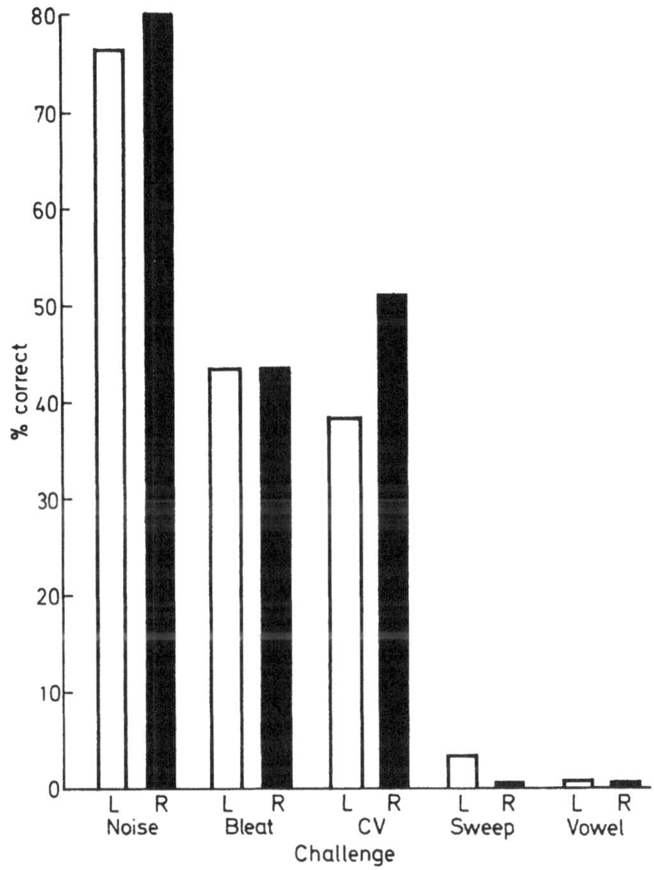

Figure 8. Monotic CV target performance for various "challenges" -- patient with unilateral medial geniculate lesion.

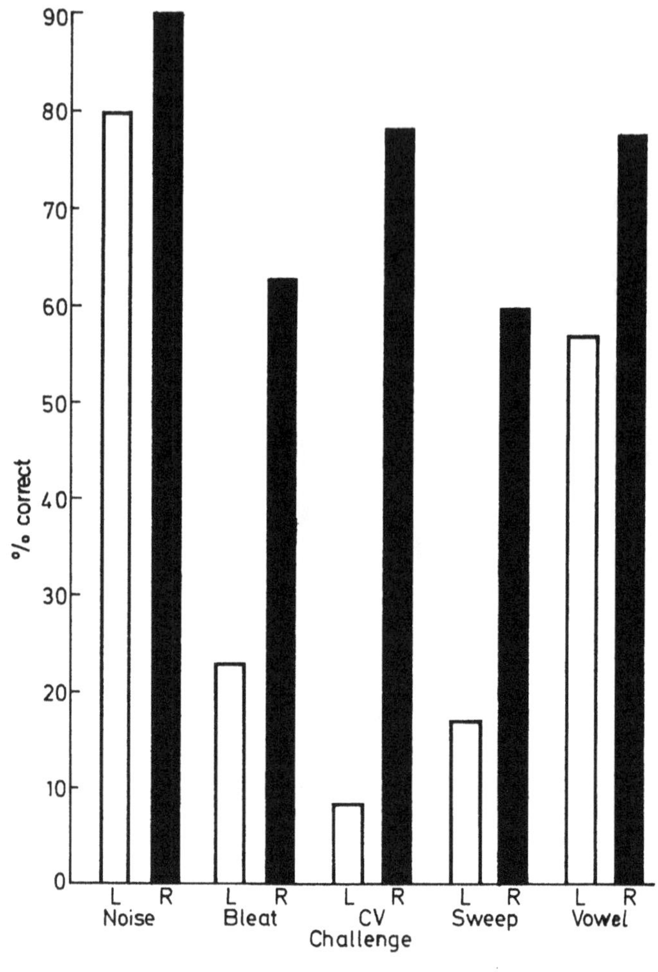

Figure 9. Dichotic CV target performance for various "challenges" -- patient with unilateral medial geniculate lesion.

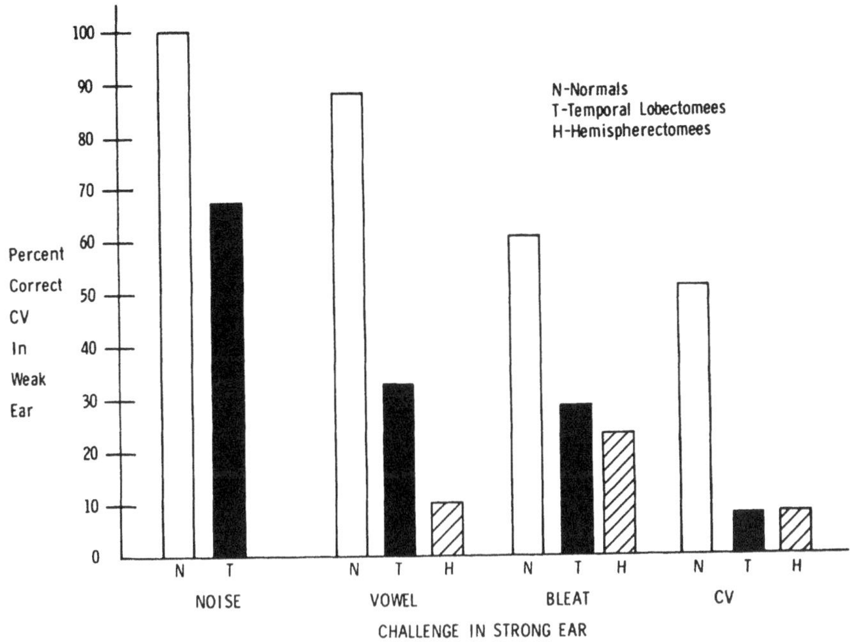

Figure 10. Performance of the weak ear (contralateral ear) of normals, temporal lobectomees, and hemispherectomees on some of our "challenges."

when contrasted to normal subject performance, may ultimately lead to a definition of what structures are involved in speech processing, and more importantly, their functional role.

CONCLUSION

Lack of isomorphism between the phoneme and the acoustic signals of speech has led to a series of theories for perception, predicated more on logical possibility than on structure and function of the human nervous system. We would like to suggest that our understanding of speech processes may be furthered by augmenting some of these logically possible theories with considerations of speech signal coding in the auditory domain. At first glance, this may appear as a step backward to the pre-spectrograph days of acoustic phonetics. This is neither our desire nor intent, for clearly, speech perception cannot be completely understood in terms of wave form analysis, or from combining the results of psychophysical experiments on the perception of duration, frequency, and intensity. However, we would remind us all of two self-evident, but often overlooked facts: (1) information which ultimately leads to (or triggers) the percept of a phoneme is part of the acoustic wave; (2) the auditory system, as the primary receiver of acoustic energy, must participate in coding this information. We would like to further suggest that the data we have obtained from dichotic listening tasks may assist us in determining how this coding occurs physiologically.

ACKNOWLEDGMENTS

This research was supported in part by National Institutes of Health, USPHS Grant No. NH 11647-01 and National Institute of General Medical Sciences, Special Fellowship 5 F03 GM52343. Necessary laboratory facilities were provided through a grant from The Kresge Foundation. Critical comments and services were supplied by L.F. Hughes, R.J. Porter, Jr., Gae Decker, Kathy Rich, and Pamela Weaver. We thank Drs. Paul Bucy, Wayne Olsen, Douglas Noffsinger, John Gilroy, and George Lynn for access to their patients, and, of course, we thank the patients themselves for their efforts.

REFERENCES

Aitkin, L. M. and Webster, W. R. Medial geniculate body of the cat: Organization and responses to tonal stimuli of neurons in ventral division. *J. Neurophysiol.*, 1972, *35*, 365.

Berlin, C. I., Porter, R. J., Jr., Lowe-Bell, Sena S., Berlin, Harriet L., Thompson, C. L., and Hughes, L. F. Dichotic signs of the recognition of speech elements in normals, temporal lobectomees, and hemispherectomees. *IEEE Trans. Audio Electroacoust.*, 1973, *AU-21*(3), 189-195.

Cullen, J. K., Jr. Tests of a model for speech information flow. Unpublished Doctoral Dissertation, Louisiana State University Medical Center, Department of Physiology, New Orleans, Louisiana, 1975.

Cullen, J. K. Jr., Berlin, C. I., Hughes, L. F., Thompson, C. L., and Samson, Diane S. Speech information flow: A model. *Proceedings of a Symposium on Central Auditory Processing Disorders*, The University of Nebraska Medical Center, Omaha, Nebraska, 1975, 108-127.

Mirabile, P. and Porter, R. J., Jr. Dichotic and monotic interactions between speech and nonspeech sounds at different stimulus-onset-asynchronies. *J. Acoust. Soc. Amer.*, 1975, *57*(1), S51(A).

DISCUSSION of "Dichotic signs of "Speech Mode" listening"
BERLIN and CULLEN

FUJISAKI:
In your functional block diagram of the central nervous system (shown by the slide), the blocks for auditory processing are located above the medial geniculate bodies (MGB), where gating of the ipsilateral signal by the contralateral signal is assumed to take place. I suspect that the nervous system below MGB may also be responsible for at least a part of the auditory processing, so that the gating at MGB may be operative between features rather than between signals. In fact, your data could be explained if we assume that the processing of quasi-stationary features is performed below MGB, while processing of transient features involves stages above MGB, and that interferences between dichotic signals occur when they both possess transient features and impinge upon the same functional block above MGB.

BERLIN:
This is a very astute point. The failure of steady-state vowels to be effective as challenges fits Dr. FUJISAKI's notion. Neurophysiological auditory feature detectors in bats, frogs and squirrel monkeys respond to time-by-frequency shifts over brief periods rather than to steady state vowel-like complexes.

LEVELT:
Your findings are analogous to facts in dichoptic vision, known as the "Law of complementary shares" (see e.g. LEVELT, 1968). If the contribution of one eye to dichoptic brightness is increased, the share of the other eye is decreased to the same amount. The share for an eye is dependent on early extracted features, especially contours. This is, therefore, in tune with Professor FUJISAKI's remark that some feature extraction must have taken place before dichotic convergence in BERLIN's model.

BERTELSON:
In your diagram of information flow in dichotic listening, the crucial element are the crossed pathways linking each ear to the contralateral hemisphere. Back to KIMURA in 1961 this is the way right ear advantage has been interpreted. There are reasons to think

this kind of interpretation is too simple. JOSE MORAIS and I have obtained what looks like rather good evidence that the critical factor in determining right ear advantage is actually not the fact that the material reaches one particular ear, but the fact that it seems to originate from a particular spatial location. The main argument is that a right side advantage is observed in a stereophonic presentation situation, where two simultaneous speech messages both reach both ears at the same intensity, an impression of lateralization respectively to the left and to the right of the median plan being created through time differences (MORAIS and BERTELSON, Journal of Experimental Psychology, Human Perception and Performances, 1975, in press). If what we are dealing with is not a relation between the stimulated ears and the hemisphere, but a relation between apparent spatial origin and the hemisphere, another, more sophisticated type of model is necessary. Perhaps the study of neurological cases can help establishing the basis for one.

BERLIN:
You have to incorporate a form of "lag effect" in interpreting your experiment, and also incorporate the work of MYERS and CHISTOVICH at the Leningrad Symposium 1973. First, in the lag effect, when the two messages reach the ears the leading message dictates the perceived source in space but the trailing message is better perceived in the 30-60 msec range. I would suspect your time differences are all on an order less than 10 msecs.
MYERS et al. also showed a separation of interaural lateralization effects vs. right ear advantages as a function of time. Thus there seem to be 2 mechanisms at work with considerably different time courses.

SUMMERFIELD:
Figure 3 in C. BERLIN's text illustrates a difference in target recognition. Performance occurs between two specific "challenges" (illustrated in top right-hard corner of Figure 1). Is this difference to be attributed to the absence of a formant transition in the P(No Trans.) +A challenge in the sense that:
a. there is no transition present; or
b. there is an absence of sound (i.e. a silent interval):
compared to the CV/Ba/ challenge?

BERLIN:

This is an important question to which I can give only a partial answer based on experimental data. CULLEN's dissertation also included a shortened challenge, which produced a release from interference of the target as the challenge was shortened. The transition alone, without a vowel, probably has little effect as a dichotic challenge. Thus the likelihood, from the evidence, points toward the silent interval and reduced total duration of stimulation as important in explaining why the transitionless CV had less effect than the full CV. If one postulates a neural model of inhibition, it could be said that the silent interval released the target from neural inhibition over a brief period of time.

HAGGARD:

DO ADES' dichotic adaptation experiments, or the fact that interhemisphere interference is diminished in hemispherectomees in your data. suggest that processing of acoustical information in a phonetically relevant way may in fact be more symmetrically located in the brain than we have recently assumed? And does it seem possible that studies on patients could ever say anything about specific relationships between the left and right hemispheres in sharing out overall language processing? This is an attractive idea to get us away from simplistic concepts of functional asymmetry but there is really no good evidence for it yet. For example the apparent left-ear (right hemisphere) superiority for processing of intonation could be due merely to the non-linguistic task involved, and the observations from clinical neurology cannot be given a very clear interpretation.

FUJISAKI:

It is certainly true that the process of speech perception is the problem of structure of mind, while the work of Drs. BERLIN and CULLEN is concerned with structure of the brain. However, they are closely related as the software and the hardware for speech perception. While the knowledge of the hardware cannot tell how the software is organized, it is still quite important to know the structural constraints imposed by the hardware, in order to obtain full understanding of the software. Therefore I consider that dichotic studies are quite relevant to studies of speech perception as long as they shed light on the structures and functional roles of the part of the brain that is directly concerned with speech processing.

SPEECH PERCEPTION IN CHILDREN

John H. V. Gilbert
The Phonetics Laboratory
University of British Columbia
Vancouver
British Columbia, V6T 1W5

INTRODUCTION

The child's ability to perceive and then produce the sounds of a language represent part of a larger general ability of the child to produce and understand an infinite number of sentences that have not been previously heard. Fundamental issues confronting workers in child language acquisition may, therefore, be represented at various linguistic, sociological, psychological and other levels. Descriptions of the *production* process have resulted in numerous accounts of phonology, syntax and semantics within psycho-, socio-, and ethno- linguistic frames, (see particularly Brown, 1973; Smith, 1973; Labov, 1970). Such descriptions give an idea of the mechanisms and strategies by which young children progressively acquire language.

Attempts to describe and account for the *perceptual* process engaged during language acquisition have been less numerous and relatively less successful. For as will be seen, a major problem encountered in discussing speech perception in the neonate-infant-child, is that it cannot be separated or divorced from, the multiple physiological-cognitive factors which will affect its effectiveness. Whether or not a *strong* form of either the motor theory or analysis-by-synthesis can meet the multiple needs of the child's developing perceptual system, must remain to be seen.

This paper presents a very preliminary and incomplete attempt to survey some of the physiological and psychological data which describe discrete parts of perceptual/productive processes of "speech" in children, a discussion of some of the results obtained, and a brief comment on the relationship between such data and existing models of speech perception in adults. The survey focuses particularly on the first year of life and, under that rubric, on some rather specific aspects of perception-production. It will become apparent that, my intuition tells me that it is not possible to separately consider phonetics and phonology.

THE CHILD AS AN ORGANIZATION

Von Bertalanffy has said "Hierarchical organization on the one hand, and the characteristics of open systems on the other, are fundamental principles of living nature" (Von Bertalanffy, 1952). Elaborating notions about ontogenetic processes in the neonate-infant-baby-child is undoubtedly contaminated by the problems of interrelating and explaining a developing, hierarchical organization (Scott, 1962; Furth, 1974). It would, therefore, appear advisable in any discussion of the language acquisition process (whether the emphasis be on perception or production), to take cognisance of this organization, neatly summarized in Piaget's (1967, 1971) words: ". . . language is mostly one particular instance of the semiotic or symbolic function, and that this function as a whole . . . and not language alone, is what causes sensori motor behavior to evolve to the level of representation or thought" (p.46). Our difficulties in discussion are therefore further compounded by the fact that feedback from the environment (which guides, or corrects, or stabilizes pre-existing patterns of behavior) must also be taken into consideration. The idea is not original. It is now apparent from the multiple works of Piaget and his co-workers, that a symbolic function exists which is broader than language and encompasses both the system of verbal signs and that of symbols in the strict sense: "Cognitive function is a more basic and primary process than language . . . the dependence-relationship of language upon cognition is incomparably stronger than v.v." (Piaget, 1971). As an appreciation of speech registers (Gleason, 1973; Snow, 1974) and speech acts (Searle, 1970) grows, so hopefully will our descriptions of the relationship between sign, symbol, environment and child become more precise.

In examining experiments with young children which seek to enlarge our understanding of one part of that complex, i.e. speech perception, our interest will be little different from that elaborated by Sven Öhman (this volume), ". . . one is interested in determining just what sorts of acoustic patterns will cause the perception of a certain phoneme." Somehow or other, the "acoustic vibrations go in, and the phonemes eventually come out", sounding (usually) like the adult language model. As I shall try to show, the distance between vibrations on the basilar membrane and vibration at the vocal fold, is enormous! Setting aside a number of very difficult questions, which numerous authors have posed for such an input-output experimental paradigm, it is apparent that early in the language acquisition process the infant and child do produce noises which are meaningful outputs of speech perception, i.e. they result in some act on the part of an observer, even though the child is faced with the very complex task of choosing from among all the audible variations in spoken language those sound relations which play the role of differentiating meanings. A major difficulty, however, is that the criteria for defining those sounds as "meaningful" in terms of the child's own system have never been agreed on by child phonologists.

GROWTH AND DEVELOPMENT

What determines the transition from "sounds" to speech sounds? If a child can semantically distinguish intonation, rhythm, and combinations of sounds in words, how and when does it begin to perceive the sounds of a language and use them as the only means of differentiating the meanings of words?

Before attempting an answer to such questions, some appreciation of the interacting complexities of the system must be obtained. It is now well known that by six months in utero the cochlea is completely formed and the child enters the world with a mechanism which is ready to begin analysing the "buzzing, blooming confusion" of the auditory world, at least in terms of frequency, intensity and time (Eisenberg, 1970). Also well documented are the facts that the hemisphere which is presumed to do the language "thing" is slightly larger than average in the neonate (Witelson and Pallie, 1973) and that right ear advantage for speech is present by 2½ years, chronological age (CA) Gilbert and Climan, 1974). Whether or not these three observations are sufficient to warrant strong forms of hypotheses about the innateness of language must await the passage of time and continued experimentation!

It bears repetition that language emerges in keeping with other aspects of physical and cognitive unfolding. It is sometimes forgotten that not all genes are at first active in growth; throughout life, previously dormant genes awaken, or at least produce new effects that they were unable to produce at an earlier age (Harrison et al., 1964). Huntington's Chorea, for example, a disease of the nervous system, is due to a single gene inherited very simply and manifesting its effects with great regularity. Unfortunately the disease does not usually appear until 30-40 years CA and, despite much searching, no clear abnormality can be demonstrated before the disease begins. It seems that a similar thing occurs in the genes governing growth. As the child grows, more of the genes show their hand; not only is the child's adult status increasingly well-predicted, but his resemblance to his parents in size becomes increasingly marked also. Since "growth" refers to things other than bones, "the strategy of the genes" (to use Waddington's (1957) famous phrase) must be carefully considered. Yakovlev and LeCours (1967), for instance, have shown that the development of the nervous system follows a sequence of morphological events which reflect, and correlate with, the changes in internal state, outward form and dynamic relations of the organism to the environment. All of these changes are part of the conception of growth and maturation of the biological systems. The concept of maturation, on the other hand, is much broader in connotation, viewing an exponential process of the progressive organization of functions and of their morphological substrata which goes on through the lifespan of the individual. Obviously, the exponential curve of maturation is constantly under pressure from the stresses and strains of life to develop parabolically!

The story of brain growth and development has been well told by Lenneberg (1967). Brain growth and weight progress rapidly up to the end of the first year, i.e. during that period of time when the substrates of language are being established.

Not only do children perform the speech act of phoneme production within six or seven months of birth but, in addition, they perform perceptual-discriminatory feats which quickly enable some children to produce single words even by eight months. Since a child usually arrives at the correct production of sounds in his immediate language environment, one can only observe that the differences which exist between its perception of the auditory events of language and an adult's perception of the same auditory events of language must converge in a short space of time. Whether the early differences in auditory perception-discrimination are only of degree will be discussed further. Although the acoustic radiation is not completely myelinated until around the fourth year of age (Yakovlev and LeCours, 1967), the phonetic inventory of the child is obviously elaborated very early and is one of the obvious pre-requisites to <u>syntactic</u> speech which often occurs by eighteen months of age, and phonological discriminative processes acquired by 3 CA (Graham and House, 1971). It would appear that, although the perceptual process has a <u>relatively</u> close correlation with development of the auditory system, persisting phonological inconsistencies would appear to indicate that production parallels a larger physiological development.

THE IN AND OUT OF PROCESSING

The linguisticality of speech acquisition would appear to hinge around six months chronological age (plus or minus one month). For it is at the break point between crying and cooing on the one hand and the onset of babbling on the other that clear perceptual and productive cues to the linguistic process are (when related to existing data) most apparent. Claims for "linguistic" processing at an even earlier age are, as yet, debatable (Butterfield and Cairns, 1974; Kuhl and Miller, 1975).

Given that the middle ear structures might act as a low-pass filter, (see below) it then becomes necessary to determine what effect, if any, the length of the external auditory canal imposes on auditory signals (including speech sounds) during infancy. Edgar Shaw (personal communication) has suggested a linear scaling solution to the problems posed by the neonate-infant's relatively short external auditory meatus. He proposes that, if it is assumed that the torso, head circumference, pinna length, concha depth and breadth, canal length, eardrum thickness, malleus, incus and stapes dimensions are scaled down by a constant factor with respect to adult dimensions, then such an hypothesis would lead to a view that the sound pressure transformation from the free field to the eardrum will have the shape and magnitude of the adult curves (Shaw, 1974), <u>but</u> shifted up the frequency scale by the appropriate factor. Given that the average head diameter at birth is 11.5 cm and the adult corresponding dimension is 17.5 cm the scaling factor is $\frac{17.5}{11.5} = 1.5$. Using Shaw's average curves for adults (Shaw, 1974) it will be seen that the gain is small up to 1.2 kHz, but rises to a peak of about 17 db at 2.7 kHz at 0° azimuth. Multiplying by 1.5 it might be inferred that the peak for the neonate would be approximately at 4 kHz. If, on the other hand the external auditory canal is absent in the infant skull it would seem likely that the gain curve is flatter than that of the adult ear, perhaps reaching a broad peak at around 5 to 7 kHz. We can assume, following Harrison et al. (1964), "That head growth is fairly rapid during the first six months and that probably by one year of age we have something approaching the adult values". The effect of this hypothetical gain, and its implications, for sound perception in the first six months, deserves further study.

Several studies have shown, none-the-less, that a simple discriminative act to auditory stimuli is possible, since the level of activity in infants and neonates is affected by stimulus intensity, complexity and gross temporal characteristics (cf Brackbill et al, 1967, for a brief review). Eisenberg (1970) has suggested that the neonatal auditory system is differentially tuned to react to the carrier frequencies of language. The support for Eisenberg's suggestion is, however, only partial. Studies by Turkewitz et al (1972) and Ling et al (1970) demonstrate that the so-called "tuning" seems to be too broad. Bench (1973) has attributed this poor tuning effect to the flaccidity of newborn middle ear structures (Keith, 1973) which may act as a low-pass filter. The implication of this would then be that low frequency sounds are more effective in eliciting response because they are more efficiently transduced through the middle ear before they become a response substrate. If such a suggestion holds, what then should be done with data from the external auditory meatus? An interesting observation from the baby-talk literature is that older speakers always raise their speaking pitch when talking to babies and infants!

There are indeed data which appear to indicate that complex sounds other than speech seem to be even more effective in eliciting responses in the neonate (Bench and Metz, 1975). It is interesting, for example, that a graded response to intensity permits the developing child to orient himself preferentially towards social sounds, hierarchical mechanisms giving a kind of time base for related input and output operations. It has been suggested by Eisenberg that these might serve as "toning up units" of the speech apparatus for articulatory tasks. During neonatal life almost every index of auditory behavior indicated previously shows range dependent effects so that, sleeping or waking, signals in the range below 4 kHz evoke responses two or three times more often than those in the higher range. There is also evidence to suggest that intensity is processed in much the same way throughout life. Thus, given such data, the observation might be made that infants' responses to complex stimuli such as, for example, mother's voice, tend to be perplexing. For apart from the well known /pa/, /da/, /ga/ (and other) discriminations of neonates evoked by Eimas (in press), Trehub (1973), etc., Horowitz (1975) has demonstrated that _2-month_-old infants are able to discriminate two different voices, and also appear capable of discriminating two different voice-tone qualities in the same voice. Even more interesting, Aronson and Rosenbloom (1971), in an experiment in space perception in early infancy, demonstrated that 30-day-old infants perceive auditory and visual information within a common space, i.e. mother's face and voice. Discrepancies within this space, i.e. when mother's voice was moved 90° right or left of her face, produced agitation and upset, noted as "tonguing" (protrusion of the tongue between the gums). For instance, 20 to 30 seconds after the onset of the displacement of a voice, greater tonguing was noted; before the switch, tonguing rate was 12.3 per sec, after the switch 60.1 per min. when mother was in view. Whether "tonguing" could be correlated with non-nutritive sucking remains to be seen.

This interesting aspect of sound localization finds a parallel in a dichotic listening task. Morais and Bertelson (1973) have shown that the strongest perceptual advantage accrues to sounds originating in the median plane, the direction of gaze. Although their subjects displayed a significant right-speaker advantage for competing CV syllables presented over left and right loudspeakers, they showed a significant front-speaker advantage if the syllables were presented over either front and left or front and right loudspeakers.

Both studies, however disparate, imply localization mechanisms and suggest that auditory perception rests, at least in part, on some low-level decision as to the _spatial_ origins of the signals, and will probably determine the proportion of incoming information that engages the infant's perceptual mechanisms.

In any event, if one holds in abeyance the "tuning to carrier frequency" suggestion, or the argument that perhaps young infants are tuned to respond to the complex stimulus of speech itself, then as far as can be seen, the better than speculative evidence for the reality of language by six months begins to achieve some importance in its correspondences with other biological aspects of development.

LARYNX, VOCAL TRACT AND CRYING

At the present time, there still exists relatively little information concerning vocal tract dimensions in the neonate. Lieberman et al. (1971) estimate a uniform tube 7.5 cms in length and open at one end. The formants of such a tube would occur at 1.1, 3.3 and 5.5 kHz, with a fundamental frequency of about 400 Hz. Hopkin (1967) has shown that the neonatal tongue is approximately half the length of the adult tongue. The larynx, according to Noback (1923), is positioned higher in the neonate vocal tract than is the case of the adult. Changes in the length of neonatal supralaryngeal vocal tract occur during cries and are the result of laryngeal movements, (Truby and Lind, 1965).

The lumen of the neonate larynx and trachea is said to be smaller in relation to the body as a whole than is the case with the adult, particularly the male adult, (Wilson, 1953), while in addition, all the laryngeal tissues, including the cartilaginous framework, the musculative, mucous and submucous tissues are looser and less fibrous than in the adult. The output of such a mechanism will very obviously be different from that of the adult. Yet, rapid increase in head size and downward displacement of the larynx within the first six months, coupled with increases in the area, thickness and length of the vocal folds, must ultimately lead to a decrease in fundamental frequency (Sheppard and Lane, 1968: Iguchi and Hirsh, 1969, and Gilbert, 1973), meaning in effect, continuing adaptations to input and output signals. As age increases, the subglottal air pressure that the infant can develop, owing to neuromuscular and anatomical development in the respiratory mechanism, also contributes to an increase in fundamental. Crying fundamental frequency is found to be higher than non-crying and there is something of a time lag between the gradual disappearance of reflexive crying and the gradual appearance of motivated crying, (Sheppard and Lane, 1968).

Given that these ongoing fluctuations in the output signal have to be "adjusted" to by the infant, what of its caretaker who provides part of the feedback from the environment, alluded to earlier? Müller et al. (1974) have shown that mothers are unable to differentiate between cries of children between 3 and 5 months, while Turnure (1971) demonstrated that there is greater motor quieting to a normal, non-distorted mothers' voice, and suggested an increase in attention to this type of verbal stimuli in the age range 3 to 9 months.

Although cry sounds, and thus the initial vocal output of the child, have by now been extensively studied (see Delack (1974); Crystal (1973) for review) the perceptual correlates both for observer (as input) and the infant proper, need further examination as the child progresses from crying to babbling.

A most perplexing problem confronting workers in child language acquisition is whether, when language emerges, it is due to some very general abilities that mature to a critical minimal stage at about the one-word level (supposing an adequate definition of that stage) and thus make many other skills possible <u>or</u> if there are in fact some speech and language-specific factors that mature and are independent from other more general processes. The evidence so far is that the emergence of speech and language is dependent on maturational change given adequate language input. The role of babbling in the transition from non-speaking to speaking human has been greatly discussed, these discussions tending to resolve into two schools of thought: continuity vs discontinuity. Whatever the disagreement, it is apparent by now that there are progressive stages of development which must be interactive, (Gilbert, 1974). Most children enter the babbling stage at between five and seven months, progressing from at first reflexive and attentional crying, through cooing (a response to an internal state, induced by sensory stimuli,) to coordinated output of transcribable (and untranscribable!) sounds.

Cruttenden (1970) relates the interesting observation, that at a very early stage in the babbling of twins he studied, the breath group did not appear to be the basic unit for utterance. He suggested that the basic unit was something which might be facetiously called the "nonsense group" at the end of which a pause occurred. He pointed out, however, that breaths are often taken within these groups. A recent study (Prescott, 1975) demonstrates a more accurate method of measuring these groupings. It is a commonplace, of course, that breathing <u>must</u> have something to do with the output from the vocal mechanism. Breathing will therefore, of course, ultimately impose its particular timing constraints. Roger Brown's (1958) notion of "babbling drift", i.e.

that the repertoire of babbling sounds of the infant drifts in the direction of the mother tongue around him, more generally speaking to more specific speech sounds, must therefore in part be associated with internal physiologic rules, imposed now by a higher centre than the lungs.

But given that breathing and laryngeal control are imposing their constraints on output, what of an observer of that same output? In relation to adult perception, Tervoort (1967) has reported that Dutch listeners could distinguish babbling per se from six months onwards, both of Dutch and non-Dutch children. Before that time, events conspire to impose limitations on adult interpretation, not the least of which is the waking state of the child.

Khrisman (1974) in the Soviet Union, working with EEG on speech perception in children aged six to eleven months, has shown that in response to a familiar word (the name of the child) interactions of the inferior parietal area of the left hemisphere are intensified primarily at the temporal and occipital cortical centers, as well as of the left temporal lobe with other areas. Such a finding, supposing accuracy of a technique about which there is debate (Friedman et al., 1975), would be a nice perceptual corroboration for the evidence of output. Pierce (1974), for instance, has shown that the range of vowel quality produced by six-month-old children includes a considerable amount of the adult vowel range. Statistically the overwhelming predominant sounds demonstrated by children at this time occur within the low-front, mid-front open and mid-central areas noted by Cruttenden. In Russian, Bel'tyukov and Salakhova (1973) have reported that "infants' babbling" contains more vowels that are inappropriate to the language than are appropriate to it, an observation which could be made about other babbling data. It is apparent, however, that as the child matures his vowel system output must rapidly approximate that of his language community. Menyuk (1967), for instance, has shown that although the child's vowels do not match the adult's at five years, his set of vowels are clearly differentiated from each other in formant frequency. Furthermore, the direction that differences take for F1 and F2 between vowels is the same for adults and children, although there is no overlap between the vowels produced by children and adults. Perceptually, the child's appreciation of vowel difference matches his productive capacity. Gilbert (1975) has shown that 2½ - 3-year-old children can identify vowel tokens in accord with their productions.

As will be seen later, the same does not hold for consonants. Acoustic characteristics other than formant frequencies must provide some cues for identification such as Kewley-Port et al. (1974) have shown; i.e. that at least for a shortlead voice onset time (VOT), the articulatory gestures used by children to produce vowels are hypothetical analogues to those used by adults.

TIMING

Ultimately, speech perception-production in the first year probably rest on temporal patterns of control which are based on and have their roots in a physiological rhythm, which is established after birth with the 24 hour circadian rhythm for body temperature, motility, sleep-wakefulness, etc. (Hellbrügge, 1960).

Prescott (1975) observed increased duration of expiratory cry utterances with maturation, although the developmental significance was less promising than pitch complexity in the cry because of the high variability in utterance durations which apparently is very normal in early life. Even among the cries of the hearing-impaired child in Prescott's study, duration was typical of the age group. It is suggested that utterance duration may therefore be primarily determined by processes other than audition. This possibility would appear reasonable in light of data on timing mechanisms. In repeated productions of a sequence if the consonant in a consonant-vowel syllable is overlong, Allen (1973) has noted, the following vowel will be over-short, making the syllable less variable. Allen hypothesized that if the mechanism is working, it knows the time allowable for each unit, so that at any given instant during production, the mechanism can compare its performance with an available model that is error sensing, and then feeds forward either to correct "durational errors" or to at least maintain whatever it can of what the model tells it ought to be durationally. A timing mechanism is undoubtedly operating at a very early stage, since for instance, Prescott has also shown that whereas in general the stops (pauses) in younger cry utterances seem to be constrictive silences, in older cries the silences are perceived by adult listeners as rhythmic patterns.

The possible significance of silence patterns in cry development is suggested by the few and relatively long intra-utterance silences which were observed in the cries of the hearing-impaired child in the

Prescott study. The duration of intra-utterance fundamental frequency variability for samples of exploratory cry utterance in the same period would, as expected, appear to increase. These developing patterns of durational change and variability in fundamental frequency in cry, would appear, therefore, to be the first signals of those suprasegmental features of intonation noted by most observers as the first signals of linguistic utterance and which are necessary physical prerequisites to "meaningful" babbling.

It is but a short step from such progressive constraints to the first auditorily observable productions of pulmonic lingual consonants and clicks at around 4 months (Cruttenden, 1970). Between the fourth and fifth month, dental, alveolar and velar plosives have been recorded (Cruttenden, 1970), as the child controls combinations of tongue movement and breathing "patterns". The breathing-tongue movement coordination would correlate well with progression of the child from a mainly prone position to an upright sitting position. It has been often suggested that at this time the infant produces all the sounds of its language environment plus many others. In his twin study, Cruttenden did not record any of the fricatives /th, s, z, sh/ and only one child showed any sign of /f/ and /v/. The absence of all fricatives from one child's babbling and all except /f/ and /v/ from the other child's babbling seem to relate to a frequently noted "order of acquisition" of English consonants. It has been hypothesized that sounds of more frequent occurrence in babbling are also of more frequent occurrence in the world's languages. In a study of 16 Yucatec infants, Huber (1970) showed that at 9 months, the phonemes /b, t, p, m, n, h/ and sundry vowels were present, which would be a very tentative corroboration for such an hypothesis. In a study of voice onset time, however, Kewley-Port et al.(1974) showed a somewhat earlier acquisition of stops, i.e. around six months of age, at which time VOT distribution has a wide range of randomly distributed values extending from voicing lead to long voicing lag. They hypothesize that short voicing lag stops are in specific ways easier for the infant to produce successfully than are apicals in the long voicing lag category, since for short voicing lag stops the child may fully close the glottis any time during apical closure provided that the velopharyngeal closure merely isolates the nasal cavities. On the other hand to produce voicing lead stops, the infant must complete glottal closure considerably before oral release and then initiate and sustain vocal fold action by the addition of other articulatory mechanisms. Precise timing constraints would not appear to be necessary for the short voicing lag stops, an observation which fits with those made earlier.

Although the VOT range at this time is about 20 msec, abduction of the vocal folds can be achieved any time during apical closure, (approximately 100 msec,) and oscillation will still begin _only_ upon apical release.

Given that such restraints as have been briefly outlined so far are indeed operating, it would appear quite clear that infants do _not_ babble all the sounds of English, in fact physiological constraints would mean that the range of sounds which are covered could _not_ be as wide as is often implied.

The intricacies of timing control assumes greater significance when we consider that in most instances physiological events precede the acoustic event of the speech sound (cf McNeilage, 1972, for review). We know that in individual speech sounds the duration of physiological events may be twice as long as the duration of acoustic events. It would therefore seem justified to claim that the sequential arrangements of muscular events require preplanning with anticipation of later events. Parenthically, it is interesting to note that there is a parallel preplanning of syntactic events (see Fromkin, 1974, for example). Thus, the occurrence of some speech events is contingent upon others which will follow. Since the output of phonemes is dependent on sequential arrangements of muscular events we must assume that timing control is implicit in language acquisition and imposed continuously through crying into babbling and the first word.

Perceptually, just what timing changes might interfere at this point have only briefly been investigated; the report of Cullen et al (1968) on the effect of delayed auditory feedback (DAF) on neonates results in longer pauses between cries. In what ways other signal distortion etc. would interfere with the infants perception is not known at this time.

DEVELOPING STORAGE

Rensch (1956) has concluded that memory retention is about proportional to brain size in animals experimented with by himself and his colleagues. The increase of some 500 cc in endocranial volume in homo-erectus would certainly represent a tremendous increase in his memory. An examination of brain maturation in children would lead one to hypothesize that it is only with increasing brain size that sufficient memory becomes available for all the linguistic utterances. However, Seckel (1960) has shown that nano-cephalic dwarfs, whose brain weight did not exceed 500 gms, are equally capable of learning language. Perhaps it would be more accurate to say that memory for speech sounds will depend not only on a _minimal_ brain weight, but most importantly on cranial connections.

Kornfeld (1971), discussing theoretical issues in child phonology, hypothesized that there are at least two possible answers to the question "What determines the childs' representational system?" These answers are _first_, that the adult system of phonological distinctions determines the child's system or _secondly_, that the child perceives and produces in his own system, which need not bear a simple relationship to that of the adult. In a number of experiments run with 1½ to 2½-year-old children, he concluded that the evidence did _not_ suggest that the children of this age level produced the same distinctions as adults. Instead, he argued, the children in his samples selectively abstracted from the adults' centre features, at least in production. From the evidence so far adduced, it would appear that the second hypothesis would have to hold at least initially, since the child's perceptual-production system could in no way allow him to perceive and produce at the same level as the adult. As has been suggested by Fry (1966) it is possible that there are two separate systems for speaking: one for perceiving and one for producing. Allowing for the prior development of the auditory system, perception will always precede production, and, moreover, in view of the normally hearing but non-verbal child's ability to perceive and understand language, it is quite possible to have perception _without_ production, although production without perception will assuredly distort the speech output.

There is little doubt that the hypothesis proposed by Öhman could be a very nice model for infant child speech perception, i.e. that in the developing organism there exists a mutual interaction between the subjective (sensory) and the objective (physical) percepts which the child is developing, through the developing activation of a private production model; the type of total cognitive process proposed by E. Gibson (1969).

J. J. Gibson has said:

> "The concrete question of innate and acquired mechanisms in perception is not a two-way issue, for we know that there are intermediates between what is intended and what is acquired. Pure genetics is one thing; pure learning is another thing; but in between there are types of development that we call growth or maturation. The anatomy and basic physiology of the organs of perception depend mainly on genetic factors as determined by evolution. The maturation of the perceptual systems depends on genetic and environmental determiners in concert. The education of the perceptual systems depends mainly on the individual's history of exposure to the environment, so there are really three questions: How much does perceiving depend on organs? How much does it depend on growth? How much does it depend on experiences?" (1966, p.268).

The organs with their receptors set limits on the kind of information that can be registered. Their ways of orienting, adjusting and exploring are partly constrained by anatomy but partly free. The basic neural circuitry for making such adjustments is built into the nervous system by the time of birth, but it continues to develop in man for a long time after birth.

It seems true that the child cannot be expected to perceive certain facts about the world until he is ready to perceive them. He is not simply an adult who doesn't have any experience but, as Piaget has shown, develops by passing through various stages. The ability to select and abstract information about the world grows as he grows.

The environment at the same time provides an inexhaustable reservoir of information. The ears are not analogous to microphones, which would imply a fixed relationship to the brain. "Listening" (and all this implies) continues to improve with experience, so that higher order variables can still be discovered even in old age. It therefore comes as no surprise that the speech in the infant's environment is a product of carefully adjusted interactional processes (Phillips, 1973).

Perceptual learning might be conceived as one of differentiation, in Gibson's terminology. The process is one of learning what to attend to both overtly and covertly. A system "hunts" so that it achieves clarity. The process can occur at more than one level. First, the pick up of information reinforces the exploratory adjustments of the organs that make it possible. And second, the registering of information reinforces whatever neural activity in the brain brings it about.

For the child who is beginning to acquire the speech code and at the same time learning to perceive the world (by developing the ability to attend to the higher order features of objects and events in graded stages), the sounds and ultimately words he hears are not simply auditory stimuli or vocal responses. These sounds embody stimulus information, especially invariant information about the irregularities of the environment. They consolidate the growing ability of the child to detect and abstract invariants while at the same time cutting across the perceptual systems.

Some of this puzzling interaction across invariant information may be seen in relatively simple perception-production tasks and provides a partial feeling for some of the complexities associated with differentiating out phonetic from phonological development. Thus, for example, Menyuk (1972), in experiments run to examine children's categorizations of /p, t, k/, /b, d, g/, /f, s, s/, /w, l, r/ in a production and perception task between 3.1 and 5.3 years (CA), showed that the task resulted in the greatest number of errors on /p, t, k/ in <u>perception</u> and in /w, l, r/ in <u>production</u>. Considering that Cruttenden (1970) reported /p, t, k/ among the first and most stable sounds in <u>production</u> during babbling, this result is somewhat interesting. Apparently the ability to produce the word does not necessarily improve perceptual performance even when the sounds have been motorically produced for some time.

The task of modifying durations of phonemes in relation to the length of utterance is learned by at least age 3 and the durations of vowels preceding voiceless sounds has been shown to be approximately the same for 3, 6 and 9 year-old children (Di Simoni, 1974). In a study with three-year-old girls, Graham and House (1971) have shown, in an analysis of perceptual discrimination errors, that the perceptual behavior of children at this age is very similar to that of adults <u>except</u> that (understandably) the children produced more errors than an adult is expected to make in a comparable task. The patterning of errors in their study did not support the Jakobsonian developmental scheme in a strict sense. They interpreted their data as compatible with a loose interpretation of that proposal, however, provided that the specific order of development of feature opposition is modified. Contrariwise, Edwards (1974) indicates that although phonemic perception develops in advance of production, children as late as 3 CA do <u>not</u> have complete phonemic perception, although the younger a child in a group of 2 to 14-year-olds, the larger the acoustic patterns must be for them to discriminate between members of a voiced/voiceless pair, (Simon, 1974, 1975). Simon suggests that young children use these features in an order of priority quite different from that used by older children and adults.

In a recent paper on word and sounds in early language acquisition of three children, Ferguson and Farwell (1975) point out that phonological development in children, like sound change in language, takes place along <u>several</u> parameters, of which the phonetic is only one. A recent model of linguistic change proposed by Wang (1969), Chen and Wang (1975), suggests that sound change takes place along 3 parameters: 1) <u>A phonetic parameter</u>: the phonetic manifestation of a sound change occurs abruptly at some point, goes through a period of variation in which some words are found in two forms, and finally approaches completion, whereupon other forms may change abruptly without going through a period of variation. 2) <u>A lexical parameter</u>: sound change starts in a subset of the relevant words, determined phonetically, socially, or by other factors, and spreads gradually through the lexicon to other forms relevant along the same dimension. 3) <u>A social parameter</u>: sound change starts with some group of people and spreads to others, or it begins in one speech style and spreads to other styles in the same individual (Ferguson, 1971). Of Wang's three parameters, the social seems least relevant at an early age, since sound changes are taking place within an individual, and children under two have very little control of different speech registers. Gleason (1973) for instance, suggests that their main register split is speech versus silence. All three children in the Ferguson and Farewell study (who were followed from 11 months to 1:2 years) showed a preference for voiced labial and alveolar stops, but voiceless velars.

IS SPEECH PERCEPTION IN CHILDREN EVER THAT NICE AND CLEAN?

Do children, in effect, have "some general techniques through which they may approach the buzzing confusion of the ambient linguistic environment?" Nelson (1973) suggests that: "<u>first</u>, the child filters out complex and extremely unfamiliar speech; <u>second</u>, he selects slightly unfamiliar stimuli for special attention and repetition. He thereby provides himself (a) with tractable input which he can use as a guide to his non-linguistic behavior and as a basis for testing his hypotheses about language, and (b) with a procedure for broadening his linguistic skills, for formulating new rules, and for adding to his lexicon." Lenneberg proposes that language begins to make its appearance when about 60% of the adult values of brain maturation are reached. Shvachkin (1973) has maintained that the basic determining factor underlying the phonemic development of child language as well as the development of articulation is the semantics of the language, the evolving need for making semantic distinctions determining the phonemic development.

> "and the moral of _that_ is - 'Take care of the sense, and the sounds will take care of themselves'." (The Duchess to Alice in '_Alice in Wonderland_')

If this is so, then even the prosodic aspects of infant utterance from crying, to babbling to word, at some point take on a semantic load which must be seen as an important part of the acquisition process. Such a notion accords well with some present findings in mothers' speech research (Snow, 1974) and would correlate with Piagetian development through the sensory motor period. It has, in fact, been suggested by MacNamara (1972) that children are able to learn to talk because they can figure out the meaning of the sentences they hear, independent of the sentences themselves, which is being borne out by current studies of language acquisition, which are pragmatically oriented.

In this very brief review there has not been occasion to mention many other factors which must of course feedback to the child in what Ferguson (1975) has termed "baby talk registers", and which are intertwined with speech perception in children.

INCONCLUSIVELY

In a far from final analysis, where might it be said that the neonate-infant in the first year of life exists vis-a-vis the adult language system? Or even perhaps, phylogenetically? If, as will undoubtedly be shown in the future, other species are capable of differentiating synthetic or natural speech stimuli of the CV variety, then it might be concluded that for a short time, at least, the perceptual system of the neonate-infant capitalizes on aspects of auditory signals which can be processed by the auditory system _at that time_, and which for the infant will ultimately be incorporated into the sounds of language. For the chinchilla, rhesus monkey or other species, however, the same auditory signals will undoubtedly be incorporated into some signal system for ethological purposes pertinent to their survival.

As a result of this all too cursory review, three points might be tentatively made. First, it is apparent that the very fact that the child IS developing means that its perceptual system cannot be accounted an exact analogue of the adults' perceptual system. Any pure form of some motor or analysis-by-synthesis speech perception theory will, in this light, be insufficient to account for non-auditory or non-motoric aspects of speech perception. The possibilities for speech perception despite nonfunctioning in either system, have been cogently argued by Lenneberg. Strong forms of motor or auditory theories will be unable to account for all the other pressures (internal and external) which are acting to bring convergence to the adult system. Apart from the purely sensory differences between child and adult, there is undoubtedly strong evidence from the work on speech registers that the neonate-infant-child IS attending to a biased sample of the adult linguistic universe: phonetic, syntactic and semantic. This subsample converges to linguistic simplicity of all kinds, the convergence being in effect forced on it by the way in which an older speaker-listener views the child's comprehension abilities, by the kind of feedback such a listener gets about how well the child is both attending and comprehending and finally by the way in which the child, for reasons already given, will filter out or turn off complex speech.

Second, it is by now also apparent that notions of a "pre-linguistic" child need re-evaluation. Whatever scepticism one might hold for the non-nutritive sucking paradigm as an indicator of speech perception, must be weighed with the abundant evidence which indicates that the neonate can perceive, discriminate and orient to, both simple and complex auditory stimuli, which must in and of themselves, bear the "seeds" of the speech signal. Whether this is in itself linguistic depends on one's definition of "language".

Third, on the basis of the physiological and psychological evidence, there is good reason for claiming that in both its productive and perceptive parts, babbling represents THE first steps of the language acquisition process.

If, perforce, the perception of speech by the child just engaging in the language game is to be properly understood, it is at the critical period of offset of reflexive crying and onset of the sensorimotor period, that the work will have to be intense.

REFERENCES

Allen, G.D., Segmental timing control in speech production, J. of Phonetics, 1, 219-237 (1973).

Aronson, E. and S. Rosenbloom, Space perception within a common auditory-visual space, Science, 172, 1161-1163 (1971).

Bel'tyukov, V.I., and A.D. Salakhova, Babbling in the hearing child, Voprosy psikhologii, 2, 105-116 (1973).

Bench, J., "Square-wave stimuli" and neonatal auditory behavior: Some comments on Ashton (1971), Hutt et al. (1968) and Lenard et al. (1969), J. Exp. Child. Psychol. 16, 521-527 (1973).

Bench, J. and L. Metz, Stimulus complexity, state and infants' auditory behavioural responses, British Journal Dis. Commun., 10, 1, April, 52-60 (1975).

Brackbill, Y., G. Adams, D.H. Crowell and M.L. Gray, Arousal level in neonates and older infants under continuous auditory stimulation, In Y. Brackbill and G. Thompson (Eds.), Behavior in Infancy and Early Childhood, New York: The Free Press, 61-69 (1967).

Brown, R., Words and Things, N.Y.: The Free Press, (1958).

Brown, R., Development of the first language in the human species, American Psychologist, 28(2), February, 97-106 (1973).

Butterfield, E.C. and G.F. Cairns, "Discussion summary - infant reception research", In R.L. Shiefelbusch and L.L. Lloyd, (Eds.), Language Perspectives-Acquisition, Retardation and Intervention, Baltimore: Univ. Park Press, 75-102 (1974).

Chen, M.Y., and W.S-Y. Wang, Sound change: Actuation and Implementation, Language, 51, 2, 255-281 (1975).

Cruttenden, A., A phonetic study of babbling, British Journal of Disorders of Communication, 5, 2, October, 110-117 (1970).

Crystal, D., Non-segmental Phonology in Language Acquisition: A Review of the Issues, Lingua, 32, 1-45 (1973).

Cullen, J.K., N. Fargo, R.A. Chase and P. Baker, The development of auditory feedback monitoring: I. Delayed auditory feedback studies on infant cry, J. Speech Hearing Res., 11, 85-93 (1968).

Delack, J.B., Prelinguistic infant vocalizations and the ontogenesis of sound-meaning correlations, Bulletin d'Audiophonologie, 4, 6, 479-499 (1974).

Di Simoni, F., Evidence for a theory of speech production based on observations of the speech of children, J. Acoust. Soc. Amer. 56, 6, 1919-1921 (1974).

Edwards, M.L., Perception and production in child phonology: the testing of four hypotheses, J. Child Language, 1, 205-219 (1974).

Eimas, P.D., "Speech perception in infancy", In L.B. Cohen and P. Salapatek, (Eds.), Infant Perception, N.Y.: Academic Press (in press).

Eisenberg, R.B., The organization of auditory behavior, J. Speech Hearing Res. 13, 453-471 (1970).

Ferguson, C.A., Short 'a' in Philadelphia English, Stanford Occasional Papers in Linguistics, 1, 2-27 (1971).

Ferguson, C.A., Baby talk as a simplified register. P.R.C.L.D., 9, April, 1-27 (1975).

Ferguson, C.A., and C.B. Farwell, Words and sounds in early language acquisition, Language, 51, 2, 419-439 (1975).

Friedman, D., R. Simson, W. Ritter, and I. Rapin, Cortical evoked potentials elicited by real speech words and human sounds, Electroencephalography and Clinical Neurophysiology, 38, 13-19 (1975).

Fromkin, V.A., Slips of the tongue, Scientific American, 110-117 (1974).

Fry, D.B., "The development of the phonological system in the normal and the deaf child," In F. Smith and G. Miller (Eds.), The Genesis of Language, Cambridge, Mass: MIT Press (1966).

Furth, H., "Two aspects of experience in ontogeny: Development and learning", In H.W. Reese, (Ed.), Advances in Child Development and Behavior, Vol. 9, N.Y.: Academic Press (1974) pp 47-66.

Gibson, E.J., Perceptual Learning and Development, N.Y.: Appleton-Century-Crofts (1969).

Gibson, J.J., The Senses Considered as Perceptual Systems, Boston: Houghton Mifflen (1966).

Gilbert, J.H., Acoustical Features of Children's Vowel Sounds: Development by CA versus Bone Age, Language and Speech, 16, 3, 218-223 (1973).

Gilbert, J.H., On babbling: Some physiological observations, PRCLD., Stanford Univ., 42-49 (1974).

Gilbert, J.H.V., Perception of four vowels by children at 2½-3 CA., (to appear) (1975).

Gilbert, J.H., and I. Climan, Dichotic studies in 2 and 3 year-olds: A Preliminary Report, Speech Communication Seminar, Stockholm Vol. 2, Upsalla: Almquist and Wiksell (1974).

Gleason, J.B., "Code switching in children's language", In T.E. Moore, (Ed.), Cognitive Development and the Acquisition of Language, N.Y.: Academic Press (1973) pp 159-167.

Graham, L.W. and A.S. House, Phonological oppositions in children: A perceptual study, J. Acoustical Soc. of Amer., 49, 2, (Part 2), 559-566 (1971).

Harrison, G.A., J.S. Weiner, J.M. Tanner and N.A. Barnicot, Human Biology, Oxford: The Univ. Press (1964).

Hellbrügge, T., "The development of circadian rhythm in infants", In Symposium on Quantitative Biology, 25, N.Y. Biological Lab: Cold Spring Harbor, N.Y., 311-323 (1960).

Hopkin, G.B., Neonatal and adult tongue dimensions, Angle Orthodont, 27, 132-133 (1967).

Horowitz, F.D. (Ed.), Visual attention, auditory stimulation and language discrimination in young infants, Monogr. Soc. Res. Child Devel. serial #158, Vol. 39, Nos. 5-6 (1975).

Huber, H., A preliminary comparison of English and Yucatec infant vocalization at nine months, CLS., 6, 114-119 (1970).

Iguchi, S., and I.J. Hirsh, Development of speech sounds in children, Acta-otolaryn., Suppl., 257 (1969).

Keith, W., Impedance audiometry with neonates, Arch. Otolaryngol., 97, June, 465-467 (1973).

Kewley-Port, D., and M. Preston, Early apical stop production: a voice onset time analysis, J. of Phonetics, 2, 195-210 (1974).

Khrisman, T.P., Relationship between the inferior parietal associative parts of the brain and speech perception in children, Acad. Pedagoc. Sci. U.S.S.R. Child and Juvenile Physiol. Res. Inst., Higher Nerv. Activ. Lab., Leningrad (1974).

Kornfeld, J., Theoretical issues in child phonology, CLS, 7, 454-468 (1971).

Kuhl, P., Speech perception by the chinchilla: The voiced-voiceless distinction in alveolar plosive consonants, Science, (in press).

Kuhl, P.D., and J.D. Miller, Speech perception by the chinchilla: the voiced-voiceless distinction in alveolar plosive consonants, Science (1975, in press).

Labov, W., The study of language in its social context, Studium Generale, 23, 30-87 (1970).

Lenneberg, E.H., Biological Foundations of Language, N.Y.: Wiley (1967).

Lieberman, P., K. Harris, P. Wolff and L. Russell, Newborn infant cry and nonhuman primate vocalization, J. Speech and Hearing Res. 14, 4, 718-727 (1971).

Ling, D., A. Ling, and D.G. Doehring, Stimulus response and observer variables in the auditory screening of newborn infants, J. Speech and Hear. Res. 13, 9-18 (1970).

MacNamara, J., Cognitive basis of language learning in infants, Psychol. Review, 79, 1-13 (1972).

McNeilage, P., "Speech Physiology," In J.H.V. Gilbert, (Ed.), Speech and Cortical Functioning, N.Y.: Academic Press (1972).

Menyuk, P., Children's perception of a set of vowels, Q.P.R., (M.I.T.), 84 Jan., 254-257 (1967).

Menyuk, P., Cues used in speech perception and production by children, Q.P.R.,(M.I.T.), No. 77, April, 310-313 (1965).

Menyuk, P., Speech sound categorization within sets by children, Q.P.R. (M.I.T.), 105, April 15, 119-123 (1972).

Morais, J., and P. Bertelson, Laterality effects in diotic listening, Perception, 2, 107-111 (1973).

Müller, E., H. Hollien and T. Murry, Perceptual responses to infant crying: identification of cry types, J. Child Lang. 1, 89-95 (1974).

Nelson, K., Structure and strategy in learning to talk, Mono. Soc. Res. Child Devel., Univ. of Chicago Press, Serial No. 149, Vol. 38, Nos. 1-2 (1973).

Noback, G.J., The developmental topography of the larynx, trachea, and lungs in fetus, new-born, infant and child, Amer. J. Dis. Child. 26, 515-533 (1923).

Phillips, J., Syntax and vocabulary of mothers' speech to young children: age and sex comparisons, Child Development, 44, 182-185 (1973).

Piaget, J., Biology and Knowledge: An Essay on the relations between organic regulations and cognitive processes, (orig. published Pans: Gallimard, 1967) Chicago: The Univ. Press (1971).

Pierce, J.E., A study of 750 Portland, Oregon children during the first year. P.R.C.L.D., 8, 19-25 (1974).

Prescott, R., Infant cry sounds; developmental features, J. Acoust. Soc. Amer. 57, 5, 1186-1191 (1975).

Rensch, B., Increase of learning capability with increase of brain size, The American Naturalist, XC, 851, March-April, 81-95 (1956).

Scott, J.P., Critical periods in behavioral development, Science, 138, 3544, 949-958 (1962).

Searle, J.R., Speech Acts: An Essay in the Philosophy of Language, Cambridge: Cambridge Univ. Press (1970).

Seckel, H., Bird Headed Dwarves: Studies in Developmental Anthropology Including Human Proportions, Springfield, Ill: Thomas (1960).

Shaw, E.A.G., "The External Ear", In W.D. Keidel and W.D. Neff, (Eds.), Handbook of Sensory Physiology, Berlin: Springer-Verlag, Vol. v/i, chp. 14 (1974).

Sheppard, W.C. and H.L. Lane, Development of the prosodic features of infant vocalizing, J. Speech Hear. Res. 11, 94-108 (1968).

Shvachkin, N.Kh., The development of phonemic speech perception in early childhood, In C.A. Ferguson and D.I. Slobin (Eds.), Studies of child language development, New York: Holt, Rinehart and Winston (1973).

Simon, C., Some measurements of the speech of children, Speech and Hearing Work in Progress, U.C.: London, November (1974).

Simon, C., Some Aspects of the Speech of Children, (Abstract), 8th Intern. Congr. Phonetic Sci., Leeds, August (1975).

Smith, N.V., The Acquisition of Phonology, Cambridge: The University Press (1973).

Snow, C.E., "Mother's Speech Research: An Overview", Paper presented at the Conference on Language Input and Acquisition, Boston, Mass., Sept., (1974).

Tervoort , B.T.M., Language development of deaf children: Some results of an 8-year project, Het gehoorgestoorde Kind, 8, 214-227 (1967).

Trehub, S.E., Infants' sensitivity to vowel and tonal contrasts, Develop. Psychol., 9, 91-96 (1973).

Truby, H.M., and J. Lind, The cry sounds of the newborn infant, Acta Paed. Scand. Suppl. 163 (1965).

Turkewitz, G., H. Birch and K. Cooper, Responsiveness to simple and complex auditory stimuli in the human newborn, Developmental Psychobiology, 5, 1, 7-19 (1972).

Turnure, C., Response to voice of mother and stranger by babies in the first year, Developmental Psychology, 4, 2, 182-190 (1971).

Von Bertalanffy, L., Problems of Life, N.Y.: Harper Torchbooks (1952).

Waddington, C.H., The Strategy of the Genes, London: Allen and Unwin (1957).

Wang, W. S-Y., Competing changes as a cause of residue, Language, 45, 9-25 (1969).

Wilson, T.G., Some observations on the anatomy of the infantile larynx, Acta. Otolaryng. 43, 95-99 (1953).

Witelson, S.F., and W. Pallie, Left hemisphere specialization for language in the newborn: neuroanatomical evidence of asymmetry, Brain, 96, 641-646 (1973).

Yakovlev, P.I. and A-R. LeCours, The myelogenetic cycles of regional maturation of the brain, In A. Minkowski, (Ed.), Regional Development of the Brain in Early Life, Oxford: Blackwell (1967).

ACKNOWLEDGMENTS

Preparation of this paper was supported in part by grant MT-4217 from the Medical Research Council of Canada. Looking over my shoulder during its writing was Carolyn Johnson, whose astute observations on form and content were extremely helpful.

DISCUSSION of "Perception of vowel sounds by 3 year olds"
GILBERT

LEHISTE:

There were two aspects of the paper I would like to comment on. First, one should really keep speech acquisition and language acquisition distinct, and they seemed to be equated in the paper. The evidence from acquisition of signs (by children of deaf parents who are native speakers of sign language) shows that infants are able to use symbols at an age much younger than the age at which spoken language is acquired.

The second comment relates to the question whether children process speech in the same way they acquire spoken language: from speech sound to word to sentence. This seems to be assumed in the paper. I would propose that children perceive speech "from the top down" and learn to react first to the suprasegmental structure of the utterance acquiring segmental discriminations somewhat later. This is supported by the general observation (experimentally untested) that when the lexical content of an utterance is contradicted by its suprasegmental structure, we believe the suprasegmental message every time. Apparently the message carried by intonation and voice quality is in a way more basic, which might well mean that ability to perceive it is acquired earlier.

HAGGARD:

I was very surprised that, when the "infant speech perception" studies first came out, people rushed into a nativist interpretation on the assumption that the performance was some close analogue of the mature perceptual performance. My reaction was that we could now say that there are acoustical patterns which from a biological point of view are intrinsically discriminable, and which many of the world's languages are likely, but not certain, to exploit. I should like to know whether, after the wide-ranging and mature survey that has gone into this paper, Dr. GILBERT concludes similarly or differently.

PISONI:

I would like to raise a general question concerning the relevance of work on infant speech perception to our understanding of the speech

perception process in adults.
It is not at all clear to me how the results of the infant studies relate to the general problems in speech perception research dealing with segmentation, invariance and stages of perceptual analysis. For the most part, these studies show that infants can make discriminations of relevant acoustic variations that are found in adult phonology. But it is another thing to claim that these infants are responding to the phonetic properties of these stimuli. I hope we can return to some of these issues later on in the symposium.

GILBERT:
Let me respond first to LEHISTE's comments. First, I entirely agree that both speech and language in ontogeny should be clearly differentiated. At the present time, however, I do not feel that we have the data at our disposal to make such discriminations. My initial comments are intended to convey the problems of differentiating speech-language from a larger symbolic process in which the infant is engaged. Second, although I have considerable sympathy with a "top down" view of speech-language acquisition, in my review I have tried to develop reservations which apply to viewing the process as exclusively "top down".

My guarded response to HAGGARD's observation is that nativist interpretations of infant speech perception studies are, to say the least, premature. Entering arguments about "innateness", (a topic which always seems to me to have been adequately taken care of by John Locke) is a waste of energy.

Finally, in response to PISONI, let me speculate that the infant perception studies appear to demonstrate some fundamental aspects of the auditory system, Pat Kuhl and Jim Miller's work on chinchillas and Philips Morse's work on the rhesus monkey indicate that we are probably seeing a phylogenetic auditory property, which can be used by homo sapiens in the development of verbal language (along with non-auditory cues) and by other animals for their own particular survival purposes.

BERLIN:
To support the notion that speech may develop "from the top down", we have data on childrens perception of the Lisker + Abramson

continuum. These were normal speaking children 3 to 7 years old who of course could articulate the 6 English stops normally. The 3 year olds, despite their normal articulation and good performance at the continua extremes, had a different slope from the other children. While we are still analyzing the VOT of the children's utterances, it is clear that the normal sounding articulation + normal discrimination of the extremes of natural speech have preceded the childrens' abilities to categorize the stimuli.

V. General Discussion and Tentative Conclusions

SPEECH UNDERSTANDING SYSTEMS AND SPEECH PERCEPTION THEORY

D.H. KLATT

The organizing committee asked me to say a few words about some of the recent efforts to build speech understanding systems. There seem to exist interesting analogies between the way in which these systems are organized and the way we perceive speech. It is certainly true that these efforts have opened my eyes to the bigger picture of the perception of spoken language. The magnitude of the problem is clearly demonstrated when one has a practical engineering goal in mind. It is difficult to study speech understanding in the abstract. That is why it can be of some benefit to observe the problems encountered in attempts to build a speech understanding system that operates in limited semantic, syntactic and lexical domains.

The first speech understanding system to be demonstrated was a program that accepted commands to move chess pieces in a game of chess played against a computer (Reddy et al., Carnegie-Mellon University). The general structure of this program is shown in Figure 1. Control of the program resides in the syntactic component which has the task of listing all possible sentence initial words. These are compared with gross acoustic characteristics in order to choose the best match. This word is sent back to the syntax component which then lists all possible second words given the first word, and the process is iterated until the input is exhausted. The resultant word string is sent to a semantic module to see if it describes a legitimate chess move. The semantics component also includes a chess-playing program that rates the quality of the move. If the move is of poor quality, semantics asks if some other word string is almost as likely based on acoustic cues.

This program was criticized by some members of the speech recognition community because it seemed that use of a chess-playing component was almost like cheating, especially since the acoustic component was very primitive indeed. A more relevant criticism for us is the use of a syntax component that generates all possible word sequences, this being rather an artificial model of human speech understanding strategies.

There presently exist three large efforts to build speech understanding systems in the United States. These systems attempt to understand spoken commands from a larger domain, and thus use all possible constraints to decode the meaning of an utterance. The

structure of these programs is suggested in Figure 2.
The figure combines ideas from three speech understanding projects -- one directed by D.R. Reddy at Carnegie-Mellon University, one directed by B. Ritea at System Development Corporation and the Stanford Research Institute, and one directed by W. WOODS at Bolt Beranek and Newman, Inc.

There exists a blackboard containing all information computed by each component. First. the acoustics module places a preliminary acoustic representation on the blackboard. Then a phonetics module places a tentative phonetic analysis on the blackboard. The remaining components read information on the blackboard and add their own conclusions, hypothesizing or rejecting words or phrases.

The box labeled "lexicon" in Figure 2 scans the preliminary phonetic analysis and proposes tentative word matches. These word hypotheses are placed on the blackboard so that the syntactic and semantic component can try to combine them into meaningful units, reject some words or word sequences. and hypothesize likely adjacent words. Information from the discourse history and a model of what the user is trying to do is also used to predict likely words.

Acceptance or rejection of the word hypotheses proposed by the higher-level components is based on a verification strategy under control of the lexical component. The abstract phonemic representation for a word or phrase is converted to a phonetic transcription using a set of phonological rules and, if necessary, syntactic and semantic information. The generated phonetic representation can be compared with the preliminary phonetic analysis in order to accept or reject the hypothesis. If it is likely that the preliminary phonetic analyzer makes errors and attempts only a partial phonetic analysis, then a better verification strategy might include a synthesis by rule module to generate the appropriate parametric representation to be compared with the raw acoustic data available on the blackboard.

Control and coordination of the activation of these modules is a very difficult problem. It is in the control strategies that the 3 systems differ the most. One system envisions a set of parallel asynchronous processors interacting with the blackboard but not under the supervision of any control program. Other systems deemphasize the blackboard as the only means of intermodule communication and create complex serial control strategies.

It might benefit all of us to look at Figure 2 and attempt to hypothesize plausible ways in which human speech perception might be structured in terms of these modules.

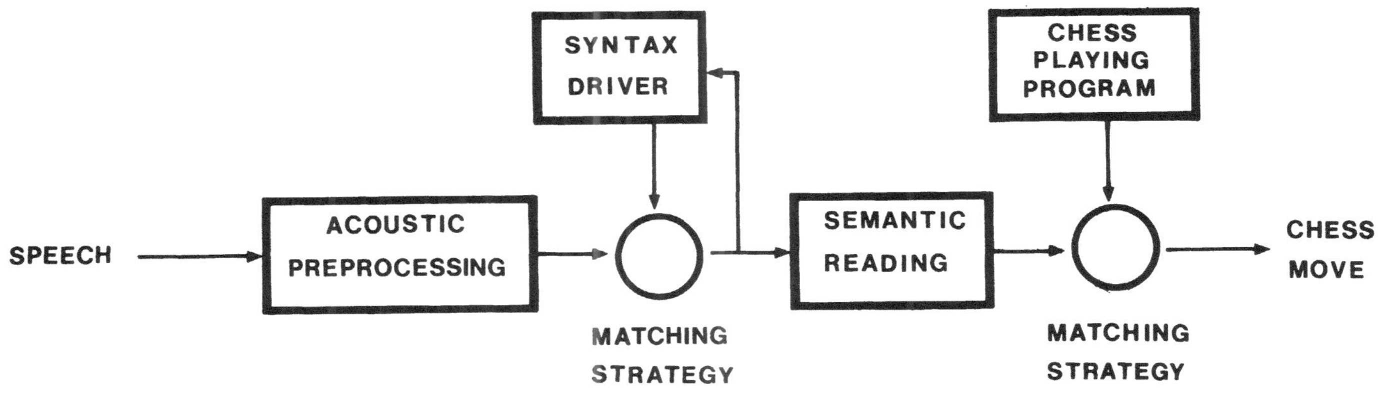

FIG. 1 FIRST SPEECH UNDERSTANDING SYSTEM (REDDY, CMU)

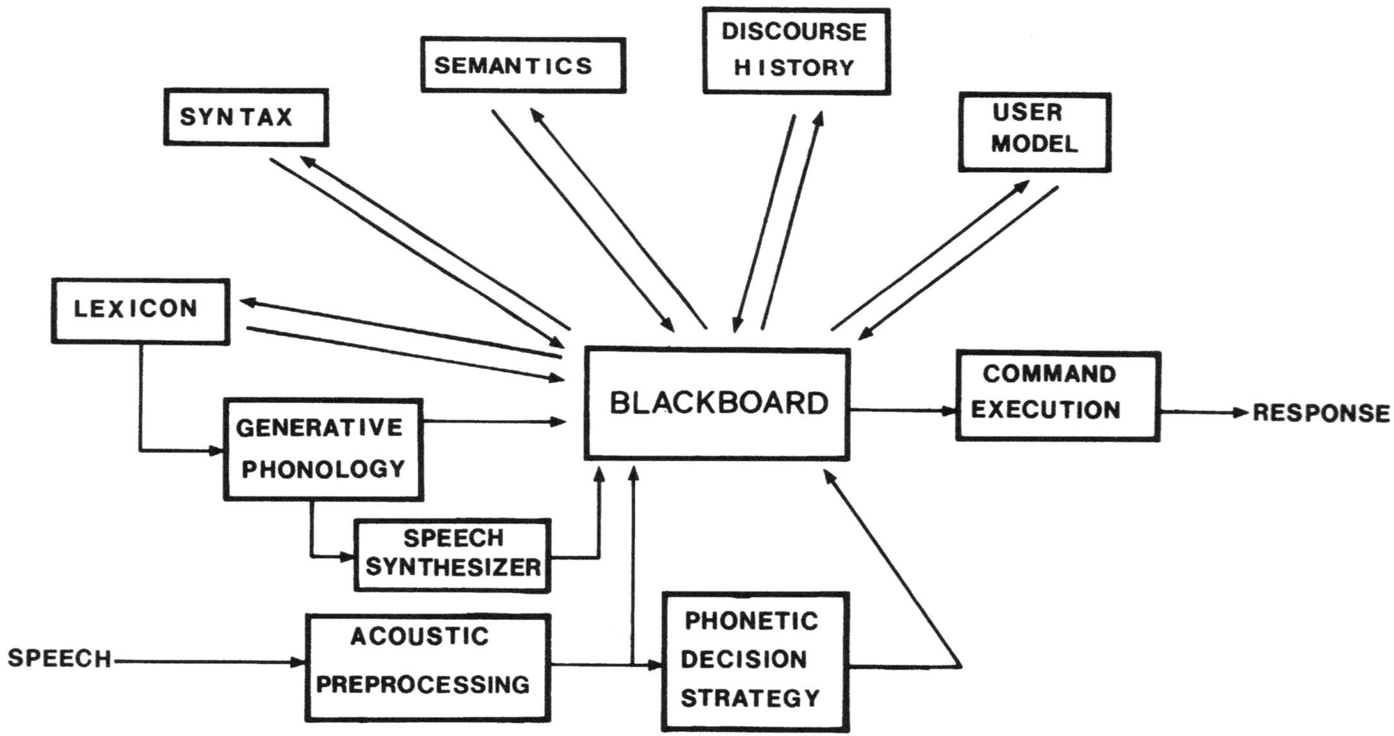

FIG. 2 STRUCTURE OF A GENERAL SPEECH UNDERSTANDING SYSTEM (CMU, BBN, SDC/SRI)

SOME PSYCHOLINGUISTS' REACTIONS TO THE SYMPOSIUM ON DYNAMIC ASPECTS OF SPEECH PERCEPTION

W.J.M. LEVELT and G.B. FLORES D'ARCAIS

The organizers of this Symposium have asked us to give some psycholinguists' reactions to what has been reported and discussed during this conference. They presumably hoped that such naive observers would be able to look objectively at the emperor's clothes. Our reaction to this is two-fold: (1) Though it may be disappointing, we have not observed any nudity. On the contrary, we were impressed by the richness of methods, phenomena and theoretical concerns among speech perception (SP) researchers. (2) We did not feel excentric to this meeting at all. Not only are many speech researchers psychologists by training, but also more generally we experienced considerable openness with respect to psychological theory, and psycholinguistic experimentation in language perception. If one compares the present situation with earlier work in SP, one finds a strongly increased interest in syntactic, semantic and even textual aspects of the speech signal. There is certainly a realistic common base now for students of "speech" and "language" perception to cooperate, and to integrate their theories. It even seems to be the case that the vigorous appeal of SP-workers to psychological models of memory, especially the early storage mechanism, is forcing psychologists into granting that their models of handling the intake and storage of continuous speech are still inadequate.

Since many of the observations we made have already been brought up by others (see especially Pisoni's paper) it will suffice for us to mention one or two issues which, in our view, still have not received sufficient attention.

- In spite of the just-mentioned convergence in the work of speech researchers and psycholinguists, there are still remarkable differences in style of theorizing. Though there are exceptions on both sides, SP-theories seem to be more structure-oriented, whereas psycholinguists show a deeper interest in process. Many of the SP-studies at this Symposium were of the following sort: here is a variable (linguistic, acoustic, etc.), does it affect a subject's reactions? Examples: Does the number of syllabes surrounding a vowel affect the subjects's identification of a vowel? Can lip-rounding be detected in vowel-preceding consonant clusters? Is voice onset time "real", i.e. influencing phoneme identification? Are syntactic variables "real" in judging segment duration? Does F_0 affect a subject's judgment with respect to the position of a sentence in a paragraph? Do intonation and syntax affect judgments of dichotic channel-switching? Though this type of question often leads to quite unexpected and nontrivial results it reminds us too much of the so-called "psychological reality" studies of the sixties. Psycholinguists have mostly passed that state now. Their interest is in the processes underlying such findings: what strategies and procedures are followed by the subject to cope with the experimental task? Which are the successive stages in the subject's decision making etc. Usually, this is called information processing approach, and we have seen too little of that during this conference. There is, for instance, not a single flow-diagram (i.e. a visual representation of a sequence of operations and decisions) in any of the Symposium papers. (There were some block-diagrams, but they are still rather "structuralist" in that the exact nature of the procedures within the blocks is left unspecified). Of course, one can do without flow-diagrams, but in our opinion one cannot do without information processing models in SP-research.

- A related point is that we have seen little concern for task structures. As in the early psychological reality studies, the objective of many SP-studies seems to be to demonstrate that a certain independent variable is effective, without much concern for the experimental task. However, since the work of Newell and Simon we know that the structure of the task is the prime determinant of the subject's decision procedure, and that there are only a few general invariant characteristics in human information processing (such as capacity-limits). Many of the SP-studies of this Symposium could have profited from a thorough task analysis: what exactly is the subject asked to do, what are the possible ways for him to cope with that task, and can we empirically determine which of these procedures is in fact chosen by him? Such an analysis might add considerably to an understanding of various interesting phenomena in SP (like those reported by Wingfield), where at present we do not even know whether they arise in early stages of stimulus encoding, or in later phases of response selection. Also, it is not unlikely that much of the rather "structuralist" controversies about "fixed" structures such as early storage systems will disappear if more attention is given to finer details of the task structure.

Finally, highly interesting theoretical issues will come up: given a certain task and the possible alternative ways to cope with it, why is it that a subject selects one procedure rather than another to arrive at an adequate response?

- An important point which has emerged several times in some of the papers presented and in the discussion, is the issue of parallel versus serial processing in speech perception. The question has been put forward by Haggard in his stimulating introduction, and has been given close attention in the last paper of the Symposium by Pisoni.

When one looks at the various models of speech perception available in the literature, it clearly appears that they are essentially, at least in most stages postulated, serial. Most models assume first some "low level" acoustic processing or preprocessing, which produces on the incoming signal certain "corrections" and equalizations. The material is then held for a brief duration in some storage of a sensory type, where some integration may take place, and then is handled over to some kind of perceptual unit where real "perceptual processing" takes place. Notice that, while some parallel processing may take place at the very first stage of acoustic processing, processing in the later stages, at least until the material reaches the level of LTM, essentially takes place sequentially. As a consequence, in most models real semantic processing would start only at a later stage. Now, it may be true that in certain experimental tasks no semantic processing takes place (for example, in a situation where the S has to search for a certain acoustic target in a series of discrete acoustic events presented in isolation) However, there is an increasing amount of evidence, even among data reported in the discussion during the Symposium, that in perception of connected speech processing takes place simultaneously in parallel, and this, very likely, from very early stages. In the attempts toward a construction of a speech perception theory, we wonder whether more attention should not be given to the possibility of very early semantic and syntactic processing, on the basis of linguistic or even non-linguistic context, of the expectations of the hearer etc. which could contribute to impose certain restrictions as to the kind of signals which may be accepted as possible candidates for perceptual synthesis, or produce certain selections or modifications on the incoming acoustic events. We do not propose to put a lexicon already on the basilar membrane, we only want to emphasize the notion that from the very early stages processing may take place at several levels simultaneously, and that the perceptual result may be the output of alternative or simultaneous activation of one or more of such levels of processing. There is a substantial amount of evidence from visual information processing studies which seems to indicate that parallel processing models can explain some critical phenomena in visual search paradigms in a more appropriate way than serial models. We would like to see more attention for a similar idea for acoustic information processing in speech perception models.

- A question which is related is whether the sensory storage which is postulated, with different labels, as an essential component in all models of speech perception, and in which the incoming information is held for a brief interval, during which some integration may take place, is to be considered a kind of "buffer" in which no recoding takes place, or whether some processing or some selection may already be at work. and that a certain amount of recoding may take place at this level. In the model proposed by Pisoni at this seminar, the S.I.S. component is functionally connected with LTM, which may mean a certain amount of interaction between SIS and LTM; in the pre-perceptual storage proposed by Massaro, on the other hand, no such connections are postulated. The question whether at the sensory mode level some recoding has already taken place, what kind of recoding is possible there, and under the influence of which factors is still, in our opinion, an open question, and specific answers in this direction are no doubt of importance in the development of a theory of speech perception.
- Another issue of major importance seems to us the series of questions one may ask about the effect of context on the perception of speech events. Various papers paid attention to several contexts, and when one compares the approach taken with the one typical of less recent work on speech perception, which was mainly concerned with very short units such as vowels or CV combinations out of any context, it seems already a real progress to have taken into account larger linguistic contexts such as words or sentences. This may have increased the ecological representativeness of the studies, and cer-

tainly may have contributed toward a theory of speech perception less limited in its explanatory, descriptive and predictive power.

Recent psycholinguistic work is taking into account larger and larger contexts. Serious attention is given not only to the larger linguistic context of a sentence, but also to the non-linguistic context, that is, to the different aspects of the communication situation - including things like the "knowledge of the world" of the two partners in the communication act, of the possible referents present, of events of which both the partners are aware, etc. Briefly, to what are often called presuppositions of a linguistic event. Several studies have shown that such presuppositions do affect perception, comprehension, storage and retrieval of linguistic material, and it is obviously an interesting question to ask, to what extent such presuppositions may affect perception of so-called "simple" linguistic events such as phonemes. We would welcome an effort in this direction.

- A question which is implicit in all theorizing about speech perception, and which has been specifically treated in Prof. Öhman's paper, is about the definition of perception. We do not think that it is possible to discuss all complex problems connected with such a definition in few words. However, a couple of comments may be appropriate here.

First, we must be aware of the different usages of the word in different experiments and in theory construction. In speech research the word perception has been used in a variety of ways. In some cases the experimenter means by perception the sensory discrimination between, say, two single acoustic events of different duration, and the paradigm adopted is the one typical of psychophysical experiments. In other cases the term is used to indicate the complex process of understanding the meaning of a word in a given context. While this variety of uses is characteristic of other fields of perception as well, one must be very alert, here as in other fields, to possible traps. An important point here is the opportunity of not confusing situations which call upon judgments on a certain speech event, with "real" perceptual situations. The interpretation of certain "click" experiments is an example of this confusion, where effects, which can clearly be assigned to response bias, are interpreted as perceptual effects.

- Another point which often has emerged in the discussion is the problem of the "perceptual unit" in speech perception. The question is whether it makes sense to talk about such units and state, for example, that "the syllable is the basic unit of speech perception." In visual perception nobody has ever found it important to decide which is the "basic perceptual unit", and research has progressed anyhow. The question could perhaps be reformulated in terms like the following;
  a. How is the speech event segmented and interpreted?
  b. Through what processing strategies do we obtain a certain perceptual result?
  c. If we want to isolate certain units such as the syllable or the word, do we know how these units are organized within the perceptual events?

- A final comment relates to the relevance of developmental studies for a theory of speech perception. There was only one developmental paper at the Symposium, but there are good reasons for SP researchers to give more attention to children's speech. Firstly, in the present literature developmental studies are rather more concerned with speech production than perception; more perceptual work is needed to achieve general understanding of speech development. Secondly, SP-studies with children may contribute to a general theory of speech perception. Specific theoretical issues where such contribution can be made are among others:
  (1) the question whether findings in SP should be interpreted as specific to language, or as special cases of more general perceptual abilities (cf. Eimas' studies relating categorical perception to neonate speech discrimination).
  (2) questions concerning the hierarchy of perceptual difficulty of various aspects of the speech signal, and concerning the hierarchy of SP skills. The order of acquisition of perceptual skills in children may not only reflect the order of complexity of perceptual tasks, but also the hierarchical skill structure of the adult. Early acquired perceptual strategies may remain basic strategies in adulthood.
  Furthermore, the child's speech perception may be used methodologically to test certain constraints in SP-theory. For instance,

parameters for certain hypothesized types of storage, will probably be different for the child, so that the functional contribution of such mechanisms to SP can be evaluated by comparing the child's and adult's behavior with respect to identical perceptual tasks.

TENTATIVE CONCLUSIONS

A. COHEN and S.G. NOOTEBOOM (editors)

In looking back to the conference we have come to the following tentative conclusions, which purport to be a reflection of those present research interests that were gathered from the discussions both during and between sessions, and in particular from the general discussion.

THEORETICAL ISSUES:

We noticed a certain reluctance among participants to commit themselves to the problem of what A theory of speech perception should look like or even which constraints such an eventual theory must be subject to.

As to the various box-diagrams discussants generally agreed that these diagrams may help researchers to concentrate on certain problems. Ultimately the exact nature of the procedures within the boxes should have to be specified (cf. above, the contribution of Levelt and d'Arcais to this general discussion).

The following specific theoretical issues came up.

- With respect to perceptual units there was a general awareness that there may be a number of different perceptual units at different processing levels. To us it seems that in looking for "perceptual units" or "decision units" one should not too easily postulate a one-to-one relationship between such units and linguistic units such as phonemes, syllables, words, phrases It might be worthwhile to transform the time-honoured question of perceptual units into a search for the scanning processes in auditory perception per se, instead of taking a lead from the units of linguistic description. It was suggested by one of the participants that there are no such things as perceptual units: speech perception may be a continuous process.
- As to the question how in the perceptual system linguistic expectancies affect the decoding process, there appeared to be a recognition that both top-down and bottom-up processes are required. Because of this, one was reluctant to get involved in an opposition between the one and the other. The question which did come up was what the lowest level is where they meet, how we can get at this, and what the general nature is of the con-

fluence of these two types of information.

- More than once during the conference the question was raised what could be gained from a study of our own immediate perceptual experiences in listening to speech. Whereas some participants favoured the idea that one should attempt to relate particular stages in models of speech perception to such perceptual experiences, others were of the opinion that it is difficult enough to infer a model from performance in a linguistic task even without trying to account for our subjective awareness as listeners.

EMPIRICAL ISSUES:

- The participants of the symposium had been assembled because they shared the belief that the field of speech research would benefit from greater attention paying paid to phenomena concerning connected speech, avoiding too great a concentration on studies of segments of phoneme or syllable size. A number of papers in the symposium demonstrate the mediating role of prosodic cues in the decoding of syntactic structures, and the importance of prosody for the perceptual coherence of longer stretches of connected speech. The perceptual processes involved remain largely unknown.
- Concern was shown with the inferences that could be drawn from specific experimental paradigms. There was a spreading realization that traditional paradigms constitute difficulties with respect to their applicability to continuous speech. In particular there were negative feelings about click experiments and the indicriminate use of dichotic listening experiments. Opinions were divided about the adequacy of phoneme monitoring and shadowing techniques. Awareness was shown of the need for new experimental methods specifically designed to get at the perceptual processes involved in the decoding of connected speech.
- There was a mild dissatisfaction with the fact that most of the experiments do not go beyond the level of demonstration and a strong feeling that one should move on from demonstrations to plotting functional relationships.

Communication and Cybernetics

Editors: K. S. Fu, W. D. Keidel, H. Wolter

Vol. 1 W. Meyer-Eppler
Grundlagen und Anwendungen der Informationstheorie
2. Auflage, neu bearbeitet und erweitert von G. Heike und K. Löhn

Vol. 2 B. Malmberg
Structural Linguistics and Human Communication
An Introduction into the Mechanism of Language and the Methodology of Linguistics
2nd revised edition

Vol. 3 J. L. Flanagan
Speech Analysis / Synthesis and Perception
2nd edition

Vol. 4 G. Herdan
The Advanced Theory of Language as Choice and Chance

Vol. 5 G. Hammarström
Linguistische Einheiten im Rahmen der modernen Sprachwissenschaft

Vol. 6 J. Peters
Einführung in die allgemeine Informationstheorie

Vol. 7 K. Weltner
The Measurement of Verbal Information in Psychology and Education
Translated from the German by B. M. Crook

Vol. 8 *Facts and Models in Hearing*
Proceedings of the Symposium on Psychophysical Models and Physiological Facts in Hearing, held at Tutzing, Oberbayern, Federal Republic of Germany, April 22–26, 1974
Edited by E. Zwicker, E. Terhardt

Some Related Titles

Communication and Cybernetics

Editors: K. S. Fu, W. D. Keidel, H. Wolter

Vol. 9 G. Hammarström
Linguistic Units and Items
Contents: Introduction. – Spoken Language. – Written Language. – Written Language in Relation to Spoken Language. – Spoken Language in Relation to Written Language. – The Tasks of Linguistics.

Vol. 10 *Pattern Recognition*
Edited by K. S. Fu
With contributions by numerous experts
Contents: Topics in Statistical Pattern Recognition. – Clustering Analysis. – Syntactic (Linguistic) Pattern Recognition. – Picture Recognition. – Speech Recognition and Understanding.

Vol. 12 J. D. Markel, A. H. Gray
Linear Prediction of Speech
In preparation

Topics in Applied Physics

Founded by H. K. V. Lotsch

Vol. 6 *Picture Processing and Digital Filtering*
Edited by T. S. Huang
With contributions by numerous experts
Contents: Two-Dimensional Transforms. – Two-Dimensional Nonrecursive Filters. – Two-Dimensional Recursive Filtering. – Image Enhancement and Restoration. – Noise Considerations in Digital Image Processing Hardware.

Vol. 11 *Digital Picture Analysis*
Edited by A. Rosenfeld
In preparation